Ajay Kumar Bhardwaj, Terenzio Zenone, Jiquan Chen (Eds.)
Sustainable Biofuels

Ecosystem Science and Applications

Editors
Jiquan Chen
Heidi Asbjornsen
Kristiina A. Vogt

Sustainable Biofuels

An Ecological Assessment of the Future Energy

Edited by
Ajay Kumar Bhardwaj
Terenzio Zenone
Jiquan Chen

DE GRUYTER

Higher
Education
Press

This work is in the *Ecosystem Science and Applications* co-published by Higher Education Press and Walter de Gruyter GmbH.

ISBN 978-3-11-055466-3
e-ISBN (PDF) 978-3-11-027589-6
e-ISBN (EPUB) 978-3-11-038133-7
ISSN 2196-6737

Library of Congress Cataloging-in-Publication Data
A CIP catalog record for this book has been applied for at the Library of Congress.

Bibliographic information published by the Deutsche Nationalbibliothek
The Deutsche Nationalbibliothek lists this publication in the Deutsche Nationalbibliografie; detailed bibliographic data are available on the Internet at http://dnb.dnb.de.

♾Printed on acid-free paper
Printed in Germany

www.degruyter.com

List of Contributors

Albaugh, Janine M.
Department of Forestry and Environmental Resources, North Carolina State University, Raleigh, NC 27695, USA
jmalbaug@ncsu.edu

Baglivi, Antonella
Department of Electronics, Information and Bioengineering, Politecnico di Milano, I-20133 Milan, Italy
antonella.baglivi@libero.it

Bandaru, Varaprasad
UC-Davis Energy Institute, University of California, Davis, CA 95616, USA
vbandaru@ucdavis.edu, bvprasad1007@gmail.com

Bar-Tal, Asher
Department of Soil Chemistry, Plant Nutrition and Microbiology, Institute of Soil, Water and Environmental Sciences, the Volcani Center, P.O. Box 6, Bet Dagan 50250, Israel
abartal@volcani.agri.gov.il

Bergante, Sara
Agricultural Research Council-Research Unit for Intensive Wood Production, St. Frassineto Po 35, 15033 Casale Monferrato, Italy
sara.bergante@entecra.it

Bhardwaj, Ajay K.
Soil and Crop Management, Central Soil Salinity Research Institute, Kachhwa Road, Karnal 132001, Haryana, India; and DOE-Great Lakes Bioenergy Research Center, Michigan State University, East Lansing, MI 48824, USA
ajaykbhardwaj@gmail.com, ajay@cssri.ernet.in, ajay@msu.edu

Ceulemans, Reinhart
Research Group of Plant and Vegetation Ecology, Department of Biology, University of Antwerp, B-2610 Wilrijk, Belgium
reinhart.Ceulemans@ua.ac.be

Chen, Jiquan
International Center for Ecology, Meteorology, and Environment (ICEME), Nanjing University of Information Science and Technology, Nanjing 210044, China; and DOE-Great Lakes Bioenergy Research Center, Michigan State University, East Lansing, MI 48824, USA
jiquan.eco@gmail.com, jqchen@msu.edu

Domec, Jean Christophe
Department of Forestry and Environmental Resources, North Carolina State University, Raleigh, NC 27695, USA; and University of Bordeaux, Bordeaux Sciences Agro UMR INRA-TCEM 1220, 33883 Villenave d'Ornon, France
jdomec@ncsu.edu

Facciotto, Gianni
Agricultural Research Council-Research Unit for Intensive Wood Production, St. Frassineto Po 35, 15033 Casale Monferrato, Italy
gianni.facciotto@entecra.it

Fichot, Régis
University of Orléans, Department of Biology, UPRES EA 1207 Laboratoire de Biologie des Ligneux et des Grandes Cultures (LBLGC), F-45067 Orléans, France; and INRA, USC 1328; and Arbres et Réponses aux Contraintes Hydriques et Environnementales (ARCHE), F-45067 Orléans, France
regis.fichot@univ-orleans.fr

Fike, John H.
Crop and Soil Environmental Sciences, Virginia Tech, Blacksburg, VA 24061, USA
jfike@vt.edu

Fiorese, Giulia
European Commission, Joint Research Centre, Institute for Environment and Sustainability, Via E. Fermi 2749, I-21027 Ispra (VA), Italy
giuliafiorese@gmail.com

Fischer, Milan
Global Change Research Centre AV CR v.v.i., Bělidla 986, 4a, 603 00 Brno, Czech Republic; and Department of Agriculture Systems and Bioclimatology, Mendel University, Zemedelska 1, 613 00 Brno, Czech Republic
fischer.milan@gmail.com

Grelle, Achim
Department of Ecology, Swedish University of Agricultural Sciences, Uppsala, Sweden
achim.grelle@slu.se

Guariso, Giorgio
Department of Electronics, Information and Bioengineering, Politecnico di Milano, I-20133 Milan, Italy
guariso@elet.polimi.it

Ioannis, Dimitriou, Swedish University of Agricultural Sciences, Department of Crop Production Ecology, Ullsväg 16, Box 7043, 756 51 Uppsala, Sweden
jannis.dimitriou@slu.se

Izaurralde, R. César
Department of Geographical Sciences, University of Maryland, College Park, MD 20742, USA; and Texas AgriLife Research & Extension, Texas A&M University, Temple, TX 76502, USA
cizaurra@umd.edu, cizaurralde@brc.tamus.edu

King, John S.
Department of Forestry and Environmental Resources, North Carolina State University, Raleigh, NC 27695, USA
john_king@ncsu.edu

Liska, Adam J.
Department of Biological Systems Engineering, University of Nebraska, Lincoln, NE 68583-0726, USA
aliska2@unl.edu

Pandey, Vimal Chandra
Eco-Auditing Group, National Botanical Research Institute, Rana Pratap Marg, Lucknow 226001, Uttar Pradesh, India
vimalcpandey@gmail.com

Parrish, David J.
Crop and Soil Environmental Sciences, Virginia Tech, Blacksburg, VA 24061, USA
dparrish@vt.edu

Raghuvanshi, Nidhi
Eco-Auditing Group, National Botanical Research Institute, Rana Pratap Marg, Lucknow 226001, Uttar Pradesh, India

Sharma, Dinesh K.
Central Soil Salinity Research Institute, Kachhwa Road, Karnal 132001, HR, India
dksharma@cssri.ernet.in, director@cssri.ernet.in

Singh, Bajrang
Restoration Ecology Group, National Botanical Research Institute, Rana Pratap Marg, Lucknow 226001, Uttar Pradesh, India
bsingh471@rediffmail.com

Singh, Gurbachan
Agricultural Scientists Recruitment Board, Krishi Anusandhan Bhavan-I, Pusa, New Delhi 110012, India; and Central Soil Salinity Research Institute, Kachhwa Road, Karnal, HR 132001, India
gurbachansingh@icar.org.in

Singh, Kripal
Division of Agronomy and Soil Science, Central Institute of Medicinal and Aromatic Plants, Picnic Spot Road, Lucknow 226015, Uttar Pradesh, India

Singh, Pushpa
Indian Institute of Sugarcane Research, Raibareli Road, P.O. Dilkusha, Lucknow 226 002, Uttar Pradesh, India
parampushpa@yahoo.com

Singh, Yashpal
Central Soil Salinity Research Institute, Regional Research Station, Lucknow 226 002, Uttar Pradesh, India
ypsingh_5@yahoo.co.in

Solomon, Sushil
Indian Institute of Sugarcane Research, Raibareli Road, P.O. Dilkusha, Lucknow 226 002, Uttar Pradesh, India
directoriisrlko@gmail.com, drsolomonsushil1952@gmail.com

Trnka, Miroslav
Global Change Research Centre AV CR v.v.i., Bělidla 986, 4a, 603 00 Brno, Czech Republic; and Department of Agriculture Systems and Bioclimatology, Mendel University, Zemedelska 1, 613 00 Brno, Czech Republic
mirek_trnka@yahoo.com

Uggè, Clara
ETA-Florence Renewable Energies, Via Giacomini 28, I-50132 Florence, Italy
clara.ugge@etaflorence.it

Vaknin, Yiftach
Department of Agronomy and Natural Resources, Institute of Plant Sciences, Agricultural Research Organization, the Volcani Center, P.O. Box 6, Bet Dagan 50250, Israel
yiftachv@volcani.agri.gov.il

Yermiyahu, Uri
Department of Soil Chemistry, Plant Nutrition and Microbiology, Institute of Soil, Water and Environmental Sciences, Gilat Research Center, Mobile Post Negev 85-280, Israel
uri4@volcani.agri.gov.il

Zenone, Terenzio
Department of Biology, University of Antwerp, 2610 Wilrijk, Belgium; and DOE-Great Lakes Bioenergy Research Center, Michigan State University, East Lansing, MI 48824, USA
terenzio.zenone@uantwerpen.be, zenone@msu.edu

Zhao, Kaiguang
School of Environment and Natural Resources, Ohio Agricultural Research and Development Center, The Ohio State University, Wooster, OH 44691, USA.
zhao.1423@osu.edu

List of Abbreviations

AB$_{wood}$	Aboveground woody biomass
AFEX	Ammonia fiber explosion
A$_{net}$	Net photosynthesis
ANPP	Aboveground net primary production
ANPP$_{wood}$	Aboveground wood net primary production
AWIFS	Advanced Wide Field Sensor
BA	Benzyladenine
BESS	Biofuel Energy System Simulator
BIG/CC	Biomass Integrated Gasification/Combined Cycle
BP	Backpressure
BREB	Bowen ratio energy balance
BTL	Biomass to liquid
C	Carbon
Ca	Calcium
Cd	Cadmium
CARB	California Air Resources Board
CDL	Cropland Data Layer
CDM	Clean development mechanism
CEST	Condensing/extraction steam turbine
CO$_2$	Carbon dioxide
CRA-CIN	Research Centre for Industrial Crops
CRP	Conservation Reserve Program
Cu	Copper
DBH	Diameter at the breast height
DEM	Digital elevation model
DGS	Distiller grains with solubles
DLUC	Direct land use change
DMC	Dry matter content
DOE	Department of Energy
EBAMM	Energy Resources Group's Biofuel Analysis MetaModel
ECCP	European Climate Change Program
ECe	Saturation extract
EISA	Energy Independence and Security Act of 2007
EPA	Environmental Protection Agency
EPIC	Environmental Policy Integrated Climate
EROI	Energy return on investment

ESP	Exchangeable sodium percent
ET	Evapotranspiration
ET_p	Potential evapotranspiration
ET_{us}	Understory evapotranspiration
EU	European Union
FAO	Food and Agriculture Organization
FAPRI	Food and Agricultural Policy Research Institute
FEAT	Farm Energy Analysis Tool
GDP	Gross domestic product
GEP	Gross ecosystem production
GHG	Greenhouse gas
GPP	Gross primary production
GREET	GHG, regulated emissions, and energy use in transportation
g_s	Stomatal conductance
GTAP	Global Trade Analysis Project
GWP	Global warming potential
HEP	Hydroelectric
IBGE	Instituto Brasileiro de Geografia e Estatìstica
ILUC	Indirect land use change
IPCC	Intergovernmental Panel on Climate Change
JCL	*Jatropha curcas* L.
K	Potassium
KBS	Kellogg Biological Station
LAI	Leaf area index
LCA	Life cycle assessment
LCC	Land Capability Classification
LCFS	Low Carbon Fuel Standard
LUC	Land use change
MAP	Mean annual precipitation
MAT	Mean annual temperature
MBC	Microbial biomass carbon
MBN	Microbial biomass nitrogen
MBP	Microbial biomass phosphorus
Mg	Megagram
N	Nitrogen
NASS	National Agricultural Statistical Services
NCCPI	National Commodity Crop Productivity Index
NEP	Net ecosystem production
N_2O	Nitrous oxide
NLDAS	North American Land Data Assimilation System
NPP	Net primary production
OLS	Ordinary least squares
ORNL	Oak Ridge National Laboratory
P	Phosphorus
PGRs	Plant growth regulators
PI	Production index

PNPB	National Program for the Production and Use of Biodiesel
P_2O_5	Phosphorus pentoxide
PP	Traditional poplar plantations distinguish from SRF
PPH	Poplar plantations under high input
PPL	Poplar plantations under low input
PV	Photovoltaic
RFP	Request for proposals
RFS2	Renewable Fuel Standard 2
RMSE	Root mean square error
SAR	Sodium adsorption ration
SB	Sugarcane bagasse
SEIMF	Spatially Explicit Modeling Framework
SEM	Scanning electronic microscopy
SOC	Soil organic carbon
SoS	System of systems
SPFIM	Surface planting and furrow irrigation method
SRC	Short-rotation coppice
SRF	Short rotation forestry
SS	Sugarcane straw
SSCF	Simultaneous saccharification and co-fermentation
SSF	Simultaneous saccharification and fermentation
T	Transpiration
USDA	US Department of Agriculture
USGS	United States Geological Survey
VOCs	Volatile organic compounds
WUE	Water use efficiency
WUE_{bfg}	Water use efficiency at the bioenergy cropping systems farm gate
WUE_{GPP}	Water use efficiency of gross primary productivity
WUE_i	Intrinsic water use efficiency
WUE_{inst}	Instantaneous water use efficiency
WUE_L	Long term water use efficiency
WUE_P	Water use efficiency of productivity
WUE_{ph}	Photosynthetic water use efficiency
Zn	Zinc
$\delta^{13}C$	Leaf carbon isotope composition

Preface

Developing sustainable biofuel energy systems has become a major effort of many governments worldwide as the demand for renewable energy increases. Many pressing issues concerning types of biofuel feedstock, genetic modification of the plants, land types and locations for selected systems, the biophysical regulation of production, resource limitations, and engineering advancement to convert biomass to ethanol are among the challenges in this regard. The knowledge generated by the scientific community in order to seek answers to the above challenges is already considerable. The idea here is not to summarize all of that knowledge but to provide a glimpse of the sustainability issues of some of the 2nd generation feedstocks by the experts in their respective fields. The scope of the book has been limited to biofuel production in the field where most of the environmental interactions and related impacts arise. We have attempted to address only some of the current research topics being used to assess the ecological sustainability of biofuel production systems for 2nd and 3rd generations. Our goal, however, was to cover most production models like perennial grasses, sugar-based feedstocks, forestry, and oilseed crops that are being perceived as frontrunners, so far, and are therefore being critically examined for their ecological and environmental impacts. The issues covered through these chapters also include materials on the socioeconomic and environmental linkages of different biofuel production systems and how these issues might hamper their long-term successful adoption. The contents of this book will also help the reader to bridge the gap of biofuel production systems from 1st generation to 2nd and 3rd generations.

The volume was revised in light of the reviewers' comments and we would like to take this opportunity to thank those voluntary reviewers for their assistance in providing useful suggestions to improve the quality of the volume. We are in debt to the following colleagues for their time and constructive suggestions and reviews: Michael Aspinwall, Ge Sun, Julie Sinistore, Werther G. Nissim, Bajrang Singh, Doug Reinemann, Kamlesh Nath, Yashpal Singh, Suresh K. Chaudhari, David Parrish, Sandeep Kumar, Poonam Jasrotia, and Lisa Delp Taylor. The editors' responsibility was to make a coherent text out of the chapter manuscripts, to ascertain that there were no major overlaps, and to take care of language and layout. Our lessons learned from research at the Great Lakes

Bioenergy Research Center (GLBRC) of the Michigan State University helped to stimulate this book. Last but not least, we thank all of the authors, as their profound knowledge and expertise have made it possible to conceptualize and produce this volume. We hope that this compilation will be informative as well as enjoyable for a broad audience. Though our main target groups are biologists, ecologists, and agriculturists, the book as a whole could be of use in courses of subjects related to resource managers, policy makers, and college students.

The Editors
September, 2013

Contents

Part I
Introduction

Chapter 1

The Sustainable Biofuels Paradigm

Ajay K. Bhardwaj, David J. Parrish, Jiquan Chen, Terenzio Zenone, and John H. Fike

1.1 Biofuels: Opportunities and Challenges

Major population increases, especially accompanied by rising standards of living in many developing nations, are inevitably leading to a heightened demand for energy. Meeting the needs of the 8.9 billion people predicted to inhabit planet Earth by 2050 [1] presents a multitude of challenges, particularly as their demands on natural resources worldwide are increasing with rising living standards. Our current pattern of energy extraction and usage, along with food and fiber production for this ever-growing population, affects the environment adversely; soil erosion, air and water pollution, deforestation, and loss of biodiversity, which are just some of the consequences. More ominously, global climate changes due to emissions of fossil carbon (C) as CO_2 and other greenhouse gases (GHGs) may present particularly challenging conditions [2] for future societies. In short, humanity's continued dependence on fossil fuels is not sustainable for the long term (or even for the short term), whether or not we have economically extractable supplies in the future [3–4]. Energy insecurity, fossil-fuel price volatility, related economic and geopolitical turmoil, and the greater issue of global climate change—all consequences of greater global demand for energy—seem to demand that we find alternative, renewable sources soon.

These issues of the resource scarcity, impacts of current sources and patterns of energy consumption are driving the quest for renewable, non-polluting energy resources. Hydropower, wind, solar, nuclear, geothermal, and similar energy options are already being explored. To varying degrees, they can provide long-term sources of heat and/or electricity based on renewable energy. However, these energy options do not notably produce liquid fuels, which are desirable for transportation. An approach that has begun to receive much attention is making liquid fuels from renewable, biological sources—so-called "biofuels".

Biofuels are energy sources derived from plants. As commonly used, the word denotes liquid fuels that are very valuable for transportation and can be fairly readily used as "drop-in" replacements for fossil fuels in the current fuel distribution infrastructure. Biomass can, of course, also be burned directly to produce heat and/or steam, a use often referred to as "biopower". Biofuels already in wide use include ethanol, made from several well-known crops such as maize (*Zea mays* L.), sugarcane (*Saccharum* spp.), sugar beet (*Beta vulgaris* L.), and sorghum (*Sorghum* spp.), and biodiesel, derived from oilseed crops such as rapeseed (*Brassica napus* L.), soybean (*Glycine max* L.), sunflower (*Helianthus annuus* L.), and oil palm (*Elaeis guineensis* Jacq.), as well as waste fats from the food industry. These readily obtained ethanol and oil sources constitute the so-called "1st generation" biofuels. Today, many other plant species with new technologies are in various stages of consideration for biofuel production (Fig. 1.1).

1.1.1   From Fossil Fuels to 1st Generation Biofuels

Using plant biomass for energy is not new, of course. Ever since the "discovery" of fire and the domestication of draft animals, wood and grasses have been used to provide energy for cooking, heating, transportation, and industry. Coal increasingly replaced wood in now-developed nations during the Industrial Revolution of the 18th and 19th centuries. Then, when internal combustion engines began to replace grass-fed draft animals in the late 19th and early 20th centuries, one of the first fuels used was ethanol. Among other things, the novel use of ethanol for fuel spurred production of "grain alcohol" and triggered the repeal of a $0.50 L^{-1} federal (USA) excise tax on it [5]. With time, however, abundant petroleum replaced fuel ethanol—just as coal had largely replaced wood—greatly facilitating human mobility and accelerating the industrial revolution.

Fossil fuels are more energy-rich than an equivalent mass of grass or wood (or ethanol) and their easily dispensed liquid forms are particularly favored for transportation. However, as noted, fossil fuels are non-renewable and their reserves are finite. Policymakers and the public are becoming increasingly aware of the unintended consequences of fossil fuel use. In attempts to mitigate fossil fuel use, some governments have moved to institutionalize the use of biofuels for a portion of their energy portfolios [6]. In an interesting repeat of history and technology, the fuel that has initially received the most attention is ethanol. Several nations have promoted and incentivized production of ethanol from grains, especially maize and sugar-rich vegetative materials such as sugarcane and sugar

Fig. 1.1 Some of the common 1st and 2nd generation biofuel crops and feed-stocks. (a) Maize (*Zea mays*), (b) Soybean (*Glycine max*), (c) Sugarcane (*Saccharum officinarum*), (d) Sorghum (*Sorghum bicolor*), (e) Agricultural crop residues, (f) Jatropha (*Jatropha curcas*), (g) Switchgrass (*Panicum virgatum*), (h) Miscanthus (*Miscanthus × giganteus*), (i) CRP Grasses, (j) Native forest, (k) Pine (*Pinus* spp.), (l) Hybrid poplar (*Populus* spp.).

beet. The technology for converting starches and sugars to ethanol is straightforward and ancient. Close to 40% of the USA's maize output is now devoted to ethanol production [7] and that percentage is expected to increase in response to mandates of the US Renewable Fuel Standard promulgated by the Energy Policy Act of 2005 and expanded by the Energy Independence and Security Act of 2007 [8].

However, much controversy surrounds the 1st generation biofuels. Some analysts suggest that nearly as much (or even more) energy may be expended in producing ethanol from maize grain as is available in the product [9–12]. Some more favorable analyses still suggest that the energy-in/energy-out ratio is much better with other potential biofuel feedstocks [10, 13]. Whether using fuel made from maize grain reduces GHG emissions is a matter of even greater controversy, primarily because of concerns about land use changes and consequences for soil C [14]. Furthermore, the conventional production of maize (for any purpose) poses some risks for soil and water resources [13, 15]. Expanded cultivation of sugarcane and oil palm for energy has raised similar concerns about land and water resources in locales where those species are grown.

Finally, ethical concerns are raised when grains or other food crops are used to make fuel—driving up food costs, especially within the developing world [13]. Concerns about food security grow when such feedstocks are used. In short, the 1st generation biofuels may be providing a valuable service in "priming the pump" for developing plant-derived drop-in fuels, but they may not provide long-term solutions—even for that small portion of our energy portfolio that they now provide.

1.1.2   A Case for 2nd and 3rd Generation Biofuels

1.1.2.1   Terminology and history

Instead of grains or other food components of plants, the whole plant's biomass (usually implying only above-ground dry matter) can serve as a starting point, or "feedstock", for producing biofuels. The so-called "lignocellulosic"—or sometimes simply "cellulosic"—or "biomass-derived" biofuels include "bioethanol", or "cellulosic ethanol", and other liquid, gaseous, and solid fuel forms. The terms introduced in quotation marks are in wide usage but are not necessarily well-defined or universally understood. "Bioethanol", for instance, could be a confusing coinage, since ethanol is commonly made from biological sources, whether it is to be used for fuel or other purposes. In this case, though, the "bio-" is presumably borrowed from "biomass", as the bioethanol term is most commonly used to describe ethanol derived from lignocellulosic materials.

"Bioethanol" made by processing lignocellulosic feedstocks is often referred to as a 2nd generation biofuel. Feedstocks can include plants grown specifically for this purpose, sometimes described as "dedicated energy crops". However, they can also include wastes or residues from agricultural and forestry operations as well as municipal solid waste and wastes from the food industry. The most frequently cited examples of lignocellulosic energy crops are perennial grasses, *e.g.*, switchgrass (*Panicum virgatum* L.) and miscanthus (*Miscanthus × giganteus*), and woody species, such as hybrid poplar (*Populus* spp.), eucalyptus (*Eucalyptus* spp.), and willow (*Salix* spp.). Energy crops are typically fast-growing, with high energy outputs per unit of time and area. Highly desirable traits in biofuel crops would include consistency of moisture, energy content, and yield per unit area [16]. Energy crops with great potential over a wide geographical range would also be desirable.

The technology for making ethanol from lignocellulosic biomass is not new. Wartime and energy "crises" have periodically spurred the production of ethanol and other fuels and chemicals from plant biomass, but the production has inevitably waned when petroleum availability and prices return to pre-war/pre-crisis levels. Higher oil prices are renewing the interest in lignocellulose-derived biofuels, but the concern for global climate changes associated with fossil fuel use is an even bigger driver for some [17].

Some consider cellulosic ethanol to be the 2nd generation biofuel and describe other biofuels as "3rd generation" or "advanced". Included in this advanced category are fatty acid methyl esters and other petroleum-like hydrocarbons that can be made by using biomass as their starting point. Some of these biofuel forms may factor favorably when solving the calculus of sustainability. However, we consider the kinetics and energetics of advanced biofuel technologies, which are undoubtedly important in determining their sustainability [18] to be beyond the scope of our analysis. Rather, we will focus largely on the widely discussed and described 2nd generation systems, or "biorefineries", where ethanol would be derived from whole-plant biomass.

1.1.2.2 Biorefinery/conversion technologies

As of the completion of this study, neither 2nd nor 3rd generation biorefineries have become proven enterprises, *i.e.*, commercially successful. Several proof-of-concept and/or pilot-scale facilities have produced ethanol and "bio-oil" from biomass feedstocks using various processes and some commercial-scale facilities are currently coming on line. While there are many promising ventures afoot—and some more may come on line before this is read—we cannot yet point to any facility that has been proven to use lignocellulosic biomass to produce drop-in fuels at prices competitive with fossil fuels.

Several different conversion technologies are being deployed in the 2nd (and 3rd) generation biorefineries that are under construction. However, we do not know what chemical processes and what efficiencies will characterize the first successful biorefineries. We can, however, generalize the technologies proposed and/or known to be in various stages of development [17, 19–20]. These include *biochemical* and/or "wet chemistry" processes, in which cellulose and hemicellulose are hydrolyzed, releasing 5- and 6-C sugars that can then be converted to ethanol or other products. A second general approach involves *thermochemical* conversions, in which biomass is exposed to high temperatures under O_2-limited conditions. Depending on the reaction type (*e.g.*, pyrolysis or gasification), the products of thermochemical processes and further steps can include bio-oil, liquid fuels (including ethanol and octane/olefins), combustible gases (H_2 and CH_4), and biochar. *Hydrothermal* processes are also drawing some favorable attention as a perhaps more efficient—and therefore more sustainable—biorefinery technology. In such systems, wet biomass would be put under high temperatures (300+°C) and pressures (200+ bars), creating conditions where water is both a solvent and a reactant and producing a range of fuel types [19].

Although no one knows which of the various technologies for converting biomass to biofuels will prove itself in the marketplace, the technology choice will likely affect whether or to what degree the system is sustainable. In a review of life cycle assessments (LCAs) evaluating several crops and systems, Fazio and Monti [21] ranked technologies in terms of their overall environmental impacts. The 1st generation systems were less desirable than the 2nd generation and, within the 2nd generation, thermochemical processing was deemed superior to biochemical in overall environmental impacts. Gaunt *et al.* [22] provided a closer look at thermochemical technologies and found very significant reductions in net CO_2 emissions when biomass was processed using "slow pyrolysis" to produce biochar—along with biofuels—and the biochar was land-applied. Again, these technologies are unproven in the marketplace. Accordingly the LCAs were limited in their ability to assess the systems employing them. Even if one of them might be more environmentally friendly, it might not be scalable or otherwise feasible as an enterprise.

1.1.2.3 Producing feedstocks for 2nd and 3rd generation biorefineries

Sustainability has been an important driver for the push to the 2nd and 3rd generation feedstocks both because of growing concerns over food and environmental security if these feedstocks are used. However, producing sufficient biomass for a new industry without adversely affecting food provisioning could prove to be a substantial challenge, even with non-food feedstocks [23]. Many have supposed that to achieve the economics of scale for successful bioenergy production will require development of multimillion-liter-per-year facilities processing perhaps hundreds of tons of biomass per day [24]. Others have argued that smaller, distributed process systems may eventually become more efficient since a larger processing facility requires more feedstock supply, land area, and enhanced transportation costs [25]. Regardless of the coming industry's nature or form, millions of hectares will be needed to grow the feedstock that will feed this new beast.

Given these needs, the reason for the interest in marginally productive lands for energy cropping is clear. Marginal lands could provide substantial opportunity, especially for the cellulosic biofuel crop production [26–28]. However, marginal lands are often environmentally susceptible to erosion and other degradation. It has been suggested that land set aside for conservation, such as the Conservation Reserve Program (CRP) lands in the USA, could be used to grow bioenergy grasses and woody plantations. These lands will likely require lower levels of fertilization and might assimilate a large amount of C [29]. Others argued that the cost of conversion in terms of C debt could be significant and, therefore, management choices will play important roles [27]. Erosion control, carbon (C) sequestration, and land reclamation could be some of the benefits directly associated with using such areas for biofuel feedstock production, along with achieving economic and energy benefits. Achieving conservation and erosion control by planting fast-growing, deep-rooted grasses or trees on marginal lands is not a new strategy; harvesting those grasses and trees for energy production could be one.

Waste materials such as crop and agro-industrial residues are also being explored as feedstocks for making biofuels. Often referred to as "opportunity fuels", these potential feedstocks are usually discarded. Worldwide, more than 300 million hectares are used for crop production [30]. The harvests of many of these crops leave residues (stalks, leaves, cobs, etc.), which are either disposed of offsite or left in the field, sometimes to be burned (Fig. 1.2). Much of this crop residue could be collected and used as an energy feedstock. Putatively sustainable removal rates of crop residues based on region, crop constitution (C:N ratio), and management might not adversely affect soil quality. However, using even a portion and retaining the rest as soil amendment and for conservation purposes has been labeled unsustainable in many studies [31].

Similar to crop residues, residues from the timber industry are of increasing interest for biofuels. After harvesting trees for timber, residues are typically either left in the forest or burned. Approximately 2.1 Mg of forest residues are available for each 28.3 m^3 of harvested timber in the western USA [16]. Woody wastes, such as residues from sawmills and the paper industry, can also provide feedstock for biofuels. Wood waste can be pelleted easily and, therefore, also offers a uniform and desirable source of energy for biopower and home heating.

Fig. 1.2 Crop residues being burned in the field to facilitate farming operations for the following season in Uttar Pradesh, India.

Compressed pellets that are uniform in size and shape have higher energy density and high calorific value. Because of this utility, markets already exist for many of these wood milling byproducts and the growth of the biofuels industries will likely create greater demand and competition for them.

In this section, we have noted some of the challenges for growing energy crops and for harvesting crop and timber residues and byproducts. However, several assessments [29, 32–34] indicate that there is significant potential to produce biofuels without affecting global food production. Such systems, particularly if paired with locally owned biorefining facilities, offer the strong potential to boost rural economies and, if carefully managed, provide positive outcomes in terms of environmental sustainability.

1.2   The Sustainability Paradigm and Biofuels

Stewart Potter, US Supreme Court Justice, famously wrote in a judicial decision that he could not adequately define pornography, stating "...but I know it when

I see it." [35]. When the topic is sustainable agriculture, many have difficulties both defining it and identifying examples of it. The phrase "sustainable agriculture" has been attributed to Wes Jackson [36–37]. In the 1990s, USDA, the US "Farm Bill", described the sustainable agriculture as a series of sociological, ecological, and economic outcomes [36]; yet, it set no standards for their measurement—or achievement. More recent attempts to define and set metrics on agricultural sustainability have come from multinational, national, and non-governmental organizations. However, none of those efforts have produced a truly consensus document.

The question of what constitutes agricultural sustainability becomes even more complicated when agriculture is asked to produce energy for transportation or electricity, especially when dealing with some of the sociological and ethical aspects of sustainability (*i.e.*, human perspectives) [38]. Many individuals and groups have worked to develop consensus definitions and metrics for sustainable energy crops [39–44], with some efforts being codified. Nations within the European Union, which has worked to institutionalize biofuel production for at least two decades, have been particular leaders in seeking sustainable production standards [45] and in developing certification methods and metrics [46–48].

When we cipher through much of the verbiage about the sustainability of production systems, we find three "universals" or fundamental attributes: economic, societal, and environmental. The sustainable production of biofuels, like any other agricultural system, demands addressing essential issues under these components. To be truly sustainable, a production system must perform well in all three categories. In that vein, "truly sustainable" agricultural systems should be characterized by their ability to be practiced essentially *ad infinitum*—to be employed for centuries without loss of productivity or diminution of life-supporting ecosystem services [49, 50]. They would adhere to a Hippocratic, first-do-no-harm standard (*i.e.*, they would preserve ecosystem health and functions). That is a very high standard and we unfortunately have to agree with [36] that agriculture, as widely practiced, often does not "measure up". Many "modern" or "industrial" agricultural production systems are consuming—not preserving—resources. We are using key finite resources, such as nutrients and fossil fuels, and slowly renewable resources, such as soil and groundwater, at clearly unsustainable rates.

Several environmental, social, and economic questions loom around biofuel production, including: Do we have enough lands to produce enough biofuels? What land covers should be converted to the bioenergy systems? Are biofuels going to help solve energy problems without creating new ones, especially environmental and social ones? Are these biorefinery systems economically viable? What are the "hidden costs"? Will biofuels help mitigate the climate dilemma only to create other ecological problems? Consequences of biofuel production systems, if significant, cannot be ignored and should be considered at an early planning stage. All biofuels and production systems are not equal in energy benefits or in environmental impacts.

A number of pressing issues related to biofuel production and its environmental footprint have been identified and examined. These include changes in surface and ground water quality and quantity, soil conservation, greenhouse gas (GHG) emissions, wildlife habitats, and species diversity, all of which are common challenges for intensive agriculture. Continuously grown crops are more susceptible to insect damage and can promote persistence of weeds. Use of nitrogen (N) fertilizers has been a principle contributor to the pollution of ground and surface water and a source of N_2O, which is a major GHG. The impact of N_2O is molecule-for-molecule over 300 times CO_2 in warming the atmosphere. Use of fertilizers to promote growth of bioenergy crops can produce the same outcomes, if efficient nutrient management strategies are not designed or implemented. Perhaps the use of these external inputs (fertilizer, water) will be more in bioenergy crops since the most probable areas for biofuel development (feedstock production) will be marginal and degraded lands [26, 51] to avoid competition with food crops for land. Broadening the natural resource base to reduce use of external inputs (fertilizers, pesticides, etc.) by efficient management of on-farm resources and optimum selection of cropping system practices could lead to better sustainability (Fig. 1.3).

Marginal lands could have higher demand for nutrients and water if pressed into use for energy crops [26] but management strategies would play a key role in maintaining the balance [27]. Of course, not using marginal lands is one option, but where is the land to be found when we have to grow food as well as fuel? Many marginal lands, such as the Conservation Reserve Program (CRP) lands in the USA, were under agriculture at some point but were either less productive or erodible. When converted to the CRP, these areas are planted with perennial grasses. Simply converting them to highly productive perennial grasses and harvesting in phases, spatially and temporally, might serve the dual purpose of conservation as well as biofuel crop production. Efficient agricultural technology based on sound ecological principles is required.

Impacts on biogeochemical cycles have been at the centers of investigations into biofuel production. Emissions of GHGs, especially, have being the center stage because of climate change concerns. A cropping system that reduces GHG emissions will be highly desirable, but other cross-cutting issues need to be weighed based on local and regional priorities (Fig. 1.4). For example, changes in plant populations, structures, and hierarchies due to changes in land use could cause changes in biogeochemical cycles. Increased evapotranspiration (ET) might reduce runoff and seepage, affecting hydrological flows. The altered hydrological flow might affect vital ecosystems in streams and rivers. Likewise, changes in the soil water profile due to changes in crops and cropping systems could also have long-term effects on soil fauna and flora.

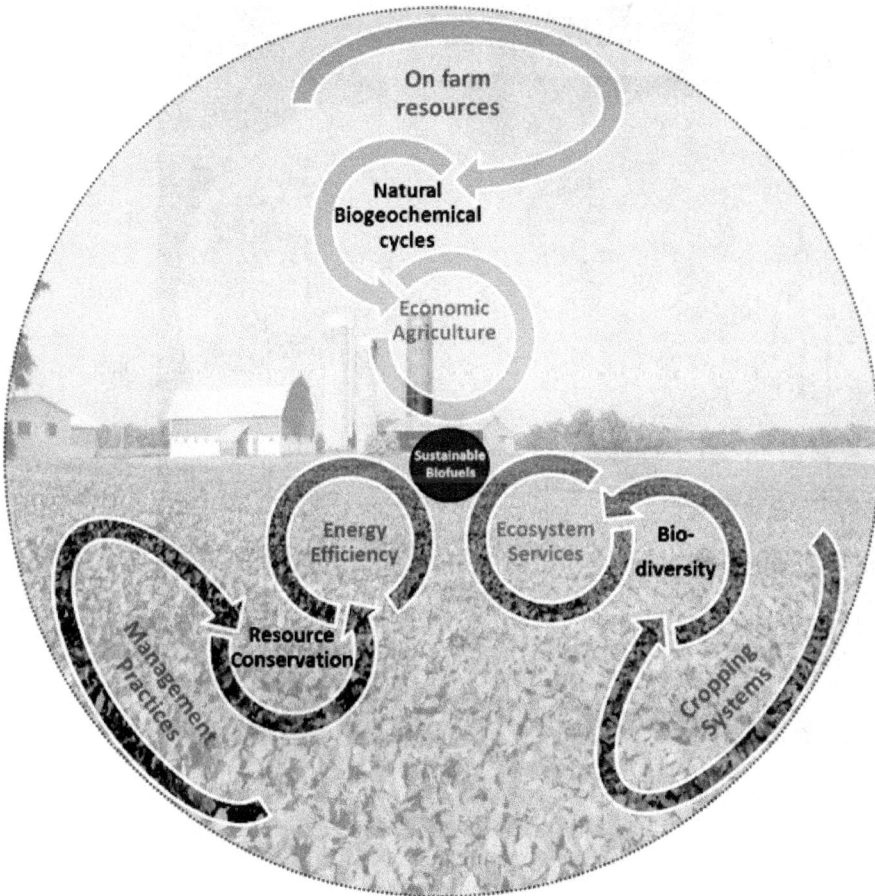

Fig. 1.3 Integration of natural biogeochemical cycles with efficient management of on-farm resources and optimum selection of cropping system practices can expand (and help conserve) the natural resource base for sustainable biofuel production.

Given the potential tradeoffs among the three essential elements of sustainability, it is important to consider whether we can actually hope to meet all of the sustainability criteria perfectly. If we cannot, what is an acceptable level of "failure" for each of the three sustainability elements? There is a need to balance the key socioeconomic drivers with ecological and environmental soundness. Thus, a better goal may be to get our biofuel production systems as close as possible to realistic, sustainable dimensions of "economically feasible", "socially beneficial", and "environmentally sound". By definition, long-term implications of new/altered production systems might be observed only after extended periods of time. A "wait and watch" strategy does not seem wise, since the potential for unintended consequences seems great. A definitive action based on creative

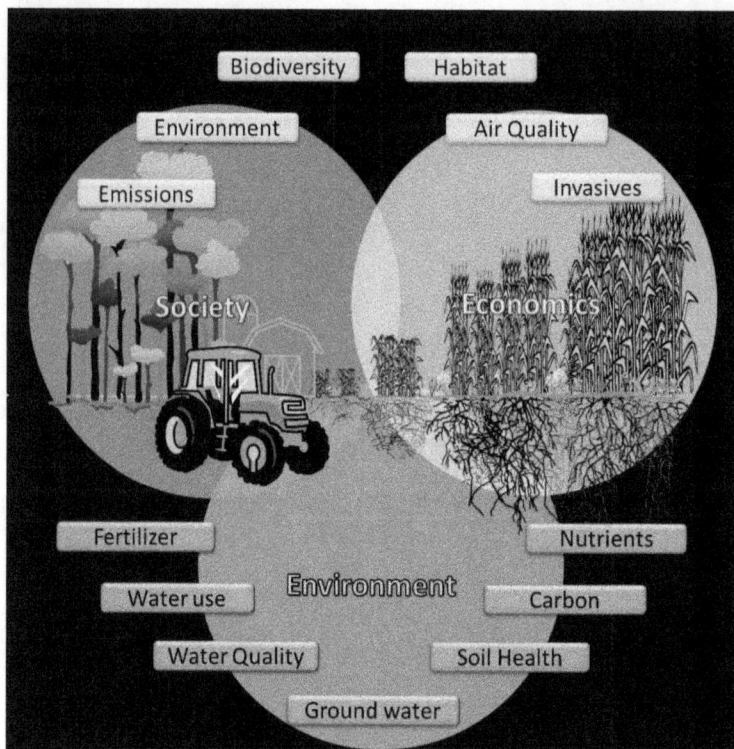

Fig. 1.4 Components in a sustainability paradigm for crop production systems. Energy cropping systems must also prove themselves to be sustainable by having suitably positive energy returns on the energy invested. Adapted from [52].

solutions to reduce environmental impacts with sustainable production will help lead us toward an energy-sufficient future. Lasting solutions must be rooted in multi-pronged approaches to achieve socioeconomic as well as environmental sustainability.

We contend that a strong case can be made for the assertion that current intensive agricultural practices—and, indeed, many other aspects of industry and society in general—are not sustainable. The resource limitations and environmental impacts associated with continued fossil energy consumption will create increasing challenges for meeting the needs of Earth's nine billion human inhabitants expected by 2050. New paradigms must be developed where agriculture is a net energy producer and a better guardian—rather than a consumer and degrader—of environmental resources. This book highlights how varied bioenergy systems—if we get them right—offer the best chances to provide positive economic, environmental, and social outcomes while meeting the needs of future societies.

References

[1] World Population to 2300. United Nations, New York, NY 10017, USA. United Nations 2004. (Accessed on Jan 13, 2013 at: www.un.org/esa/population/publication/long range 2/worldpop2300final.pdf).

[2] Contribution of Working Groups I, II and III to the Fourth Assessment Report of the Intergovernmental Panel on Climate Change [Core Writing Team: Pachauri RK, Reisinger A (Eds.)]. IPCC. Climate Change 2007: Synthesis Report. IPCC Geneva Switzerland 2007.

[3] Biello D. 2012. Has petroleum production peaked, ending the era of easy oil? Scientific American. (Accessed on Jan 31, 2013, at http://www.scientificamerican. com/artical.cfm?id=has-peak-oil-already-happened).

[4] Hirsch RL. Packing of World Oil Production: Recent Forecasts. A report by DDE National Energy Technology Laboratory, Morgantown, WV 26507, USA, 2007.

[5] Payne WA. Are biofuels antithetic to longterm sustainbility of soil and water resources? Advances in Agronomy 2010; 105: 1–46.

[6] IPCC. IPCC Special Report on Renewable Energy Sources and Climate Change Mitigation. Prepared by Working Group III of the Intergovernmental Panel on Climate Change [Edenhofer O, Pichs-Madruga R, Sokona Y K, Seyboth K, Matschoss P, Kadner S, Zwickel T, Eickemeier P, Hansen G, Schlömer S, von Stechow C (Eds.)]. Cambridge University Press, Cambridge, United Kingdom and New York, NY USA, 2011.

[7] Bioenergy Statistics. USDA ERS (Accessed on March 25, 2013 at: http://www. ers.usda.gov/data-products/us-bioenergy-statistics.aspx).

[8] Alternative Fuels Data Center, US Department of Energy. USDOE ADFC. (Accessed on March 25, March 2013 at: http://www.afdc.energy.gov/laws/RFS).

[9] Pimentel D. Ethanol fuels: energy security economics and the environment. Journal of Agricultural and Environmental Ethics 1991; 4: 1–13.

[10] Hall CAS, Dale BE, Pimentel D. Seeking to understand the reasons for differnent energy return on investment (EROI) estimates for biofuels. Sustainability 2011; 3: 2413–2432.

[11] Mamani-Pati F, Clay DE, Carlson G, Clay SA. Production economic and energy life cycle analysis can produce contrary results for corn used in ethanol production. Journal of Plant Nutrition 2011; 34: 1278–1289.

[12] Zhang J, Palmer S, Pimentel D. Energy production from corn. Environment Development and Sustainability 2012; 14: 221–223.

[13] Payne WA. Are biofuels antithetic to long-term sustainability of soil and water resources? Advances in Agronomy 2010; 105: 1–46.

[14] Searchinger T, Heimlich R, Houghton RA, et al. Use of US croplands for biofuels increases greenhouse gases through emissions from land-use change. Science (Washington) 2008; 319: 1238–1240.

[15] Wright CK, Wimberly MC. Recent land use change in the Western Corn Belt threatens grasslands and wetlands. Proceedings of the National Academy of Science (USA) 2013;110:4134–4139.

[16] Biomass Combined Heat and Power Catalog of Technologies: U.S. EPA, Washington, DC, USA. US EPA, 2007. (Accessed on Jan 10, 2013 at: www.epa.gov/chp/documents/biomass_chp_catalog.pdf).

[17] Sims REH, Mabee W, Saddler JN, Taylor M. An overview of second generation biofuel technologies. Bioresour Technol. 2010; 101: 1570–1580.

[18] Cruse RM, Herndl CG, Polush EY, Shelley MC. An assessment of cellulosic ethanol industry sustainability based on industry configurations. Journal of Soil and Water Conservation (Ankeny) 2012; 67: 67–74.

[19] Balan V, Kumar S, Bals B, Chundawat S, Jin M, Dale B. Biochemical and thermochemical conversion of switchgrass to biofuels. In: Monti A (Ed.) Switchgrass: A Valuable Biomass Crop for Energy. Springer 2012; 153–185.

[20] Drapcho CM, Nhuan NP, Walker TH. Biofuels Engineering Process Technology. McGraw-Hill Inc, New York, 2008.

[21] Fazio S, Monti A. Life cycle assessment of different bioenergy production systems including perennial and annual crops. Biomass and Bioenergy 2011; 35: 4868–4878.

[22] Gaunt JL, Lehmann J. Energy balance and emissions associated with biochar sequestration and pyrolysis bioenergy production. Environmental Science and Technology 2008; 42: 4152–4158.

[23] Fike JH, Parrish DJ, Alwang J, Cundiff JS. Challenges for deploying dedicated, large-scale, bioenergy systems in the USA. CAB Reviews: Perspectives in Agriculture, Veterinary Science, Nutrition and Natural Resources 2007; 2: 28.

[24] Cundiff JS, Fike JH, Parrish DJ, Alwang J. Logistic constraints in developing dedicated large-scale bioenergy systems in the Southeastern United States. Journal of Environmental Engineering 2009; 135: 1086–1096.

[25] INEEL Bioenergy Initiative. INEEL External Website Document. INL 2001. (Accessed on July 22, 2013 at: www.inl.gov/bioenergy/docs/bioenergy-strategic-plan.pdf).

[26] Bhardwaj AK, Zenone T, Jasrotia P, Robertson GP, Chen J, Hamilton SK. Water and energy footprints of bioenergy production on marginal lands. Global Change Biology Bioenergy 2011; 3: 208–222.

[27] Zenone T, Chen J, Deal M, et al. CO_2 fluxes of transitional bioenergy crops: effect of land conversion during the first year of cultivation. Global Change Biology Bioenergy 2011; 3: 401–412.

[28] Gelfand I, Sahajpal R, Zhang X, Izaurralde RC, Gross KL, Robertson GP. Sustainable bioenergy production from marginal lands in the US Midwest. Nature 2013; 493: 514–517.

[29] Tilman D, Jason H, Clarence L. Carbon negative biofuel from low input high diversity grassland biomass. Science 2006; 314: 1598–1600.

[30] US Department of Agriculture, 2013. World agricultural production. Circular Series. USDA Foreign Agricultural Service (Accessed on Feb. 9, 2013 at: http://www.fas.usda.gov/psdonline/circulars/production.pdf).

[31] Lal R. World crop residues production and implications of its use as a biofuel. Environment International 2005; 31: 575–584.

[32] Robertson GP, Dale VH, Doering OC, et al. Sustainable biofuels redux. Science 2008; 322: 49–50.

[33] Eisentraut A. Sustainable Production of Second Generation Biofuels: Potential and Perspective in Major Economies and Developing Countries. Information paper. International Energy Agency, Paris France, 2010.

[34] Biomass supply for a bioenergy and bioproducts industry. Perlack RD and Stokes LB (Leads), ORNL/TM-2011/224. US Department of Energy. Billion ton update: Oak Ridge National Laboratory, Oak Ridge, TN, 2011; 227.

[35] Gerwitz P. On "I know it when I see it." Yale Law Journal 1996; 105: 1023–1047.

[36] Gomiero T, Pimentel D, Paoletti MG. Is there a need for a more sustainable agriculture? Critical Reviews in Plant Sciences 2011; 30: 6–23.

[37] Jackson W. New Roots for Agriculture. University of Nebraska Press, 1980.

[38] Dufey A, Grieg-Gran M. (Eds.) Biofuels Production, Trade and Sustainable Development. International Institute for Environment and Development, London, 2010.

[39] Buchholz T, Luzadis VA, Volk TA. Sustainability criteria for bioenergy systems: results from an expert survey. Journal of Cleaner Production 2009; 17: S86-S98.

[40] Lora EES, Palacio JCE, Rocha MH, Reno MLG, Venturini OJ, del Olmo OA. Issues to consider existing tools and constraints in biofuels sustainability assessments. Energy 2011; 36: 2097–2110.

[41] Paine LK, Peterson TL, Undersander DJ, et al. Some ecological and socio-economic considerations for biomass energy crop production. Biomass & Bioenergy 1996; 10: 231–242.

[42] Dale VH, Efroymson RA, Kline KL, Langholtz MH, Leiby PN, Oladosu GA, Davis MR, Downing ME, Hilliard MR. Indicators for assessing socioeconomic sustainability of bioenergy systems: a short list of practical measures. Ecological Indicators 2013; 26: 87–102.

[43] Solomon BD, Limburg K, Costanza R. Biofuels and sustainability. Annals of the New York Academy of Sciences 2010; 1185: 119–134.

[44] Khanna M, Hochman G, Rajagopal D, Sexton S, Zilberman D. Sustainability of food energy and environment with biofuels. CAB Reviews: Perspectives in Agriculture Veterinary Science Nutrition and Natural Resources 2009; 4(028): 1–10.

[45] Certification of sustainable biofuels. European Biofuels Technology Platform (Accessed on March 20, 2013 at http://www.biofuelstp.eu/certification.html).

[46] Lewandowski I, Faaij APC. Steps towards the development of a certification system for sustainable bio-energy trade. Biomass and Bioenergy 2006; 30: 83–104.

[47] Van Dam J, Junginger M, Faaij A, Jurgens I, Best G, Fritsche U. Overview of recent developments in sustainable biomass certification. Biomass and Bioenergy 2008; 32: 749–780.

[48] Ahlgren S, Röös E, Di Lucia L, Sundberg C, Hansson PA. EU sustainability criteria for biofuels: uncertainties in GHG emissions from cultivation. Biofuels 2012; 3: 399–411.

[49] Fike JH, Parrish DJ. Switchgrass. In: Singh BP (Ed.) Biofuel Crops: Production, Physiology and Genetics. CAB International 2013; 199–230.

[50] Fike JH, Parrish DJ, Fike WB. Sustainable cellulosic grass crop production. In: Singh BP (Ed.). Biofuel Crop Sustainability. John Wiley and Sons, Hoboken, NJ, USA, 2013.

[51] Spiertz H. Challenges for crop production research in improving land use, productivity and sustainability. Sustainability 2013; 5: 1632–1644.

[52] Adams WM. The Future of Sustainability: Re-thinking Environment and Development in the Twenty-first Century. Report of the IUCN Renowned Thinkers Meeting, 29–31 January, 2006. (Accessed on May 25, 2013 at: http://cmsdata. iucn.org/downloads/iucn_future_of_sustanability.pdf).

Part II
Biofuel Crop Models

Chapter 2

Switchgrass for Bioenergy: Agro-ecological Sustainability

David J. Parrish and John H. Fike

2.1 Introduction

2.1.1 Switchgrass—A Short History of and the Case for Its Use as a Biofuel Feedstock

Switchgrass has perhaps gained more attention than any other species among the suite of potential perennial herbaceous feedstock "contenders". In the span of about two decades, switchgrass has gone from relative scientific obscurity to becoming the "poster child" for bioenergy. It raises the question that a common prairie grass of little distinction could so rapidly gain wide attention as a potential feedstock for biorefineries: How did it achieve such status? We have discussed the history of switchgrass as a crop and as a biofuel feedstock in some detail elsewhere [1, 2], and we shall summarize those accounts here.

"In the beginning", switchgrass was just one of the native species commonly found on the tall-grass prairies of North America. It was not considered as a crop per se, and it received scant scientific attention through the first half of the 20th century (Fig. 2.1). The first reports of it being grown in monocultures came in the 1940s; the average number of research publications dealing specifically with switchgrass was only about six per year going into the 1980s. However, there has been an explosion of interest (as indicated by the indexed citations) in the species as an energy crop since then.

In 1984, the US Department of Energy (DOE), through its Oak Ridge National Laboratory (ORNL), issued a request for proposals (RFP) to screen herbaceous, lignocellulosic species as energy crops [3]. Seven subcontracts were awarded,

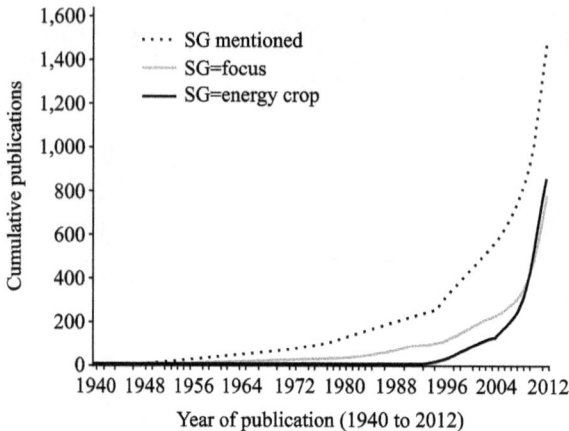

Fig. 2.1 A graphical history of publications that mention switchgrass as captured from the CAB Direct indexing service of CAB Abstracts. The upper curve is the cumulative total of publications with "switchgrass" appearing in any indexing field. The middle curve tallies publications where switchgrass was a major focus of the work. The lower curve, which exceeds the middle one beginning in 2010, is the total number of publications discussing switchgrass as an energy crop. Essentially, all of the increase in the last 10 years can be attributed to bioenergy studies.

representing seven eastern and central states in the USA. Each of the subcontractors proposed a region-appropriate list of species to be screened, but there was no benchmark species to allow cross-region comparisons of biomass productivity. At a meeting with subcontractors, it was determined that switchgrass would be included in all studies based largely on its known wide adaption.

When the final reports of the five-year DOE screening studies were compiled, switchgrass proved to be one of the best biomass producers across all seven states. As a result, subsequent DOE RFPs called for work on switchgrass as a "model" biofuel species [3–4]. A second and then a third five-year round of DOE/ORNL-funded subcontract work included long-term cultivar and management studies as well as breeding and tissue culture work [4].

DOE funding for extramural switchgrass research largely ended in 2002 [4]. Thereafter, the US Department of Agriculture (USDA) began slowly and then more vigorously to assume leadership of energy cropping work to include developing its Bioenergy and Energy Alternatives program that has spawned major studies with switchgrass [1]. Also stepping into the biofuels arena increasingly in the first decade of the 21st century was the private sector. Major petroleum and chemical companies have invested in biofuels research, as have traditional members of the agricultural sector; a number of new companies are developing the bioenergy potential of switchgrass and other species. Over this same time period, interest in switchgrass has "gone global". In a tally taken in 2005, the species had been investigated for its energy production potential in 11 countries [5]. The list now stands at more than 20 [1].

Switchgrass has moved in just a few decades from being an obscure prairie grass to being one of the most widely cited energy crop candidates. After a limited screening effort, it was identified as a model species and received significant, well-orchestrated funding from DOE for 10 years, at a time when little work was being done on any other herbaceous biomass species. Some other species are now beginning to have significant amounts of effort devoted to their study, but the lead of switchgrass on the field has been substantial. At a minimum, it continues to be used as the reference species against which other energy-crop candidates are often compared. As we elaborate below, some salient features of the species (productivity, adaptation, stress tolerance, etc.) make it a very good candidate as a biomass producer [4, 5]. As with any potential energy crop, though, we must ask, "Would its use be sustainable?"

2.2 Energetic and Economic Considerations in Sustainability

2.2.1 Energy In: Energy Out (Is Making Biofuel from Switchgrass Energetically Feasible?)

Calculating the energy balance of biofuel production requires models based on a comprehensive analysis of system inputs and outputs. Typically, life cycle assessment (LCA) models are used. LCAs assign values to the various components of the system being studied in order to interpret their "goodness of fit" in terms of environmental effects. In the case of bioenergy systems, LCAs are used to model energy use, GHG emissions, sequestered C, water balance, and land use change among other impacts. Here, we interject an observation attributed to statistician George Cox, "essentially, all models are wrong, but some are useful". Thus, conclusions about bioenergy system sustainability based on LCA models will be strongly affected by (sometimes seemingly arbitrary) decisions regarding system boundaries, crop production inputs, logistics, biorefinery efficiencies, end products, "byproducts" that can have significant economic and environmental implications, and product uses.

System boundaries vary widely among LCA studies, and published analyses span the range from "well-to-tank"—or "field to tank"—to "cradle-to-grave" [6]. Such divergent scopes arise from the fact that LCA studies are often conducted for very different reasons—thus making useful comparisons of LCA studies quite challenging. The assessments that are most comprehensive in terms of system boundaries—and thus more useful as evaluation tools—often minimize differences among fuels, which is often the issue of greatest interest [7]. Useful comparisons of LCAs are further hampered by the lack of uniform unit currencies among studies. For example, the functional units of system outputs in various studies have included fuel energy, fuel weight, and fuel output per unit area; these different "accounting" practices can produce large differences in assessment outcomes [6].

When LCAs are used to model the energy balance of bioenergy systems, their output is often reported as the "energy return on investment" (EROI), a ratio of the energy produced relative to the energy invested in its production. Conceptually computing an EROI is simple, but capturing and assessing all of the inputs and outputs and finding agreement about assumptions and boundaries make the task daunting.

The preponderance of this section's discussion of energy balance will deal with cellulosic ethanol production systems. This is not because of bias toward or preference for biochemical processes. Rather, it reflects the dearth of EROI data on thermochemical or hydrothermal biorefinery technologies (see Chapter 1 for expanded discussion of likely technologies for converting biomass to biofuels). Some argue that bioenergy systems that burn biomass for heat are more efficient and have lower GHG emissions than biorefinery systems [8–9]. However, again, the supporting data are limited. Developing liquid fuels that can be "dropped into" the existing fuel distribution infrastructure is a major driver for deploying biomass-to-biofuel systems. "Bioethanol" from maize has been proven as a drop-in fuel, clearing the way—or perhaps priming the pump—for cellulosic ethanol.

Calculating the EROI requires estimating energy embodied in inputs and outputs for feedstock species, fertilizer requirements and sources, harvest systems, hauling, storage, and conversion processes as well as products and byproducts (and their end uses). The array of system components, their potential configurations, and their interactions throughout bioenergy supply, production, and delivery make difficult direct comparisons of different bioenergy systems' EROIs. Moreover, the values used for such estimates are not always easily agreed upon, and this represents a source of much academic ferment among scientists ruminating on these issues [10].

Several studies have looked at the energy balance of ethanol production (whether from maize grain or cellulosic biomass); and, aside from some sharp critics [11–12], most have found these systems to have positive EROIs [13–14]. However, having a positive (or greater than unity, really) EROI is not sufficient to assure sustainability. Hall et al. [15] suggested that a minimum sustainable EROI for a biofuel system must be about 3:1. If the ratio is less than 3:1, the system is likely being powered too heavily by fossil fuel inputs. In the case of ethanol made from maize grain, EROI values typically range from 1.3 to 1.7; a more recent analysis suggests the value may be closer to 1 [16]. These estimates, coupled with environmental costs associated with maize production [17], reaffirm our view that maize-to-energy schemes are not the best path forward. We can hope, though, that they have helped to pave the way for cellulosic ethanol and other biofuel enterprises.

Estimates of EROI for cellulosic ethanol, as might be produced from switchgrass, are generally much higher than for starch-based ethanol systems. However, because we lack sufficient performance parameters from mature biomass-to-biofuel enterprises, a certain amount of good faith estimation has to go into their generation. In an interesting discussion of widely divergent EROI values for cellulosic ethanol [10], one author calculated an EROI of almost 18 : 1, compared with a second author's estimate of 0.72 : 1 [18]. The authors "debated" within their articles' pages their rationales for more or less conservative approaches [10], and we will summarize some of their unresolved points in the next two paragraphs.

There is general agreement that the largest energy costs for producing and delivering switchgrass to a biorefinery are N fertilizer and fuel for field operations and transport. These resources accounted for 93% of energy inputs in a large, multi-site field study in the Great Plains [19] and over 95% in a Canadian study [20]. Disagreement can arise over the fertilizer costs for these systems [10], which would vary as a function of productivity and management; we will discuss issues related to N fertility later.

A key source of disagreement—and differences in computed EROI values— comes from the LCA modelers' assumptions about energy returns per unit of biomass—the degree to which biomass energy will be conserved in bioethanol— and the energy that will be required for biomass processing [10]. For example, large amounts of energy are required for distilling ethanol, and a chief assumption of biofuel proponents is that substantial amounts of that energy will come from combustion of biomass residues. The heart of the counterargument seems to be that it will not be technologically or thermodynamically feasible to capture all (or even a sufficient amount) of the heat energy in the biorefinery's waste residues [11]. However, the successful use of sugarcane bagasse to power ethanol production of Brazil gives some support for the notion that residues are a feasible heat-processing source [10].

Our discussion of energy balance has largely been centered on farm production and delivery and the subsequent processing/conversion operations. However, a number of other variables affect a system's EROI (as well as GHG emissions and economic returns). Larger, fully integrated systems have been promoted for their economies of scale, but such systems can have certain physical limitations. Diseconomies of scale can arise in bioenergy systems due to the logistics required to handle and deliver very large quantities of material to a central facility [21]—

and perhaps redistribute processed byproducts to the farmshed [22–23]. Thus, on-farm or near-farm processing or pre-processing systems may offer opportunities to reduce energy requirements for transportation. Densifying biomass via pelleting, cubing, fiber expansion, torrefaction, pyrolysis, or other methods at pre-processing sites could further reduce the energy required for transportation and conversion [23–26].

In summation of this section, the critical question of whether biomass-to-biofuel systems will be feasible energetically and have a significantly positive net energy balance remains to be answered satisfactorily. Greater refinement (and agreement) is needed in setting boundaries for such analyses, and we definitely need real-world values for energy use and conversion efficiencies at successful cellulosic biorefineries.

2.2.2 Economic Tipping Points (Is Making Biofuel from Switchgrass Economically Feasible?)

A bioenergy system's EROI has obvious implications for its economic value. Positive economic returns on investment are essential components of the sustainability of market-driven enterprises. The low EROI of maize-based biofuel systems has meant that government interventions, *i.e.*, subsidies, have been required to drive the industry forward, at least through the early stages of industry development [27].

Government policies will likely have to play a key role in the development of liquid biofuel systems, at least until their products are cost competitive with those from fossil fuels. We make the "liquid" distinction here because biomass is already competitive with fossil fuels in some markets. As just one example, a hospital in Virginia, USA, is using grass biomass as boiler fuel. At USD $187 per Mg (ground and delivered), this fuel is cost-competitive with heating oil, which is the only other fuel available to the hospital. A case for phasing out heating oil more generally has been made by Wilson *et al.* [8]. Energy prices will be critical for biomass-based systems because of the effect they have on production costs. The "field-to-wheel" production of biofuel will presumably, at some point, be competitive with "well-to-wheel" production of gasoline/petrol—ideally at a not-too-high price point.

One policy change that would very likely promote rapid expansion and long-term survival of a lignocellulosic bioenergy industry in the USA is a requirement for GHG emission offsets [28]. Drop-in fuels produced from switchgrass or other perennial crops can far outdistance petroleum and grain-based fuels in reducing GHG emissions [29]. We will present more information on GHG emissions below, where it is considered under the heading of ecological/environmental concerns. We simply observe here that institutionalizing GHG emission offsets could provide great incentive to a nascent cellulosic biofuels industry.

2.2.3 Using Value-added Products to Shift the Tipping Point

Although producing biofuels at prices cost-competitive with petroleum has been a goal of many biofuel companies, it may not be a successful business model for the fledgling industry. Refineries that can produce multiple products—not just fuel—will have greater opportunity for financial success [30]. Thus, producing high-value, high-margin chemicals and value-added products—even with biofuel as a byproduct—may be a more viable business model. Although the products would be very different, producing coproducts of equal or greater value than the biofuel itself could boost the potential for the economic success of both biochemical and thermochemical platforms.

Chemicals, polymers, fibers, and food and feed ingredients are among the products being viewed as profit centers for a biorefinery industry. The maize-ethanol biorefinery already takes this approach, producing gluten feed and distillers' grains through different milling processes. Cellulosic biorefineries also might be able to produce feed products [31]. Recent research suggests that nutraceuticals are a potential product for these systems as well [32]. Interestingly, the debate over using biomass residues for in-plant fuel [10] may prove moot; advances in process chemistry could make lignin residues more valuable as sources of high-value chemicals [33] than sources of process energy.

Some researchers are working to convert switchgrass and other plants into "biofactories" that can produce high-value chemistries. For example, scientists working with switchgrass have been able to incorporate genes that synthesize polyhydroxyalkanoate, a constituent of bio-plastics [34–35]. While recognizing the concerns associated with transgenic plants, some have argued that their benefits will outweigh their risks or limitations [36]. For example, unlike petroleum-based plastics, plant-based plastics are biodegradable and thus represent an "environmentally benign and carbon-neutral source of polymers" [36].

2.2.4 Farmer and Factory Relationships: Getting the Ball Rolling

Just as the development of a bioenergy and bioproducts industry will require sufficient economic justification, so too will farmers and land managers need adequate financial justification to begin producing switchgrass or other feedstocks, and there may be some conservative resistance to such change [37]. Without economic incentives, would-be biomass growers will have no reason to "get in the game". Without biomass, captains of industry will have no reason to build biorefineries. Others and we have described this predicament as a "chicken or the egg" conundrum [24]; but, in reality, the "captains" have the initial role in getting it underway.

Industry development can occur and flourish where the processing facility will pay producers a fair price to grow switchgrass. Exactly what a "fair price" is may be a moving target, though, and studies suggest that this will vary by region. Regional price (or value) variation will be driven by differences in productivity, land rent, and the value of switchgrass or other cellulosic feedstocks relative to all potential crops [38–40].

It is supposed that the arrangement between switchgrass producers and biorefineries will be formalized by long-term contracts [21]. The contract structure will be crucial for both farmers and biorefiners. Long-term arrangements will be required by industry in order to obtain sufficient feedstock for continuous operations. However, long-term (5- or 10-year) contracts will be problematic for farmers who rent the land they farm; and the amount of farmland rented on a year-to-year basis can be great in some areas. Contracts that provide dividends for both quantity and quality of biomass will likely prove more appealing to growers than contracts based on a single fixed farm-gate price. Also, as we noted above, whether contracts can provide sufficient incentive (profit) to stimulate producer interest will, in part, be a function of external factors such as the price of energy and government policies.

2.2.5 Ethical/Social/Fairness Dimensions of the Sustainability

We include this heading only to remind us that, besides its economic and ecological dimensions, sustainability has "soul". Essentially, all sustainability standards for biofuel enterprises include criteria dealing with social justice. The welfare of the industry's workers and the consequences of the biofuel enterprise on the wellbeing of all segments of society must be taken into consideration. Probably the biggest ethical/social/fairness issue is the concern of "food vs. fuel". It is a very legitimate—even crucial—concern. We will not argue that switchgrass-based systems can avoid the concern because the feedstock is essentially hay; rather, any incentives to produce switchgrass for biofuels create the likelihood that at least some food-producing land will be diverted to biomass production. We strongly hope that unambiguous solutions can be found and that scientists, producers, and policymakers will find ways to avoid conflicts between land's food provisioning and fuel provisioning services.

2.3 Ecological/Environmental/Resource Considerations of the Sustainability

In a chapter on agroecological sustainability of switchgrass-based bioenergy systems, this and the next section finally get at the "heart of the matter". However, we felt compelled to develop the points made in the previous section. Any truly sustainable biofuel-producing system must simultaneously meet economic, energetic, ethical, and environmental/ecological targets. In a simplified analysis, the economic target is profitability. The energetic target is substantive net energy production. The ethical target is providing societal gains/benefits for all segments of society. The ecological target is to conserve—or, better yet, improve—our natural resources and to produce bioenergy in ways that can be practiced ad infinitum, *i.e.*, without doing any harm to the provisioning services of all ecosystems—not just agroecosystems.

In our consideration of the potential of switchgrass-based bioenergy systems to meet ecological sustainability targets, we have simplistically dissected ecosystems into four key components: soil, air, water, and life forms. We acknowledge that this reductionist approach beggars the complexity of ecosystems, but it serves as a logical way of analyzing at least some key sustainability criteria.

2.3.1 Sustaining the Soil Resource

Grasses—especially perennial grasses, such as switchgrass, that grow in swards (Fig. 2.2)—can provide great soil-conserving (and even soil-building) benefits. Their dense foliage and mass of fibrous roots can protect against wind and water erosion [41–42]. The perennial nature of switchgrass combined with intelligent harvest management could be a plus in this regard. Because senescent switchgrass is rather resistant to biomass losses [43–44], delaying harvest could (among other benefits) reduce the time during which fields might be essentially bare during the dormant season. Although some concerns have been raised about increased erosion potential after biomass is removed [45], the strong stubble and harvest residues can buffer against erosive losses. Without doubt, growing switchgrass on potentially erodible or highly erodible land would pose less threat to the soil resources than does maize or other row crops [45–48]. However, we must address concerns about agricultural intensification and land use changes—caused by likely increased demand for land to grow energy crops and food—driving energy crop production onto land that is more sensitive to erosion [7, 46, 49].

Fig. 2.2 A stand of Cave-in-Rock switchgrass growing in Pittsylvania County, VA, USA. Switchgrass is primarily used for forages, and uses for bioenergy are growing; but many early plantings have been for conservation purposes such as erosion mitigation and wildlife habitats.

Protection of the soil resource can be aided by using minimum- or no-till technology, and switchgrass lends itself readily to no-till planting [5]. With good management, soil erosion could likely be held to sustainable levels in switchgrass-for-energy systems, *i.e.*, soil losses would not outstrip soil-formation processes. Indeed, perennial grasses such as switchgrass may improve soils through addition of organic matter [50] and facilitation of soil-forming processes. Carbon sequestered in the soil (and held out of the atmosphere) for scores or even hundreds of years would be a very positive byproduct of the increased soil organic matter [51]. However, the issue of C sequestration—especially as it relates to GHG emissions vis-à-vis any land use changes—will be seen to be more complex in the next section.

2.3.2 Sustaining the Air Resource: GHGs and Climate

The air resource must likewise be protected from potential negative impacts of energy cropping systems. Better yet, atmospheric and climatic consequences of large-scale production of crops such as switchgrass might be positive [29, 52]. Ideally, biofuels would be at least CO_2-neutral, releasing into the atmosphere only the CO_2 recently sequestered in biomass; although, at least for the time being, fossil fuels will inevitably be used in biofuel production. Life cycle assessments suggest that using E85 fuel made from switchgrass would release up to 65% less CO_2 compared to gasoline/petrol [53].

Deploying switchgrass-for-energy on a large scale could result in reductions in CO_2 emissions when C is sequestered in root biomass and then "turns over" as roots die and become soil organic matter [54]. The effect would be short-term because soil organic matter can accumulate only to some upper limit, which varies with location, soil, and a variety of other factors [50]. However, a soil's natural capacity for C sequestration might be surpassed if biofuel systems develop around pyrolitic conversion technologies (see Chapter 1). Such systems may produce large quantities of recalcitrant, non-reactive (in the soil) biochar, which can be added to the soil and provide several benefits besides hightened levels of C sequestration [55–56].

While the production of biofuels is inherently intended to be protective of the atmosphere, concerns have been raised that energy cropping systems might not result in net reductions of GHG [57]. LCAs of some systems have suggested that more fossil fuel is used in the production of biofuels than is foregone in their consumption [58]. Arguments and roots of the different values for the same systems are similar to those seen when analyzing energy balances [10] (see above). In one rather encompassing LCA study [59], perennial crops such as switchgrass had much lower CO_2 emissions than maize, and their thermochemical processing resulted in lower CO_2 emissions than did biochemical processing. Another recent report on the biogeochemistry of bioenergy suggests that bioenergy systems based on perennial species could be GHG sinks [48]. Gaunt et al. [56] opined that the volume of GHGs avoided could be even greater if biochar-producing technologies (slow pyrolysis) were used and biochar was applied to the land. Slow pyrolysis results in a 30% reduction in biofuel output, but GHG savings would more than compensate for that in their argument. Again, though, we cannot overemphasize that some of the factors used to calculate such estimates are based on theoretical—not real-world—biorefinery operating conditions.

Introducing the issue of land use change and its consequences on GHG emissions [49, 57, 60] has brought a new, complicating factor into play—one that has many reexamining energy crops' potentials for sustainability [7, 45, 61]. The concern is that allocating or diverting land to energy crop use will cause cascading direct and indirect effects on land use that will increase net CO_2 emissions. The concern is now widely recognized and is being taken into consideration in sustainable biofuels standards [62], but sustainability formulators are finding it difficult to assess how much CO_2 (and other GHGs) flux to attribute to this unintended consequence of biofuel production, i.e., bringing new or sensitive land into production. The net effect will depend in part on the previous use of the land brought into production (either for biofuel or for "displaced" crops) and the nature of the production on that newly purposed land [61].

One recent modeling study compared the biofuel systems of first- and second-generations and included calculations for GHG emissions due to land use changes [63]. In this study, the land base that is currently used to produce ethanol from maize was reallocated to the switchgrass-for-ethanol production, and the consequences to various processes—including GHG emissions—were examined. The study's analysis suggests a 2nd-generation-only biofuel economy—using the land base that ethanol-from-maize does today—would result in 82% more ethanol production, nearly 20% less N runoff or leaching, and a reduction of 29% to 473% in GHG emissions, making the system a net GHG sink.

CO_2 is not the only GHG of concern in biofuels systems; N_2O and CH_4, two very potent GHGs, can also be released during the production and conversion of biomass, negating any positive CO_2 benefits [64]. The potential for the release of N_2O following fertilization of crops can be significant [45, 57, 65]. In a field study done on emissions of CO_2, N_2O, and CH_4 from switchgrass stands, N fertilization at a fairly low rate (67 kg N ha^{-1}) doubled N_2O emissions (as well as yield), but CO_2 and CH_4 did not change [66]. Another recent study brought to analysts' attention the potential for GHG releases from switchgrass biomass (or other bioenergy feedstocks) while in storage [67]. That work suggests that GHG reductions attributed to bioenergy systems may be overestimated by 10% or more if GHG emissions while in storage are not taken into consideration.

Besides GHG, which can have climate-altering effects, crops may also emit other gases (in addition to their well-understood fluxes of H_2O, O_2, and CO_2). In a very recent report, Graus et al. [68] quantified and described the field emissions of volatile organic compounds (VOCs) from maize and switchgrass. Both crops emitted VOCs in the same low nmol range, but switchgrass VOC emissions were slightly less. Methanol was the major component, but various aromatics and terpenoids were also present. The authors concluded that VOC emissions (when expressed as functions of liters of fuel produced) were equivalent to the VOCs associated with the use of a liter of gasoline/petrol [68].

Employing perennial grasses, such as switchgrass, on a large scale might have positive effects on the climate in ways that do not directly relate to GHG [69]. Shifting from annual to perennial crops, such as switchgrass, could have the biogeophysical effect of cooling localities and regions where such shifts occurred. The predicted cooling is related to changes in both transpiration and albedo over the cropland. The estimates of cooling could be as much as 1°C over major portions of the Corn Belt. The authors [69] surmised that the cooling effect would be equivalent to a reduction of 78 Mg C ha^{-1} in C emissions, more than six times the amount of C that might be sequestered and/or foregone as CO_2 emissions. There could be positive consequences on water use as well, and we will mention those in the next section.

2.3.3 Sustaining the Water Resource: Depletion and Pollution Concerns

With regard to the water resource, we do not suppose that irrigating to produce biomass for energy production would be sustainable. Where there is not enough rainfall to produce a crop of switchgrass, the region is likely to be water-limited. Currently, in many agricultural settings, groundwater removal exceeds recharge; and, even if the annual recharge-minus-withdrawal deficit is very small, the practice is, by definition, unsustainable. Similarly, when surface water is being siphoned to the extent that streams no longer provide or sustain key ecosystem services, the practice is unsustainable. Fisheries and navigation are but two of the ecosystem services that might be lost in order to produce fuel for cars and airplanes; this is neither sustainable nor justifiable.

Some research suggests that switchgrass culture may adversely affect hydrologic cycles because it draws more and deeper water due to its longer roots and extended growing season (relative to maize) [70–71]. Similar but even more adverse observations have been made about miscanthus [70, 72–73]. Tall-grass prairie ecosystems, dominated by deep-rooted and warm-season (C4) grasses such as switchgrass [74], are prolific biomass producers in regions that receive > 50 cm of annual rainfall. They are climax ecosystems that—a priori—meet all of the criteria for sustainability. It is mildly ironic, then, that a native prairie species might be considered too profligate in its water use. Also, in contradiction to the suggestion that bioenergy production using natural rainfall might adversely affect the hydrologic cycle, the biogeophysical report mentioned above (about the possible cooling effects of the stands of warm-season grasses) suggests that transpirational demand would be reduced by the localized cooling [69].

By any sustainability standard, energy cropping systems must, in addition to not depleting the water resource, not do any damage to water quality. Again, perennial grasses have an inherent advantage over annual rowcrop species in this regard [75], and switchgrass may be among the best [5]. It is often the species of choice for buffer strips between croplands and bodies of water, where it can physically and biologically filter chemicals and sediments out of runoff. This ability would seemingly accrue to even greater advantage where the entire field was planted in switchgrass, partially because switchgrass can be productive with much lower levels of N (and P) fertilizer than can maize. Avoiding soil runoff has clear downstream consequences. Reduced sediment and nutrient runoff will translate to improved aquatic habitats and reduced eutrophication. So also would be the case for reducing runoff of pesticides [75]. However, to the degree that any energy cropping system might adversely affect water quality (to include groundwater contamination), that system's sustainability would have to be questioned.

The environmental impacts of converting land currently in row crops to bioenergy crops such as switchgrass should be profoundly sanguine [47, 59]. More pointedly, a move from 1st generation maize biofuels to 2nd generation biofuels would seem to be a big step toward sustainability [24]. Several studies suggest reduced soil losses and improved water quality due to the lower rates of N and P fertilization that will likely to be used on biomass crops (relative to row crops such as maize). However, although perennial energy crops are typically expected to reduce erosion and off-site nutrient impacts [47], land use changes could again cloud the picture. Some have modeled greater N loading to surface waters based on the assumption that fertilized hectares will expand as energy crops displace native/natural grasslands [72]. In contrast, Ng et al. [76] reported that converting land to miscanthus, another frequently mentioned biomass species, should be expected to reduce NO_3 loading, and Wu et al. [71] projected similar reductions for N and P loading when land is switched to switchgrass production (either from row crop production or from pasture).

Issues of clean and abundant water are not of concern solely during the production of feedstock. Depending on the technology adopted and stringency of in-plant recycling/reuse, biorefineries could be water gluttons, and processing wastes might pose threats to water (and air) quality. Water demands have, at times, constrained the growth of the ethanol-from-maize industry [77]. Water supply limitations have the potential to limit development of 2nd generation biorefineries as well, and the constraint will likely be greatest for biorefineries situated near urban centers and/or in regions with limited precipitation. Pyrolytic technologies would be water-sparing, given that they utilize combustion-like processes—not water-based chemistries. Furthermore their biochar coproduct could improve both water and nutrient retention in soils [78–79].

2.3.4   Sustaining Biological Resources: Biodiversity

For its first 2 million years, switchgrass presumably occurred only in natural mixtures of other grasses and forbs, i.e., it was part of the community of organisms making up the living components of various ecosystems. Its productivity was its contribution to the overall primary productivity of those ecosystems. Since the mid-20th century, though, it has been collected from natural settings and planted and managed in monocultures, i.e., it has become a crop [1]. That summary of the evolutionary and agronomic history of switchgrass suggests why the species is still essentially a wild plant. It has not undergone the degree of conscious and unconscious selection by its cultivators for traits that many other crops have—traits that have made them more amenable to cultivation and more valuable to their cultivators. Switchgrass' essential wild status presents challenges and opportunities to those who would manage it for biomass production.

That summary also suggests some ignorance or audacity—or perhaps optimism—inherent in planting and growing monocultures of still essentially wild species that evolved to grow in mixtures. Evolution has fitted them for interaction with many species. Indeed, they often grow better in symbiotic associations, e.g., mycorrhizal fungi and higher plants [80]. Furthermore, all species express "preferences" sometimes for subtle variations in their environment—proclivities that become or define their niche and often minimize inter-species competition for resources. However, rather than allowing multiple species to take advantage of their ability to grow well together and achieve high levels of annual primary productivity, the general agricultural model is to act as if a single niche occurs over tens or hundreds of hectares and then hope that the identical plants established there will not compete too seriously for the set of resources that they all equally need. For most crops, the ability to grow well in monocultures and to tolerate intra-species competition was presumably selected for during domestication just by virtue of being grown in fields planted only as maize, wheat, or soybean. But an essentially wild species, such as switchgrass, that has not yet struck such a Faustian bargain would at least hypothetically be more of a challenge to grow in monoculture.

So, might there be an advantage to mimicking/mirroring natural ecosystems and planting polycultures for biomass production? (NB: biomass composed of mixed species would not presumably be as suitable/uniform a feedstock as biomass from monocultures.) In the best-known and most widely cited study, monocultures and mixtures of a number of prairie species were planted, provided minimal inputs, and then monitored for productivity for several years [81–82]. The authors concluded that mixtures under these low-input conditions were more productive, i.e., produced more biomass, than monocultures. We quibble with their protocol and therefore their interpretation, however. Their protocol called for taking a yield sample from the experimental plots annually, but the remainder of the plot was burned—not harvested. Therefore, they did not truly simulate biomass production systems. By burning the plots, they "recycled" key nutrients. Their approach more closely mirrors the prehistoric pattern of burning off the prairie periodically and/or returning nutrients in the excrements of grazers. We do not know how these mixtures would have fared had they truly been harvested annually with no nutrient inputs. Furthermore, there is a cost for low productivity. For example, low-input, high-diversity plantings may be more productive than low-input switchgrass monocultures, but the net energy yields are about half of those in well-managed switchgrass production systems [19], and economic returns would also be lower. Higher intensity management may also be a better way to achieve environmental goals where it spares land providing valuable ecosystem services [83]. A study similar to Tilman et al. [81–82] has recently been reported [84]. In this latter case, all biomass was removed annually and nutrients were applied. The results [84] suggested that switchgrass is equally productive in monoculture as in the mixtures tested and mixtures provided no biomass yield advantage.

Of course, even the best-managed field does not contain only one species. While there may be only one species of a plant evident, there will be hundreds, probably thousands (if not hundreds of thousands), of other species still living there. Indeed, the growth and success of the planted species often depends on the presence and activity of many organisms. Nevertheless, there is an inevitable loss of biodiversity—biological richness—in monocultures. Many "higher" life forms (both plants and animals) are excluded from the field to the degree that the manager can control them. Managers often take a rather blunderbuss approach to controlling insects, fungi, bacteria, nematodes, etc. when they threaten their crops, with the effect of reducing biodiversity even more.

We immediately concede that growing switchgrass (or any species) in monoculture will reduce biodiversity compared to natural ecosystems. The question we pose for this review is how does switchgrass compare with row crops and other energy crops in their effects on biodiversity? We will answer that question somewhat anecdotally using examples of a few life forms that may serve as "canaries in the mine". For example, butterflies have been suggested as possible indicator species for looking at the health and biodiversity of biofuels crop systems [85].

Energy cropping systems that provide wildlife habitats gain an advantage in the search for sustainable systems. Grassland fauna could be favored in suitably managed energy cropping systems [86]. Multiple-species plantings—mixed intercrops—might add niches for greater biodiversity as well as increase yield potential in some cases [81, 87].

Several studies have examined avian fauna in switchgrass plantings [86, 88–89]. Some have looked at other vertebrates, insects, and soil organisms [90–91]. In all such cases, one finds a less diverse fauna in switchgrass stands than might be present in a more diverse plant community. Species that prefer tall grasses may be more abundant, but many other species may be essentially excluded. An exception might include birds in migration. One study suggests that switchgrass fields may make as good or better spring stopover points as multi-species plantings [89].

Planting switchgrass might increase grassland bird habitat potential. However, practices that might be used to maximize biomass production might also reduce species richness and/or abundance of grassland-dependent birds [92]. There are concerns that highly productive upland switchgrass ecotypes may not make good habitats for ground-nesting birds. Stands that are suitably dense for good biomass production may prove impenetrable, especially for fledglings—an issue that might be addressed with varied harvest timings [86]. In fields sown to lowland switchgrass ecotypes, habitats for ground-nesting birds may be improved through management. Lowland types exhibit a bunch-grass habit and can grow quite productively in row widths of up to 90 cm [93]; so, wide-row plantings could meet the needs of both wildlife and growers.

Sometimes, too much biodiversity—or at least the occurrence of undesirable organisms—can be problematic. Weeds can seriously limit the success of establishing switchgrass stands; however, once established at sufficient densities, switchgrass plants grow so vigorously that they tend to exclude invasive species [5]. Of more concern could be whether switchgrass might become a weed itself, invading other crops or ecosystems. Since switchgrass is a native species and occurs naturally (but not abundantly) throughout much of the eastern USA, it has already shown that it is not aggressively invasive. It "knows its place" and appears primarily as a member of communities of mixed, tall-growing grasses. Studies that try to anticipate the weediness of introduced species suggest that switchgrass poses minimal risk [94].

Insect pests and diseases have generally been considered of limited concern with switchgrass culture [5], but they could become serious problems as plantings for energy crops expand [95]. Catalogued fungal diseases for which switchgrass is a host now total more than 150 [96]. The growth of that list could be a function of more diseased fields of switchgrass or of more effort directed toward monitoring diseases in switchgrass fields—or both. We have recently documented elsewhere a history of some of the "outbreaks" of fungal and viral diseases in switchgrass [97]. Diseases could pose serious concerns for sustainability, both as they affect yield and as control measures might affect biodiversity. It is hoped that this wild species might carry good resistance to many pathogens. Some clone- or ecotype-specific resistance has been shown for both fungal and viral pathogens, and breeders can use those traits to minimize the need for chemical control [95, 98].

Insect pests may be of less concern than diseases; although, again, pressure may mount as switchgrass-for-energy systems scale up [5]. Older reports suggested that switchgrass was not a preferred host for many insect species in its very early days as a crop [99], and it still appears to be an inferior host for insects compared to other warm-season crops [100–101]. A baseline study of insects in Nebraska switchgrass stands found that about 60% of arthropods collected were of the orders Thysanoptera and Hymenoptera [102]. Leafhoppers, grasshoppers, grass flies, and wireworms were the most abundant of potential pest species.

As already mentioned, the switchgrass gene pool appears to contain genes for resistance to various pests and diseases and presumably also to stresses such as heat and drought. It is hoped that switchgrass breeders will be able to use those factors to develop higher-biomass cultivars. But what about using genes from other species, i.e., transgenes? Transgenic modification of energy crops introduces major concerns about biodiversity and sustainability [36, 103]. Some of those concerns are so great that transgenic approaches should probably be rejected out of hand for a naturally occurring species such as switchgrass until (or if) the concerns can be allayed. The introgression of transgenes into native populations, sometimes called "genetic pollution", would be essentially inevitable with an open-pollinating, self-incompatible species such as switchgrass.

In sum, growing switchgrass monocultures on a large scale, when done by bringing previously uncropped land into production, would reduce biodiversity over that same scale. On the other hand, switchgrass culture will likely restore and conserve biodiversity better than many row crops, to include maize used for biofuel purposes. As a minimum, it should be no more harmful than other large-scale monocultures and potentially much less harmful than some.

2.4 Managing Switchgrass for Bioenergy and Sustainability

2.4.1 Description, Adaptations, and Selection

In the past two and a half decades, much effort has been expended examining the biology, genetics, physiology, and agronomy of switchgrass [5]. Knowledge of these factors can inform sustainable management and use of the crop. In the next few sections, we will discuss the biology and agronomy of switchgrass used as an energy crop. Numerous comprehensive production guides are available elsewhere. We primarily want to provide readers an appreciation for some of the ways that growers and biorefiners can use the species' characteristics to an advantage in developing sustainable systems built around it.

In North America, switchgrass occurs naturally from Mexico to Canada and from the Atlantic coast to the Sierra Nevada Mountains [104]. Wide geographic distribution and broad adaptability reflect the species' great genetic diversity. This range of adaptation was a key reason why switchgrass was selected as a model species in early energy cropping studies [1, 3–4].

Within geographic regions, switchgrass' natural suitability to an array of soil and fertility environments is largely a function of ecotypic adaptation. The species is categorized into two ecotypes: upland and lowland (Fig. 2.3). These ecotypes are sometimes also referred to as cytotypes because their chloroplastic DNAs have distinctive sequences [105]. The ecotypic names describe typical niches for each type. Mesic, upland sites are more often occupied by upland ecotypes, with their lower sensitivity to moisture stress; whereas hydric bottomlands are the typical habitats for lowland ecotypes [74]. Lowland ecotypes are taller, often reaching over 3 m, while uplands are generally shorter, with some being

Fig. 2.3 Upland and lowland switchgrass ecotypes. Uplands are shorter and finer-stemmed. In this October photo in Virginia, USA, the lowland ecotypes (extreme left and center) clearly have not senesced as much as the uplands.

no more than 0.5 m [74, 106]. Lowlands are generally more robust plants with thicker stems and have a bunchgrass growth habit. In contrast, uplands have finer stems and may form sod due to their more active rhizomes [107]. Despite considerable overlap in their regional distribution, lowlands are the principal ecotypes in the southern USA, while upland types are more typically found in the drier, colder portions of the American Great Plains.

Soil type, while it is often very different by landscape position, does not seem to play a particularly strong role in switchgrass production [108]. However, slope within a soil type may be an important variable affecting switchgrass production due to differences in infiltration and drainage [109]. In a similar manner, soil texture may also be an important factor in switchgrass establishment and yield. This is largely a function of soil water holding capacity, and both excessively and poorly drained soils can limit switchgrass productivity [110].

Switchgrass production guides often recommend amending soil to a pH of 6.0 or greater for planting, but switchgrass tolerates a pH from 5 to 8 for germination [111]; it is rare that soil acidity limits yields in production settings [112]. Switchgrass root growth has been reported at pH as low as 3.7 [113]; and, while this is not typical for production fields, it does demonstrate the species' adaptability to harsh soil conditions.

Switchgrass plants (both ecotypes) produce multiple tillers in a determinate fashion. The tillers stay vegetative and continue to produce stems and leaves, *i.e.*, biomass, until the plants are exposed to a night length that triggers flowering and causes biomass production to essentially cease. Although the photoperiod is the primary environmental cue, it is perhaps not the only one [114].

Photoperiod response is a genetically determined trait tied to a cultivar's latitude of origin. Differences in photoperiodicity among cultivars affect both yield and survival. When cultivars are moved to higher latitudes from their region of origin/adaptation, reproductive growth is delayed, because the plant will not experience the photoperiod signal (appropriate night length) until later in the growing season. This results in continued vegetative development and greater biomass yields. Conversely, a plant moved from its provenance to lower latitudes will initiate reproductive development earlier and produce a lower yield.

Switchgrass cultivars also vary morphologically and adaptively along a longitudinal gradient in North America [115]. Taller plants adapted to greater humidity are common to the humid east, while shorter, compact plants tolerant of more mesic and windy conditions are found in the Great Plains. Both types can be negatively affected when moved from their longitude of origin. Abiotic stress is the likely issue for eastern plant materials moved westward in the USA; however, in the case of western plants brought to the more humid east, the dissonance is perhaps more likely due to pathogens for which the western genotypes have developed no resistance.

As of this publication, no switchgrass cultivars specifically bred for biomass systems have been registered or released. This should not be surprising, as cultivars historically have been selected on the basis of forage traits and for conservation purposes. The breeding and selection of lines for biomass production are occurring, but there is considerable lag time built into the process before releases might occur.

Because of their genetic/evolutionary match to particular provenances and latitudes, switchgrass lines should be selected on the basis of their adaptation to local climatic conditions and photoperiods. Seasonal moisture and humidity and maximum and minimum temperatures are primary considerations for cultivar selection. As noted, we can move southern-adapted cultivars to higher latitudes to good effect for biomass yield, but this strategy has its limits. Switchgrass requires sufficient time to "harden off" for winter [116–117], and moving cultivars to higher latitudes where growing seasons are compressed and winters are colder can limit the plant's preparation for dormancy and/or expose it to temperatures for which it is not physiologically adapted. A general rule of thumb is to choose cultivars that are within 500 km of their latitude of origin [106].

2.4.2 Establishment

Among bioenergy crops, switchgrass presents certain advantages from an agronomic perspective. High productivity, broad adaptation, and tolerance to marginal sites helped it become a "model" energy crop. However, not all marginal sites will be suitable for energy production systems. Environmentally fragile sites with limited potential productivity and challenging logistics should not be targets for the production of switchgrass—or any other crop.

Compared with some other bioenergy crops, the ability to establish switch-grass from seed also presents some practical and economic advantages. The species has a reputation for being difficult to establish [5]; and, from an environmental perspective, stand failure presents a greater risk of erosion and nutrient runoff. Replanting means more expense and lost productivity. "Getting it right" will be important to achieve high yields quickly following establishment and to create greater probability of economic success for these systems [40, 119].

We have previously written, in some detail, about switchgrass establishment and some of the causes for stand failure [97, 110, 119], so we will summarize and generalize here. Many producers are accustomed to planting crops that have larger seeds with greater vigor and less sensitivity to planting depth or to weeds. Thus, when working with switchgrass, they may not be suitably equipped—literally and figuratively—for the exacting conditions needed. High levels of seed dormancy, poor seed placement (too deep or too shallow), poor seed-soil contact due to loose soil or high amounts of residue, poor timing, weed or pest competition, and poor weed management are among the several factors that can cause stand failure [97, 110, 120–122].

In the context of promoting switchgrass for sustainability purposes, we note that no-till planting methods would typically be recommended. However, no-till practices and equipment are not familiar to all (maybe most) potential growers. Ultimately, to have successful establishment, stand frequencies should approach a minimum of 40% soil coverage by the end of the first growing season [118]. This is possible with a wide range of seeding rates [123–124] depending on how well weeds and field conditions are managed and if seeds have low dormancy [5]. Low seeding rates coupled with high stand success obviously will translate into more economical production [40, 119].

2.4.3 Fertility in an Agroecological and Sustainability Context

Switchgrass seedling growth is slow relative to many crops, and the new plants often compete poorly with weeds during establishment. Because weeds often respond more vigorously to nutrient inputs than do switchgrass seedlings, the general recommendation is to apply no N until the second growing season [126–127]. Outside of first-year growth [113], few studies support P and potassium (K) fertilizer inputs at seeding; generally, soil levels are adequate if within "medium" range on standard soil tests. Responses to lime applications are also limited and most likely to be observed when applied in combination with nutrients [113, 128].

The perception of switchgrass' tolerance to marginal environments and pro-ductivity with limited inputs largely put the crop "on the map" with early bioenergy research efforts. This issue is not trivial from a sustainability standpoint, given the energy embedded in the production of lime and chemical fertilizers and the potential for runoff of nutrients [48]. Estimates of the direct energy inputs needed for switchgrass production suggest that just about a half is used for N fertilizer [10]. However, views on crop N needs may have moderated somewhat given the broad range of the response to N fertility found in the literature [5].

2.4.3.1 Fertility in a sustainability context: A case for N

We give N greater focus because, among the nutrients commonly used, it is the most costly both economically and environmentally. N is primarily derived from fossil fuel sources; and, because it can escape the cropping system via runoff, leaching, or as N_2O or NH_3, it has tremendous potential to negatively affect environmental sustainability metrics [47–48, 128]. From a biorefiner's perspective, N (and other fertilizer applications) has the potential to negatively affect fuel yield by changing plant morphology and composition [129].

The mixed data on switchgrass' yield response to N fertility [5] may reflect variations in plant demand, soil type and N status, atmospheric deposition, plant capacity to capture soil N, plant-microbial relations, plant capacity to recycle N, and harvest management factors (e.g., timing and cutting height), among others. Greater yield responses to applied N are likely in coarse-textured soils with low nutrient holding capacity [93, 130], while little response to N is likely to be seen on soils with high amounts of organic N [131]. Both older and newer evidence suggests that biological N fixation is a factor in the low apparent N requirements of switchgrass [132–134]. Also, at least one research program in the USA is working to develop microbial endophytes that can stimulate switchgrass production [135].

Switchgrass has a great ability to internally translocate nutrients. N is mobilized from storage organs in spring, and N concentrations in biomass decline across the growing season as plants grow and mature [136]. Aboveground N stocks are then remobilized and translocated to roots and rhizomes at the end of the growing season [107, 137–138]. This capacity greatly reduces the need for exogenous fertilizer applications under typical biomass cropping system management.

Achieving sustainability in switchgrass-based biofuel systems will require using management practices that both minimize and balance fertility imports and exports. To this end, most management guides call for a single, end-of-season harvest in order to minimize nutrient removal in biomass. Post-season N concentrations in switchgrass biomass will often be in the range of 5 to 8 g kg^{-1} [109, 139–141]. Assuming typical switchgrass yields in the range of 10 to 15 Mg ha^{-1} [4]

and 6 g N kg^{-1} biomass, replacement N input rates would be on the order of 90 kg ha^{-1}. However, the reader is again cautioned about taking such generalizations too far, given all of the variations in plant productivity that we have discussed.

Incorporating alternative nutrient sources may also improve long-term productivity and economic sustainability of biomass production systems. Animal manures may be useful in this regard [142], although sustainability of the production systems that generate these waste products may be in question. Legumes have also been incorporated into switchgrass production systems as N sources, although their success has been variable [143–144].

There is some evidence that switchgrass reduces N partitioning to roots under higher fertility management [138, 145], and fewer tillers have been observed with higher N inputs, particularly under one-cut management [130, 140]. Though this may not affect aboveground yields [130], C sequestration and GHG emissions may be affected [146]. Thus, N fertility's effects on overall system sustainability relative to GHG footprints bear further investigation [147].

2.4.3.2   *Fertility in a sustainability context: P and K*

Responses to P and K have generally been limited in switchgrass production systems [148], and single, end-of-season harvests will reduce losses of these nutrients from the system. While these nutrients may seem to be "bit" players in switchgrass production, as they seldom seem limiting, there is some question about their long-term availability [149]. There will likely be opportunities to improve nutrient capture and to tighten nutrient cycles (particularly for P) in the field, *e.g.*, by using mycorrhizae [80, 150], and at the biorefinery. We would anticipate greater development of such biological and engineering systems when scarcity and a rising cost of nutrients force the agricultural community to address such concerns.

2.4.4   Mechanization, Storage, and Hauling

The goal of bioenergy systems will be to optimize feedstock yield and quality while minimizing, to the degree possible, inputs such as fertilizers, water, fuel, equipment, and labor. We have already touched on the issues of harvest timing and nutrient removal interactions, but they bear further exploration in terms of the harvest methods themselves. Delaying harvest until the plants have naturally dried down at the end of the season not only reduces nutrient losses from the system, but also reduces feedstock moisture content—and feedstock moisture is a critical consideration in handling, hauling, and storage [24].

The two primary systems suggested for harvesting switchgrass biomass are "chopping" and baling. Each approach has variations and has advantages and shortcomings as well. There is also evidence that harvest system energy requirements may be improved through modifications of existing harvest equipment [151–152].

The major advantage of chopping at harvest is that size reduction occurs in the field, providing a flowable material that will move readily through first stages at a biorefinery. However, this typically requires two or more persons: one to run the chopper and one or more to haul away the chopped material (Fig. 2.4). The chopped material is typically blown into a trailer that operates in tandem with the chopper, and then the filled trailers must be hauled away and others brought in their place. It would be preferable to move this chopped material directly to a biorefinery, but this will be impractical due to the system scale. Thus, densification and storage facilities will likely be required, along with efforts to minimize storage losses.

Fig. 2.4 Field chopping "pre-processes" switchgrass, producing a flowable material that is easier to handle by the process systems within a biorefinery. However, such systems require additional labor to collect the chopped material in the field and haul to a central collection point (as in the insert) and specialized storage.

Photos by Bobby Grisso.

In contrast to chopping, mowing and baling may require only one person, as newer tractor designs will allow for a front-end mower attachment coupled with a baler behind the tractor (Fig. 2.5). Even if cutting and baling are done in two passes through the field, baling severs the tie between in-field harvest and hauling operations [21]. After harvest, multiple-bale collection and removal systems can increase the efficiency of handling. Also, at least with round bales, storage losses may be minimal ($\sim 5\%$), given that net-wrapped switchgrass bales can shed water [153]. Large square bales may be preferred where large acreages are available for bioenergy production, and they will allow greater densification. However, the square bale systems may be less appropriate in regions with small, irregularly shaped fields of uneven terrain; and these bales will add to storage costs, since they do not shed moisture. For both round bales and large square bale systems, however, biomass size reduction issues must still be addressed. The material in the bales must, at some point, be chopped or ground to a size that will allow for further processing.

Fig. 2.5 A combined mower and baler designed for one-pass harvesting of herbaceous biomass such as switchgrass. Such systems will reduce labor required to harvest and bundle biomass.

Photo by John Cundiff and courtesy of FDC Enterprises.

2.4.5 Demands of a Bioenergy Industry

The previous discussion of harvests and handling logistics should give the reader some small flavor of these issues facing the bioenergy industry. Indeed, they are sufficient to warrant their own chapter and will be an important aspect of developing a bioenergy industry "at scale". Building a bioenergy industry that makes a significant contribution to the energy needs will be a vast enterprise.

Some biorefinery scenarios envision that biomass will not be stored on site more than three days before being processed [154]. Providing a continuous supply thus will require copious off-site storage (and perhaps pre-processing) facilities. These off-site storage areas must be able to collect biomass over a relatively short period of time, maintain it in a suitable state throughout storage, and be readily accessible in inclement weather to allow for on-time delivery to the refinery [155].

The costs of building this infrastructure will require careful evaluation of the economic and environmental costs and tradeoffs, and this may lead to some not-so-obvious conclusions. For example, the infrastructure and energy demands for a system that collects biomass for a few months each year may be much greater than for a year-round harvest and storage system. While expanding the harvest window and collecting "green" biomass may be resisted by some agronomists, such evaluations will be needed when looking to create a large-scale refinery [21, 24]. The value of such management practices must be weighed in terms of biorefinery outputs as well. The nature of the conversion system and the quality of the feedstocks will affect outputs of fuel, coproducts or byproducts, and waste materials. The sizes of the systems will also need evaluation, as some suggest that larger, centralized processing points will improve economic outcomes, while distributed refineries will have greater environmental sustainability [156].

This is perhaps a good point to ask questions of purpose. Are these systems being developed to strengthen rural economies? To minimize exposure to national security issues associated with foreign oil consumption? To help address global (or local) environmental issues? While it is far from a comprehensive treatment of this issue, we hope that the reader can recognize the complicated and sometimes conflicting values and demands being placed on these systems in hopes of meeting environmental, economic, social, and policy goals [24, 157].

2.5 Conclusions

We must find sustainable alternatives to fossil fuels, and many arguments suggest that making ethanol from grain is not a sustainable approach. Using lignocellu-

losic biomass, such as switchgrass, as a feedstock for the production of ethanol
or other biofuels holds greater promise. In various analyses, using switchgrass as
a bioenergy feedstock appears to provide advantages in GHG emissions relative
both to fossil fuels and to ethanol from maize. Furthermore, switchgrass-based
energy systems show promise in meeting several other criteria that are essential
for sustainability. Some major challenges must be met before making biofuels
from any biomass source can be clearly established as sustainable. Concerns include: bringing to maturity biorefinery technologies that can produce billions of
liters of biofuel annually, scaling up agronomically and economically to the millions of hectares required to produce sufficient biofuel feedstock, and developing
logistics for handling and storing millions of megagrams (Mg) of biomass. These
goals must be met while conserving natural resources and minimizing ecosystem service disruptions. Furthermore, we must develop clear, usable metrics for
assessing sustainability.

Biorefineries that will be using switchgrass or other biomass feedstocks will be
in operation shortly. Lessons learned from and data generated by those facilities
will help to establish whether such systems can, indeed, be sustainable. Switchgrass would appear to offer advantages over some other biomass candidates—
advantages that could ultimately redound to establishing its sustainability as a
feedstock for a suitably designed system.

References

[1] Parrish DJ, Casler MD, Monti A. The evolution of switchgrass as an energy
    crop. In: Monti A (Ed.). Switchgrass: A Valuable Biomass Crop for Energy.
    Springer-Verlag, 2012; 1–28.
[2] Fike JH, Parrish DJ. Switchgrass. In: Singh BP (Ed.). Biofuel Crops: Produc-
    tion, Physiology and Genetics. CAB International 2013: 199–230.
[3] Wright L, Turhollow A. Switchgrass selection as a "model" bioenergy crop: a
    history of the process. Biomass and Bioenergy 2010; 34: 851–868.
[4] McLaughlin SB, Kszos LA. Development of switchgrass (*Panicum virgatum*) as
    a bioenergy feedstock in the United States. Biomass and Bioenergy 2005; 28:
    515–535.
[5] Parrish DJ, Fike JH. The biology and agronomy of switchgrass for biofuels.
    Critical Reviews in Plant Sciences 2005; 24: 423–459.
[6] Van der Voet E, Lifset RJ, Luo L. Life-cycle assessment of biofuels convergence
    and divergence. Biofuels 2010; 435–449.
[7] Hart Energy Consulting/CABI Head Office. Land Use Change: A Science and
    Policy Review. Houston TX USA and Wallingford Oxfordshire UK, 2010.

[8] Wilson TO, McNeal FM, Spatari S, Abler DG, Adler PR. Densified biomass can cost-effectivley mitigate greenhouse gas emissions and address energy security in thermal applications. Environmental Science & Technology 2012; 46: 1270–1277.

[9] Samson R, Mani S, Boddey R, *et al.* The potential of C4 perennial grasses for developing a global BIOHEAT industry. Critical Reviews in Plant Sciences 2005; 24: 461–495.

[10] Hall CAS, Dale BE, Pimentel D. Seeking to understand the reasons for different energy return on investment (EROI) estimates for biofuels. Sustainability 2011; 3: 2413–2432.

[11] Patzek TW. A probabilistic analysis of the switchgrass ethanol cycle. Sustainability 2010; 2: 3158–3194.

[12] Pimentel D. Ethanol fuels: energy security economics and the environment. Journal of Agricultural and Environmental Ethics 1991; 4: 1–13.

[13] Farrell AE, Plevin RJ, Turner BT, Jones AD, O'Hare M, Kammen DM. Ethanol can contribute to energy and environmental goals. Science 2006; 311: 506–506.

[14] Hammerschlag R. Ethanol's energy return on investment: a survey of the literature 1990-present. Environmental Science & Technology 2006; 40: 1744–1750.

[15] Hall CAS, Balogh S, Murphy DJR. What is the minimum EROI that a sustainable society must have? Energies 2009; 2: 25–47.

[16] Murphy DJ, Hall CAS, Powers B. New perspectives on the energy return on (energy) investment (EROI) of corn ethanol. Environment Development and Sustainability 2011; 13: 179–202.

[17] Yang Y, Bae JH, Kim JB, Suh SW. Replacing gasoline with corn ethanol results in significant environmental problem-shifting. Environmental Science & Technology 2012; 46: 3671–3678.

[18] Patzek TW, Pimentel D. Thermodynamics of energy production from biomass. Critical Reviews in Plant Sciences 2005; 24: 327–364.

[19] Schmer MR, Vogel KP, Mitchell RB, Perrin RK. Net energy of cellulosic ethanol from switchgrass. Proceedings of the National Academy of Sciences (US) 2008; 105: 464–469.

[20] Samson R, Duxbury P, Drisdelle M, Lapointe C. Assessment of pelletized biofuels. 2000; 41. (Accessed 24 March 2013 at: http://wwwreapcanadacom/online_library/feedstock_biomass/15%20Assessment%20ofPDF).

[21] Cundiff JS, Fike JH, Parrish DJ, Alwang J. Logistic constraints in developing dedicated large-scale bioenergy systems in the Southeastern United States. Journal of Environmental Engineering 2009; 135: 1086–1096.

[22] Bruins ME, Sanders JPM. Small-scale processing of biomass for biorefinery. Biofuels, Bioproducts & Biorefining 2012; 6: 135–145.

[23] Eranki PL, Dale BE. Comparative life cycle assessment of centralized and distributed biomass processing systems combined with mixed feedstock landscapes. Global Change Biology Bioenergy 2011; 3: 427–438.

[24] Fike JH, Parrish DJ, Alwang J, Cundiff JS. Challenges for deploying dedicated, large-scale, bioenergy systems in the USA. CAB Reviews: Perspectives in Agriculture, Veterinary Science, Nutrition and Natural Resources 2007; 2: 1–28.

[25] Stephen JD, Mabee WE, Saddler JN. Biomass logistics as a determinant of second generation biofuel facility scale location and technology selection. Biofuels Bioproductions & Biorefining 2010; 4: 503–518.

[26] Wright MM, Brown RC, Boateng AA. Distributed processing of biomass to bio-oil for subsequent production of Fischer-Tropsch liquids. Biofuels Bioproducts & Biorefining 2008; 3: 229–238.

[27] Gardner B. Fuel ethanol subsidies and farm price support. Journal of Agricultural & Food Industrial Organization 2007; 5: 1–20.

[28] Popp M, Nalley LL. Modeling interactions of a carbon offset policy and biomass markets on crop allocations. Journal of Agricultural and Applied Economics 2011; 43: 399–411.

[29] Hill J, Polasky S, Nelson E, et al. Climate change and health costs of air emissions from biofuels and gasoline. Proceedings of the National Academy of Sciences (US) 2009; 106: 2077–2082.

[30] Arato C, Pye EK, Gjennestad G. The lignol approach to biorefining of woody biomass to produce ethanol and chemicals. Applied Biochemistry and Biotechnology 2005; 123: 871–882.

[31] Kammes KL, Bals BD, Dale BE, Allen MS. Grass leaf protein, a coproduct of cellulosic ethanol production, as a source of protein for livestock. Animal Feed Science and Technology 2011; 164: 79–88.

[32] Hernández-Soto R, Sandoval-Fabian G, Cardador-Martínez A, Estarrón-Espinoza M. Quantification of phenolic content, tocopherols and phytosterols in different morphological fractions obtained as byproducts of the corn wet milling. In: Nutraceuticals and Functional Foods: Conventional and Non-conventional Sources 2011; 187–201.

[33] Varanasi P, Singh P, Auer M, Adams PD, Simmons BA, Singh S. Survey of renewable chemicals produced from lignocellulosic biomass during ionic liquid pretreatment. Biotechnology for Biofuels 2013; 6: 14–22.

[34] Snell KD, Peoples OP. PHA bioplastic: a value-added coproduct for biomass biorefineries. Biofuels Bioproducts & Biorefining 2009; 3: 456–467.

[35] Somleva MN, Snell KD, Beaulieu JJ, Peoples OP, Garrison BR, Patterson NA. Production of polyhydroxybutyrate in switchgrass a value-added co-product in an important lignocellulosic biomass crop. Plant Biotechnology Journal 2008; 6: 663–678.

[36] Mooney BP. The second green revolution? Production of plant-based biodegradable plastics. Biochemical Journal 2009; 418: 219–232.

[37] Wen Z, Ignosh J, Parrish D, Stowe J, Jones B. Identifying farmers' interest in growing switchgrass for bioenergy in Southern Virginia. Journal of Extension 2009; 47: 5RIB7.

[38] Bransby DI, Samson R, Parrish DJ, Fike JH. Harvest and conversion systems for producing energy from switchgrass: logistic and economic considerations. Forage and Grazinglands 2008; FG-2008-0722-01-RV.

[39] Jain AK, Khanna M, Erickson M, Huang HX. An integrated biogeochemical and economic analysis of bioenergy crops in the Midwestern United States. Global Change Biology Bioenergy 2010; 2: 217–234.

[40] Perrin R, Vogel K, Schmer M, Mitchell R. Farm-scale production cost of switch-grass for biomass. BioEnergy Research 2008; 1: 91–97.

[41] Kort J, Collins M, Ditsch D. A review of soil erosion potential associated with biomass crops. Biomass and Bioenergy 1998; 14: 351–359.

[42] Vaughan D, Cundiff J, Parrish D. 1989. Herbaceous crops on marginal sites—Erosion and economics. Biomass 1989; 20: 199–208.

[43] Adler PR, Sanderson MA, Akwasi A, Boateng A, Weimer PJ, Jung HJG. Biomass yield and biofuel quality of switchgrass harvested in fall or spring. Agronomy Journal 2006; 98: 1518–1525.

[44] Lee KD, Owens VN, Boe A, Koo BC. Biomass and seed yields of big bluestem, switchgrass, and intermediate wheatgrass in response to manure and harvest timing at two topographic positions. GCBCB Bioenergy 2009; 1: 171–179.

[45] Payne WA. Are biofuels antithetic to long-term sustainability of soil and water resources? Advances in Agronomy 2010; 105: 1–46.

[46] Wright CK, Wimberly MC. Recent land use change in the Western Corn Belt threatens grasslands and wetlands. Proceediings of the National Academy of Science (US) 2013; 110: 4134–4139.

[47] Powers SE, Ascough JC II, Nelson RG, Larocque GR. Modeling water and soil quality environmental impacts associated with bioenergy crop production and biomass removal in the Midwest USA. Ecological Modelling 2011; 222: 2430–2447.

[48] Robertson GP, Hamilton SK, Grosso SJ, Ddel, Parton WJ. The biogeochemistry of bioenergy landscapes: carbon, nitrogen, and water considerations. Ecological Applications 2011; 21: 1055–1067.

[49] Searchinger T, Heimlich R, Houghton RA, et al. Dong FX, Elobeid A, Fabiosa J, Tokgoz S, Hayes D, Yu TH. Use of US croplands for biofuels increases greenhouse gases through emissions from land-use change. Science (Washington) 2008; 319: 1238–1240.

[50] Sartori F, Lal R, Ebinger MH, Parrish DJ. Potential soil carbon sequestration and CO_2 offset by dedicated energy crops in the USA. Critical Reviews in Plant Sciences 2006; 25: 441–472.

[51] Lal R. Carbon sequestration in dryland ecosystems. Environmental Management 2004; 33: 528–544.

[52] Bessou C, Ferchaud F, Gabrielle B, Mary B. Biofuels, greenhouse gases and climate change: a review. Agronomy for Sustainable Development 2011; 31: 1–79.

[53] Yu B, Lin L, Vvan Dder Voit E. Life cycle assessment of switchgrass-derived ethanol as transport fuel. International Journal of Life Cycle Assessment 2010; 15: 468–477.

[54] Gu Y, Wylie BK, Zhang L, Gilmanov TG. Evaluation of carbon fluxes and trends (2000-2008) in the Greater Platte River Basin: a sustainability study for potential biofuel feedstock development. Biomass and Bioenergy 2012; 47: 145–152.

[55] Laird DA. The charcoal vision: a win-win-win scenario for simultaneously pro-ducing bioenergy permanently sequestering carbon while improving soil and wa-ter quality. Agronomy Journal 2008; 100: 178–181.

[56] Gaunt JL, Lehmann J. Energy balance and emissions associated with biochar sequestration and pyrolysis bioenergy production. Environmental Science & Technology 2008; 42: 4152–4158.

[57] Lora EES. Palacio JCE, Rocha MH, Reno MLG, Venturini OJ, del Olmo OA. Issues to consider existing tools and constraints in biofuels sustainability assessments. Energy (2011); 36: 2097–2110.

[58] IPCC. IPCC Special Report on Renewable Energy Sources and Climate Change Mitigation. Prepared by Working Group III of the Intergovernmental Panel on Climate Change [Edenhofer O, Pichs-Madruga R, Sokona Y, Seyboth K, Matschoss P, Kadner S, Zwickel T, Eickemeier P, Hansen G, Schlömer S, von Stechow C (Eds)]. Cambridge University Press Cambridge United Kingdom and New York NY USA, 2011.

[59] Fazio S, Monti A. Life cycle assessment of different bioenergy production systems including perennial and annual crops. Biomass and Bioenergy, 2011; 35: 4868–4878.

[60] Fargione J, Hill J, Tilman D, Polasky S, Hawthorne P. Land clearing and the biofuel carbon debt. Science 2008; 319: 1235–1238.

[61] Curtright AE, Johnson DR, Willis HH, Skone T. Scenario uncertainties in estimating direct land-use change emissions in biomass-to-energy life cycle assessment. Biomass and Bioenergy 2012; 47: 240–249.

[62] EC. Certification of sustainable biofuels. European Biofuels Technology Platform. (Accessed on March 20, 2013 at http://www.biofuelstp.eu/certification.html).

[63] Davis SC, Parton WJ, Ddel Grosso SJ, et al. Impact of second-generation biofuel agriculture on greenhouse-gas emissions in the corn-growing regions of the US. Frontiers in Ecology and the Environment 2012; 10: 69–74.

[64] Crutzen PJ, Mosier AR, Smith KA, Winiwarter W. N_2O release from agrobiofuel production negates global warming reduction by replacing fossil fuels. Atmospheric Chemistry and Physics Discussions 2007; 7: 11191–11205.

[65] Smith KA, Searchinger TD. Crop-based biofuels and associated environmental concerns. GCB Bioenergy 2012; 4: 479–484.

[66] Schmer MR, Liebig MA, Hendrickson JR, Tanaka DL, Phillips RL. Growing season greenhouse gas flux from switchgrass in the Northern Great Plains. Biomass and Bioenergy 2012; 45: 315–319.

[67] Emery IR, Mosier NS, Liu SJ, Abrahamson LP, Scott GM. The impact of dry matter loss during herbaceous biomass storage on net greenhouse gas emissions from biofuels production. Biomass and Bioenergy 2012; 39: 237–246.

[68] Graus M, Eller ASD, Fall R, et al. Biosphere-atmosphere exchange of volatile organic compounds over C4 biofuel crops. Atmospheric Environment 2013; 66: 161–168.

[69] Georgescu M, Lobell DB, Field CB. Direct climate effects of perennial bioenergy crops in the United States. Proceedings of the National Academy of Science (US) 2011; 108: 4307–4312.

[70] Le PVV, Kumar P, Drewry DT. Implications for the hydrologic cycle under climate change due to the expansion of bioenergy crops in the Midwestern United States. Proceedings of the National Academy of Sciences (US) 2011; 108: 15085–15090.

[71] Wu M, Demissie Y, Yan E. Simulated impact of future biofuel production on water quality and water cycle dynamics in the Upper Mississippi river basin. Biomass and Bioenergy 2012; 41: 44–56.

[72] Wu YP, Liu SG. Impacts of biofuels production alternatives on water quantity and quality in the Iowa River Basin. Biomass and Bioenergy 2012; 36: 182–191.

[73] Van Loocke A, Bernacchi CJ, Twine TE. The impacts of *Miscanthus × giganteus* production on the Midwest US hydrologic cycle. GCBCB Bioenergy 2010; 2: 180–191.

[74] Porter CL Jr. An anlysis of variation between upland and lowland switchgrass *Panicum virgatum* L. in central Oklahoma. Ecology 1966; 47: 980–992.

[75] Love BJ, Einheuser MD, Nejadhashemi AP. Effects on aquatic and human health due to large scale bioenergy crop expansion. Science of the Total Environment 2011; 409: 3215–3229.

[76] Ng TL, Eheart JW, Cai XM, Miguez F. Modeling miscanthus in the Soil and Water Assessment Tool (SWAT) to simulate its water quality effects as a bioenergy crop. Environmental Science & Technology 2010; 44: 7138–144.

[77] Keeney D, Muller M. Water use by ethanol plants: potential challenges. 2006. Institute for Agriculture and Trade Policy. Minneapolis, MN, USA, 2006.

[78] Uzoma KC, Inoue M, Andry H, Zahoor A, Nishihara E. Influence of biochar application on sandy soil hydraulic properties and nutrient retention. Journal of Food, Agriculture & Environment 2011; 9: 1137–1143.

[79] Jeffery S, Verheijen FGA, van der Velde M, Bastos AC. A quantitative review of the effects of biochar application to soils on crop productivity using meta-analysis. Agriculture, Ecosystems & Environment 2011; 144: 175–187.

[80] Clark RB. Differences among mycorrhizal fungi for mineral uptake per root length of switchgrass grown in acidic soil. Journal of Plant Nutrition 2002; 25: 1753–1772.

[81] Tilman D, Reich PB, Knops JMH. Biodiversity and ecosystem stability in a decade-long grassland experiment Nature (London) 2006; 441: 629–632.

[82] Tilman D, Hill J, Lehman C. Carbon-negative biofuels from low-input high-diversity grassland biomass Science 2006; 314: 1598–1600.

[83] Anderson-Teixera KJ, Duval BD, Long SP, DeLucia EH. Biofuels on the landscape: is "land sharing" preferable to "land sparing"? Ecological Applications 2012; 22: 2035–2048.

[84] Hong CO, Owens VN, Lee DK, Boe A. Switchgrass, big bluestem and Indiangrass monocultures and their two- and three-way mixtures for bioenergy in the Northern Great Plains. BioEnergy Research 2013; 6: 229–239.

[85] Haughton AJ, Bond AJ, Lovett AA, *et al.* A novel integrated approach to assessing social economic and environmental implications of changing rural land-use: a case study of perennial biomass crops. Journal of Applied Ecology 2009; 46: 315–322.

[86] Roth AM, Sample DW, Ribic CA, Paine L, Undersander DJ, Bartelt GA. Grassland bird response to harvesting switchgrass as a biomass energy crop Biomass and Bioenergy 2005; 28: 490–498.

[87] Bonin CL, Tracy BF. Diversity influences forage yield and stability in perennial prairie plant mixtures. Agriculture, Ecosystems & Environment 2012; 162: 1–7.

[88] Robertson BA, Doran PJ, Loomis LR, Robertson JR, Schemske DW. Perennial biomass feedstocks enhance avian diversity. GCB CB Bioenergy 2011; 3: 235–246.

[89] Robertson BA, Landis DA, Sillett TS, Loomis ER, Rice RA. Perennial agroenergy feedstocks as en route habitat for spring migratory birds. BioEnergy Research 2013; 6: 311–320.

[90] Werling BP, Meehan TD, Robertson BA, Gratton C, Landis DA. Biocontrol potential varies with changes in biofuel-crop plant communities and landscape perenniality. GCBCB Bioenergy 2011; 3: 347–359.

[91] Robertson BA, Porter C, Landis DA, Schemske DW. Agroenergy crops influence the diversity biomass and guild structure of terrestrial arthropod communities. BioEnergy Research 2012; 5: 179–188.

[92] Robertson BA, Rice RA, Sillett TS, et al. Are agrofuels a conservation threat or opportunity for grassland birds in the United States? Condor 2012; 114: 679–688.

[93] Ma Z, Wood CW, Bransby DI. Impact of row spacing nitrogen rate and time on carbon partitioning of switchgrass. Biomass and Bioenergy 2001; 20: 413–419.

[94] Barney JN, DiTomaso JM. Bioclimatic predictions of habitat suitability for the biofuel switchgrass in North America under current and future climate scenarios. Biomass and Bioenergy 2010; 34: 124–133.

[95] Thomson LJ, Hoffmann AA. Pest management challenges for biofuel crop production. Current Opinion in Environmental Sustainability 2011; 3: 95–99.

[96] Farr DF, Rossman AY. Fungal Databases, Systematic Mycology and Microbiology Laboratory, ARS, USDA. (Retrieved March 22, 2013, from http://nt.ars-grin.gov/fungaldatabases/.)

[97] Fike JH, Butler TJ, Mitchell RB. The agronomy of switchgrass for biomass. In: Luo H and Wu Y (Ed.). Compendium of Bioenergy Plants: Switchgrass. CRC Press 2013.

[98] Schrotenboer AC, Allen MS, Malmstrom CM. Modification of native grasses for biofuel production may increase virus susceptibility. GCBCB Bioenergy 2011; 3: 360–374.

[99] Walkden HH. Cutworm and armyworm populations in pasture grasses waste lands and forage crops. Journal of Economic Entomology 1943; 36: 376–381.

[100] Nabity PD, Zangerl AR, Berenbaum MR, DeLucia EH. Bioenergy crops Miscanthus × giganteus and Panicum virgatum reduce growth and survivorship of Spodoptera frugiperda (Lepidoptera: Noctuidae). Journal of Economic Entomology 2011; 104: 459–464.

[101] Prasifka JR, Bradshaw JD, Lee ST, Gray ME. Relative feeding and development of armyworm on switchgrass and corn and its potential effects on switchgrass grown for biomass. Journal of Economic Entomology 2011; 104: 1561–1567.

[102] Schaeffer S, Baxendale F, Heng-Moss T, *et al.* Characterization of the arthropod community associated with switchgrass (Poales: Poaceae) in Nebraska. Journal of the Kansas Entomological Society 2011; 84: 87–104.

[103] Carpenter JE. Impact of GM crops on biodiversity. GM Crops 2011; 2: 7–23.

[104] Hitchcock AS. Manual of the Grasses of the United States. Washington DC: United States Department of Agriculture. 1935.

[105] Hultquist SJ, Vogel KP, Lee DJ, Arumuganathan K, Kaeppler S. Chloroplast DNA and nuclear DNA content variations among cultivars of switchgrass *Panicum virgatum* L. Crop Science 1996; 36: 1049–1052.

[106] Moser LE, Vogel KP. Switchgrass, big bluestem and indiangrass. In: Barnes RF, Miller DA, Nelson CJ (Eds.). Forages: An Introduction to Grassland Agriculture. Iowa State University Press Ames Iowa, 1995.

[107] Beaty ER, Engel JL, Powell JD. Tiller development and growth in switchgrass. Journal of Range Management 1978; 31: 361–365.

[108] Sanderson MA, Reed RL, Ocumpaugh WR, *et al.* Switchgrass cultivars and germplasm for biomass feedstock production in Texas Bioresource Technology 1999; 67: 209–219.

[109] Fike JH, Parrish DJ, Wolf DD, *et al.* 2006. Long-term yield potential of switchgrass-for-biofuel systems. Biomass and Bioenergy 2006; 30: 198–206.

[110] Parrish DJ, Fike JH. Selecting, establishing, and managing switchgrass (*Panicum virgatum*) for biofuels. In: Mielenz JR (Ed.). Methods in Molecular Biology: Biofuels, Methods and Protocols. Springer/Humana Press, New York. 2009: 27–40.

[111] Hanson JD, Johnson HA. Germination of switchgrass under various temperature and pH regimes. Seed Technology 2005; 27: 203–210.

[112] McKenna JR, Wolf DD. No-till switchgrass establishment as affected by limestone phosphorus and carbofuran. Journal of Production Agriculture 1990; 3: 475–479.

[113] Stucky DJ, Bauer JH, Lindsey TC. Restoration of acidic mine spoils with sewage sludge: I Revegetation Reclamation Review 1980; 3: 129–139.

[114] Esbroeck GAV, Hussey MA, Sanderson MA. Variation between Alamo and Cave-in-Rock switchgrass in response to photoperiod extension. Crop Science 2003; 43: 639–643.

[115] Casler MD, Boe AR. Cultivar x environment interactions in switchgrass. Crop Science 2003; 43: 2226–2233.

[116] Hope HJ, McElroy A. Low-temperature tolerance of switchgrass (*Panicum virgatum* L.). Canadian Journal of Plant Science 1990; 70: 1091–1096.

[117] Casler MD, Vogel KP, Taliaferro CM, Wynia RL. Latitudinal adaptation of switchgrass populations. Crop Science 2004; 44: 293–303.

[118] Schmer MR, Vogel KP, Mitchell RB, Moser LE, Eskridge KM, Perrin RK. Establishment stand thresholds for switchgrass grown as a bioenergy crop. Crop Science 2006; 46: 157–161.

[119] Parrish DJ, Fike JH, Bransby DI, Samson R. Establishing and managing switchgrass as an energy crop. Forage and Grazinglands 2008; FG-2008-0220-01-RV.

[120] Fike JH, Parrish DJ, Fike WB. Sustainable cellulosic grass crop production. In: Singh BP (Ed.). Biofuel Crop Sustainability. John Wiley & Sons, Hoboken, NJ, USA. 2013.

[121] Berti MT, Johnson BL. Switchgrass establishment as affected by seeding depth and soil type. Industrial Crops and Products 2012; 41: 289–293.

[122] Curran WS, Ryan MR, Myers MW, Adler PR. Effects of seeding date and weed control on switchgrass establishment. Weed Technology, 2012; 26: 248–255.

[123] Mooney DF, Roberts RK, English BC, Tyler DD, Larson JA. Yield and breakeven price of "Alamo" switchgrass for biofuels in Tennessee. Agronomy Journal 2009; 101: 1234–1242.

[124] Foster JL, Guretzky JA, Huo C, Kering MK, and Butler TJ. Effects of row spacing, seeding rates, and planting date on establishment and biomass of switchgrass. Crop Science 2013; 53: 309–314.

[125] Jung GA, Shaffer JA, Stout WL. Switchgrass and big bluestem responses to amendments on strongly acid soil. Agronomy Journal 1988; 80: 669–676.

[126] Sanderson MA, Reed RL. Switchgrass growth and development: water nitrogen and plant density effects Journal of Range Management 2000; 53: 221–227.

[127] Taylor RW, Allinson DW. Response of three warm-season grasses to varying fertility levels on five soils. Canadian Journal of Plant Science 1982; 62: 657–665.

[128] Mitchell RB, Schmer MR. Switchgrass harvest and storage. In: Monti A (Ed.). Switchgrass Green Energy and Technology. Springer-Verlag London, 2012; 113–127.

[129] Allison GG, Morris C, Lister SJ, et al. Effect of nitrogen fertiliser application on cell wall composition in switchgrass and reed canary grass. Biomass and Bioenergy 2012; 40: 19–26.

[130] Muir JP, Sanderson MA, Ocumpaugh WR, Jones RM, Reed RL. Biomass production of "Alamo" switchgrass in response to nitrogen, phosphorus, and row spacing. Agronomy Journal 2001; 93: 896–901.

[131] Stout WL, Jung GA. Biomass and nitrogen accumulation in switchgrass: effects of soil and environment. Agronomy Journal 1995; 87: 663–669.

[132] Tjepkema J. Nitrogenase activity in the rhizosphere of Panicum virgatum. Soil Biology & Biochemistry 1975; 7: 179–180.

[133] Riggs PJ, Moritz RL, Chelius MK, et al. Isolation and characterization of diazotrophic endophytes from grasses and their effects on plant growth. In: Finan TM, O'Brian MR, Layzell DB, Vassey JK, Newton W (Eds.). Nitrogen Fixation: Global Perspectives. Proceedings of the 13th International Congress on Nitrogen Fixation, Hamilton, Ontario, Canada, 2-7 July 2001. CABI Press, London, 2002: 263–267.

[134] Ker K, Seguin P, Driscoll BT, Fyles JW, Smith DL. Switchgrass establishment and seeding year production can be improved by inoculation with rhizosphere endophytes. Biomass and Bioenergy 2012; 47: 295–301.

[135] Kim SH, Lowman S, Hou GC, Nowak J, Flinn B, Mei CS. Growth promotion and colonization of switchgrass (Panicum virgatum) cv. Alamo by bacterial endophyte Burkholderia phytofirmans strain PsJN. Biotechnology for Biofuels 2012; 5: 37. doi: 10.1186/1754-6834-5-37.

[136] Waramit N, Moore KJ, Heggenstaller AH. Composition of native warm-season grasses for bioenergy production in response to nitrogen fertilization rate and harvest date. Agronomy Journal 2011; 103: 655–662.

[137] Lemus R, Parrish DJ, Abaye O. Nitrogen-use dynamics in switchgrass grown for biomass. BioEnergy Research 2008; 1: 153–162

[138] Garten CT Jr, Brice DJ, Castro HF, et al. Response of "Alamo" switchgrass tissue chemistry and biomass to nitrogen fertilization in West Tennessee USA. Agriculture Ecosystems & Environment. 2011; 40: 289–297.

[139] Madakadze IC, Stewart KA, Peterson PR, Coulman BE, Smith DL. Cutting frequency and nitrogen fertilization effects on yield and nitrogen concentration of switchgrass in a short season area. Crop Science 1999; 39: 552–557.

[140] Fike JH, Parrish DJ, Wolf DD, et al. Switchgrass production for the upper southeastern USA: influence of cultivar and cutting frequency on biomass yields. Biomass and Bioenergy 2006; 30: 207–213.

[141] Guretzky JA, Biermacher JT, Cook BJ, Kering MK, Mosali J. Switchgrass for forage and bioenergy: harvest and nitrogen rate effects on biomass yields and nutrient composition. Plant and Soil 2011; 339: 69–81.

[142] Lee KD, Owens VN, Doolittle JJ. Switchgrass and soil carbon sequestration response to ammonium nitrate manure and harvest frequency on conservation reserve program land. Agronomy Journal 2007; 99(2): 462–468.

[143] Springer TL, Aiken GE, McNew RW. Combining ability of binary mixtures of native, warm-season grasses and legumes. Crop Science 2001; 41: 818–823.

[144] Bow JR, Muir JP, Weindorf DC, Rosiere RE, Butler TJ. Integration of cool-season annual legumes and dairy manure compost with switchgrass. Crop Science 2008; 48: 1621–1628.

[145] Heggenstaller AH, Moore KJ, Liebman M, Anex RP. Nitrogen influences biomass and nutrient partitioning by perennial, warm-season grasses. Agronomy Journal 2009; 101: 1363–1371.

[146] Garten CT Jr, Smith JL, Tyler DD, et al. Intra-annual changes in biomass, carbon, and nitrogen dynamics at 4-year old switchgrass field trials in west Tennessee, USA. Agriculture, Ecosystems & Environment 2010; 136: 177–184.

[147] Jung JY, Lal R. Impacts of nitrogen fertilization on biomass production of switchgrass (Panicum virgatum L.) and changes in soil organic carbon in Ohio. Geoderma 2011; 166: 145–152.

[148] Kering MK, Biermacher JT, Butler TJ, Mosali J, Guretzky JA. Biomass yield and nutrient responses of switchgrass to phosphorus application. BioEnergy Research 2012; 5: 71–78.

[149] Neset TSS, Cordell D. Global phosphorus scarcity: identifying synergies for a sustainable future. Journal of the Science of Food and Agriculture 2012; 92: 2–6.

[150] Brejda JJ, Yocom DH, Moser LE, Waller SS. Dependence of 3 Nebraska Sandhills warm-season grasses on vesicular-arbuscular mycorrhizae. Journal of Range Management 1993; 46: 14–20.

[151] Lewandowski I, Clifton-Brown JC, Scurlock JMO, Huisman W. Miscanthus: European experience with a novel energy crop. Biomass and Bioenergy 2000; 19: 209–227.

[152] Johnson PC, Clementson CL, Mathanker SK, Grift TE, Hansen AC. Cutting energy characteristics of *Miscanthus × giganteus* stems with varying oblique angle and cutting speed. Biosystems Engineering 2012; 112: 42–48.

[153] Cundiff JS, Marsh LS. Harvest and storage costs for bales of switchgrass in the southeastern United States. Bioresource Technology. 1996; 56: 95–101.

[154] Resop JP, Cundiff JS, Heatwole CD. Spatial analysis to site satellite storage locations for herbaceous biomass in the piedmont of the Southeast. Applied Engineering in Agriculture 2011; 27: 25–32.

[155] Hess JR, Wright CT, Kenney KL. Cellulosic biomass feedstocks and logistics for ethanol production. Biofuels Bioproducts & Biorefining 2007; 3: 181–190.

[156] Egbendewe-Mondzozo A, Swinton SM, Bals BD, Dale BE. Can dispersed biomass processing protect the environment and cover the bottom line for biofuel? Environmental Science & Technology 2013: 47: 1695–1783.

[157] Ng TL, Cai XM, Ouyang YF. Some implications of biofuel development for engineering infrastructures in the United States. Biofuels Bioproducts & Biorefining 2011; 5: 581–592.

Chapter 3

Sugarcane as an Alternative Source of Sustainable Energy

Sushil Solomon and Pushpa Singh

3.1 Introduction

Projections on the doubling of the world's population and increasing aspirations
of people to have better qualities of life have resulted in massive increase in demand for energy in all its forms. The world currently derives about 60% of its
energy from fossil fuels, the supplies of which are limited and, at the present rate
of consumption, these reserves are likely to last less than 30 years. Inevitably,
fuel prices have begun to rise faster than the average rate of inflation. An equally
worrying consequence of the projected increase in fossil-fuel consumption is the
emission of the greenhouse gas (GHG) carbon dioxide, whose rate of accumulation in the atmosphere is also expected to double. It is thus in the direct global
interest that the renewable energy transition is immediate, rapid, and orderly
and the only natural renewable carbon resource known that is large enough to
be used as a sustainable substitute for fossil fuels is biomass. In addition to
substantially reducing carbon emissions, the production of bioethanol, biodiesel,
bio-oil, and other modern biofuels from biomass also promotes the generation
of jobs and income, especially in rural regions where the largest concentrations
of poverty and extreme poverty are found around the world. It is in this context that the rationality of sugarcane as an alternative source of sustainable
energy shall require more emphasis in planning energy requirements in the coming decades.

Sugarcane (*Saccharum* spp.), a C4 photosynthetic large-stature perennial grass,
is a worldwide crop cultivated in more than 100 countries in tropical, semi-
tropical, and subtropical regions. It has one of the most efficient photosynthetic
mechanisms among commercial crops, which allows it to fix almost 2%–3% of
radiant solar energy and transforms it into chemical energy that is usable as a

food and fuel source. The high photosynthetic capability also allows it to show a high coefficient of CO_2 fixation, comparable to the moderate climate zone woods, thus contributing to the decrease of the greenhouse effect. A hectare of sugarcane produces about 100 tonnes of green matter every year, which is more than twice the agricultural yield of most other commercial crops. A crop of 100 tonnes leaves around 60–65 tonnes of dry matter (8–10 tonnes of trash consisting of dry leaves) in the field, while, after the extraction of sucrose, nearly 50–60 tonnes of bagasse is generated in sugar factory zones. This dry matter, when burned, has the potential to produce 4000 kcal per kg (7200 Btu/lb). The total dry matter content thus has a fuel equivalent of about 10–20 tonnes of oil and an efficient use of the energy potential of sugarcane can result into approximately 1 tonne of oil equivalent for every tonne of sugar produced. Bagasse is generally used as a raw material for heat energy to run the sugar mills and also for cogenerating electricity, whereas trash, with the exception of its use as mulch, is yet to be exploited for its fuel potential. As a metabolic energy carrier for animal feeding, each cultivated hectare delivers 75, 000 million calories each year, which is equivalent to more than eight times the yield of other fodder (60 to 120 kWh of electric power per tonne of cane). From sugarcane harvest and processing, it is possible to obtain more than eight products and byproducts, which are potential raw materials for the extractive, chemical, biochemical, and fuel industries, leading to the production of more than fifty commercial products (Fig. 3.1). Practically all products and byproducts obtained from sugarcane have the potential to serve as substrates for liquid or solid-state fermentation processes and by usage of the available 2nd and 3rd generation technologies, a significant number of production processes and products can be developed.

The 1st-generation biofuels have been dependent on food crops such as oilseeds (rapeseed, palm oil), starch crops (cereals, maize), or sugar crops (sugar beet and sugarcane). Conversion technologies, though, as far as commercial production costs, have been typically high due to high feedstock costs and the net overall avoided GHG emissions range between 20%–50% compared to conventional gasoline or diesel [1]. Another constraint is that such food crops need to be produced on better quality land and increased demand directly competes with food markets. However, sugarcane-based energy production is a notable exception to these key concerns as overall production costs are competitive and net GHG balance achieves 80%–90% reduction; sugar prices have remained constant or have decreased slightly over the past years, despite strong increases in bioethanol production from sugarcane.

Sugarcane, thus, has a significant advantage as a renewable raw material, in the production of basic chemicals, with a yield not equaled by any other plant. Since the energy-delivering capacity of sugarcane is equivalent to five times that used by the crop, the energy produced by sugarcane plants and variations among different varieties or germplasm makes it an extremely important renewable and

Filter mud — Fertilizer, Animal feed, Cane wax

Bagasse — Chemical, Mechanical, Microbial processes

Flue gases

Sugar

Cane tops & Leaves(Trash) — Sugar cane

Jaggery

Furnace ash

Molasses — Chemical, Mechanical, Microbial processes

Proteins from juice

Fuel — Electricity, Charcoal, briquettes, Producer gas

Fibrous products — Pulp industry, Fiber board, Paper(Writing Newsprint), Writing board

Miscellaneous — Furfural, Alpha cellulose, Xylitol, Plastics, Poultry litter, Animal feed, Concrete, Soil amendment (Microbial processes)

Direct utilization — Fertilizer, Animal feed

Distillery industry — Rum, Ethanol, Rectified spirit, Anhydrous alcohol, Ethanol derivatives

Fermentation Industry — Vinegar, Citric acid, Butanol-acetone, Citric acid Lactic acid, Glycerol, Yeast(Bakers, Feed) (Microbial processes)

Miscellaneous — Aconitic acid, Mono sodium glutamate, Dextran, Lysine, Xanthum gum, Itaconic acid

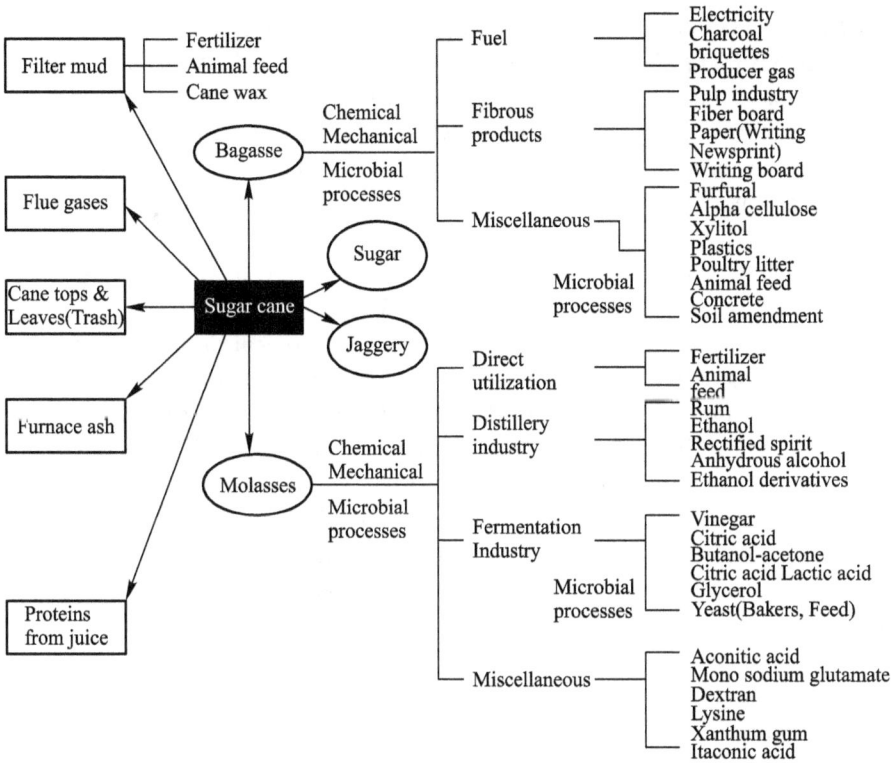

Fig. 3.1 Products and byproducts of sugarcane.

sustainable energy bio-resource. All of the above factors taken together along with the possibilities offered by further genetic improvement turn sugarcane into an ideal energy crop for the next century. This chapter analyzes the present energy potential of sugarcane and its conversion efficiencies and tries to project the data for the future, considering the gains in agricultural productivity, fuel efficiency in light of emerging energy needs, and the phasing in of emerging technologies, such as gasification, hydrolysis, and biomass to liquid (BTL), especially to achieve the conversion of sugarcane to ethanol, electricity, and liquid fuels.

3.2 Energy Expenses in Sugarcane Production

The total energy expenditure for growing one hectare of sugarcane has been computed to be 148.02 GJ ha^{-1} for the plant crop and the energy output is about 112.22 GJ ha^{-1} for plant crops. Electricity is the main energy input accounting for 43% in the plant crop while the second single largest energy input in the plant crop is diesel fuel used in the farm machinery and transport accounting for

23.0% (34.04 GJ), followed by nitrogen (N) fertilizer 14.4% (21.32 GJ), sugarcane cuttings 8.3% (12.25 GJ), and machinery 6.0% (8.93 GJ). Chemicals are the smallest of all inputs with 1.3% (1.92 GJ). Energy outputs in sugarcane farms with about 90 t ha^{-1} yield are 112.22 GJ ha^{-1} for plant crops. The output to input energy ratio has been reported to be 0.76. Energy productivity, specific energy, and net energy gain are 0.63 kg MJ^{-1}, 1.59 MJ kg, and 35.8 GJ ha^{-1}, respectively [2].

Worldwide sugarcane occupies an area of 22.00 million ha, approximately 0.5% of the total world area used for agriculture [3], lying between the latitude 36.7° north and 31.0° south of the equator, extends from tropical to subtropical zones, and its average yield is 70.9 t ha^{-1}. Sugarcane produces the world's greatest crop tonnage [3] and is the most efficient collection of solar energy in the plant kingdom, fixing 2% of the available solar energy into chemical bonds of organic compounds, chiefly composed of carbohydrates (sugar and lignocellulose) that have an energy content of $\sim$ 15.9 MJ kg^{-1}. The high photosynthetic capability also allows it to show a high coefficient of CO_2 fixation, thus contributing to the decrease of the greenhouse effect. Since it is an efficient assimilator, it can produce more than 200 tonnes of biomass (in fresh weight) per ha in the best experimental conditions and a huge figure of 381 t ha^{-1} has been estimated as a theoretical maximum annual yield of sugarcane in the most favorable conditions. Typical values in informal sources for average sugarcane yield may range between 50–150 t ha^{-1} and, in wet tropics, good rain-fed sugarcane yield is 70–100 t ha^{-1}, whereas in dry tropics and subtropics, good sugarcane yield using irrigation may often be 110–150 t ha^{-1}.

The sugarcane plant basically consists of stem and straw composed of 73%–76% water, 10%–16% soluble solids, and 11%–16% fibers. Physically, sugarcane is constituted by four major fractions, whose relative magnitudes depend on the sugar agro-industrial process: fiber, non-soluble solids, soluble solids, and water (Fig. 3.2). The fiber is composed of the whole organic solid fraction, originally found in the cane's stem, and is characterized by its marked heterogeneity. The non-soluble solids, or the fractions that cannot be dissolved in water, are constituted mainly by inorganic substances (rocks, soil, and extraneous materials) are greatly influenced by the conditions of the agricultural sugarcane processing, types of cutting, and harvesting. Soluble solid fractions that can be dissolved in water are composed primarily of sucrose as well as other chemical components such as waxes, in a smaller proportion. The solids are primarily sugars and non-sugars while the fiber is composed of many cellulosic and non-cellulosic products. The sugarcane straw (or trash) is divided in three principal components; that is, fresh leaves, dry leaves, and tops. Chemically, the leaves, stalks, stubbles, and roots (2/3) of sugarcane serve to be the richest sources of lignocelluloses, while the juice produces sucrose as the chief product and several byproducts as wastes, molasses being the prime waste. Sugarcane has a high harvest index of

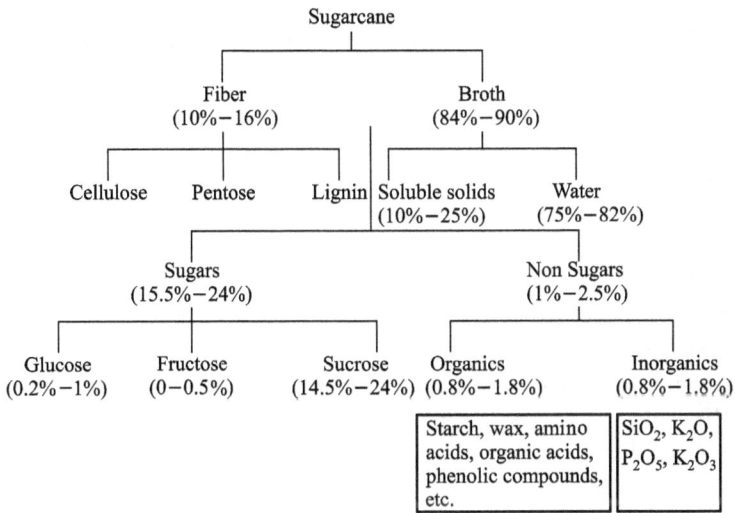

Fig. 3.2 General composition of sugarcane. Adapted from [4].

0.8 because a majority of the plant's organs are harvested, except for a fraction (~ 0.2) of plant material that remains in the stubble, roots, and trash consisting of dead stalks and leaves. The potential theoretical yield of aboveground biomass is 177 t ha^{-1} yr^{-1} or a fresh weight cane yield of 381 t ha^{-1} yr^{-1}. Sugarcane thus happens to be a highly valuable economic renewable, natural agricultural resource as every part of the plant can be put to creative utilization and thus serves to be a rich reservoir of sugar along with biofuel, fiber, fertilizer, and a myriad of byproducts/co-products with ecological sustainability.

A hectare of sugarcane produces about 100 tonnes of green matter every year, which is more than twice the agricultural yield of most other commercial crops. A crop of 100 tonnes leaves around 60–65 tonnes of dry matter (8–10 tonnes of trash) in the field, while after extraction of sucrose, nearly 50–60 tonnes of bagasse is generated in sugar factory zones. This dry matter, when burned, has the potential of producing 4,000 kcal per kg (7200 Btu/lb). The total dry matter content thus has a fuel equivalent of about 10–20 tonnes of oil and an efficient use of the energy potential of sugarcane can result in approximately 1 tonne of oil equivalent for every tonne of sugar produced (Tab. 3.1).

Tab. 3.1 Primary energy of sugarcane per tonne of clean stalks (higher heating value).

Component	Energy (MJ)
146 kg of sugars	2,400
130 kg of stalk fibers	2,300
140 kg of straw fibers	2,500
Total	7,200 (1.2 boe)*

*boe = barrel of oil equivalent.

3.3 Nutrient and Fertilizer Expenditures of Sugarcane

Sugarcane crop needs about $1,500$–$2,500$ mm of water evenly distributed during the growth phases. About 37–330 kg of water is used for producing one kg of cane and $1,000$–$2,000$ kg of water for producing one kg of sucrose, respectively. The crop needs dry, sunny and cool conditions in order to ripen to harvest state and boost its sugar content to 10%–12%. Rooting and sprouting of the planted stem pieces occurs best at 32–38°C and stalk growth reaches its optimum at 22–30°C, but ripening of stems and their sugar enhancement proceeds most successfully at 10–20°C. Optimum soil pH for sugarcane is 6.5 but the plant can be grown in soils with pHs of 5–8.5. Sugarcane grows best in more than a one-meter-deep layer of soil and parts of its root system may extend into a depth of five meters. However, the bulk of its roots (85%) typically are harbored in the uppermost 60 cm of soil, especially if the plant is irrigated often and with small doses of water at a time. Deeper root systems are generated by irrigating the plants less frequently and with greater doses each time. Sugarcane needs nitrogen (100–200 kg ha^{-1}, referring to a yield level of 100 t ha^{-1}) as well as potassium (125–160 kg ha^{-1}), but rather little phosphorus (20–90 kg ha^{-1}) is sufficient. In wet tropical areas, only about 6% of fertilizer nitrogen is utilized by the cane plant, whereas in temperate regions, 20%–40% of N fertilizer is being utilized by sugarcane for yield production. However, in the ripening period, the N content in the soil should be as low as possible in order to reach high sucrose content in the stems, especially in hot and wet conditions.

3.4 Sugarcane Bagasse: A Sustainable Energy Resource

The sugarcane stems are milled to obtain the cane juice, which is subsequently used for sugar (sucrose) or alcohol (ethanol) production. The residual fraction from the sugarcane stem milling is called bagasse. Sugarcane bagasse (SB) is chemically composed of cellulose, hemicellulose, and lignin. Cellulose and hemicellulose fractions are composed of a mixture of carbohydrate polymers. The cellulosic fraction is solid and rich in glucose and the hemicellulose fraction is liquid and rich in xylose, glucose, and arabinose. Each tonne of sugarcane yields about 250 kg of bagasse. The chemical composition varies according to the variety of the cane, maturity status, and efficiency of the milling plant. When it comes to the milling plant, bagasse has about 50% moisture and contains 46%–

52% fiber, 43%–52% moisture, 3% sugar, and 0.55% minor constituents (Tab. 3.2). As bagasse is a rich source of lignocellulose and it has been responsible for its use in all the fiber-based industries. Being rich in cellulose, it is utilized in all of the industries where cellulose serves as the base material. As the internal pith is of no use, it is used in the manufacture of pulp, paper, and other cellulosic products. While the cellulosic wastes are converted into paper, the pentosan contents are used for manufacturing furfural.

Tab. 3.2 Chemical properties of sugarcane bagasse.

Property	Value
Water content	43%–52%
Fiber Content	46%–52%
Soluble solids	2%–6%
Average density	150 kg m^{-3}
Low heat value	1,780 kcal kg^{-1}
High heat value	4,000 kcal kg^{-1}

3.4.1   Electricity Generation from Bagasse

The burning of bagasse for steam generation in sugarcane factories and utilization of this medium or high pressure steam for prime movers and power generation is a standard practice of the sugarcane factories throughout the world. The fuel value of bagasse is mainly on account of its fiber content. It is used as captive fuel in sugar factories because the sugar industry is a seasonal industry that deals with highly perishable raw material like sugarcane and its intermediate products *viz*, cane juice, massecuites, molasses and syrup, etc. Therefore, it cannot depend on traditional and extraneous fuels like coal, furnace oil, or natural gas. In these circumstances, the sugar industry depends solely on the readily available supply of bagasse produced within the same premisis. Now a days, many fuel-efficient factories are coming up in countries such as Brazil, Cuba, Mauritius, USA, China, Indonesia, Australia, India, and other places where surplus bagasse is being used for generating electric power to their grids. The process of bagasse cogeneration and the cost of electricity generation are sketched in Figure 3.3. According to Paturau [5–6], a modern sugarcane factory, producing raw sugar and designed for fuel economy, would require 30 kWh of power and 300 kg of exhaust steam per tonne of cane. Under these conditions, 50% of the bagasse produced will be surplus and can be used for electricity generation (Tab. 3.3). In this direction, some Indian state electricity boards, such as Tamil Nadu State Electricity Board, have agreed to accept surplus power from cogenerators

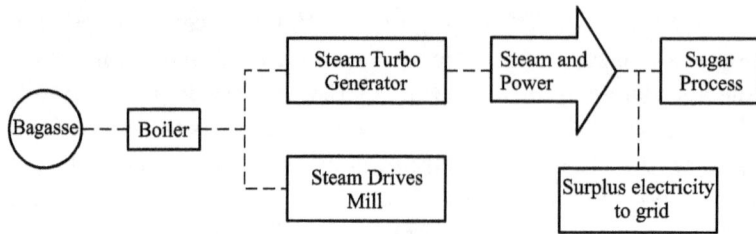

Fig. 3.3 The bagasse cogeneration process.

Tab. 3.3 Global market potential of bagasse.

Country	Sugarcane production (t yr^{-1})	Potential for electricity production (GWh yr^{-1})	Bagasse potential as % of electricity demand
Brazil	386, 232, 000	38, 623	11.50
India	290, 000, 000	29, 000	5.83
China	–	9, 390	0.72
Thailand	74, 071, 952	7, 407	8.15
Pakistan	52, 055, 800	5, 206	8.36
Mexico	45, 126, 500	4, 513	2.42
Columbia	36, 600, 000	3, 660	9.19
Australia	36, 012, 000	3, 601	1.95
Cuba	34, 700, 000	3, 470	25.93
USA	31, 178, 130	3, 118	0.09
Phillipines	25, 835, 000	2, 584	6.16
Others	244, 581, 738	24, 458	0.32
Total	1, 350, 293, 120	135, 029	0.97
Total (excl. China, Australia, USA, etc.)	944, 621	94, 462	7.45

on a banking basis. Similar projects are being set up in other states of India like Maharashtra, Punjab, and Uttar Pradesh. However, a major hurdle in a cogeneration project is to fix the price at which excess electricity would be sold by the sugar mills to the state electricity departments. It is necessary that sugar mills get an adequate return on the investment so that they will have to make on the cogeneration of power from bagasse.

Cogeneration from sugarcane waste (bagasse) provides one of the best examples of renewable-based cogeneration, yet it remains largely unexplored. The advantages of bagasse as a fuel for cogeneration are numerous, ranging from environmental to social and economic. From a financial point of view, bagasse cogeneration is a classic win-win for the sugar industry, as it boasts numerous ad-

vantages over traditional generation. Cogeneration of energy from bagasse is attractive as it combines low-cost, efficient, and social benefits with the provision of
clean, renewable energy. Bagasse cogeneration, especially in high-temperatures
and pressure configurations, thus plays an important role in encouraging much
more efficient use of resources and ensuring widespread access to electricity services (Tab. 3.4). However, insufficient incentive to supply electricity to the grid
because of low or non-existent buyback rates has meant that around two-thirds
of harvested bagasse was wasted. This situation is now set to improve with the
introduction of more effective biomass feed-in tariffs in countries such as Brazil
and India.

Tab. 3.4 Electricity from bagasse.

Characteristics and costs	Best conditions	Moderate conditions
Characteristics		
Boiler (46 Bar A, 440°C) capacity tonnes steam per hour	90	90
Turbo-alternator (condensing at 0.10 Bar A) capacity (MW)	20	20
Total capital investment for generating station in working order (USD, million)	9	11
Electricity generated yearly (GWh)	150	120
Weight of mill-run bagasse utilized (t)	333,000	266,000
Acquisition cost of mill-run bagasse (USD per tonne)	15	20
Average transport cost per tonne of bagasse (USD)	4	5
Cost of electricity generated (in USD cents per kWh)		
Depreciation and maintenance (10%)	0.60	0.92
Annuity repayment (0.16275 for 10 years at 10% interest)	0.98	1.49
Labor and administration (USD 100 000 yearly)	0.07	0.08
Transport cost of bagasse	0.89	1.11
Acquisition cost of bagasse	3.33	4.48
Total generation cost per kWh	5.87	8.08
	US cents	US cents
	6.00 kWh	8.00 kWh

Source: [7]

Cogeneration is a highly efficient energy conversion process. The same amount
of bagasse yields more power (heat as well as electricity) in cogeneration mode
than in conventional combustion processes that do not recover heat. More efficient fuel uses can thus also be further countries' sustainable development goals.
The potential to make a meaningful contribution to the energy balance is especially great in Cuba, Brazil, India, Thailand, Pakistan, Colombia, Mexico, and

the Philippines. Overall, the potential among these countries (which account for 70% of the global cane production) reaches as high as 25% in Cuba and, as an average, is a significant 7.45% of the total demand. The potential, in absolute terms, is also high in China. As a decentralized mode of electricity generation, bagasse cogeneration reduces transmission and distribution (T&D) losses by supplying electricity near its generation point whilst reducing loads on grid wires. This could be most significant in large countries such as India and Brazil, where average T&D losses account for around 23% and 16% of centrally generated electricity, respectively, mainly due to long distances between generation and end users. In countries such as Brazil and India, where peak power can be up to ten times costlier than off-peak power, sugar mills can benefit immensely from the opportunity to sell electricity to the grid.

The advantages of bagasse cogeneration in increasing the security of power supply issues also include the capacity to generate during the dry season. Sugar mills that produce and export electricity also increase grid stability and reliability as well as decrease the need for costly capital investments in grid upgrades in these areas. In Brazil, for instance, Sao Paolo State has already developed all of its large economically viable HEP sites. Thus, the promotion of electricity generation from bagasse cogeneration is seen as a means of avoiding electricity imports from other regions to meet the State's demand. The capital costs of bagasse cogeneration plants are the lowest of all renewable forms of power generation, equal to those of biomass gasification projects, whilst generation costs, despite being higher than biomass gasification projects, such as small hydroelectric (HEP) and photovoltaic (PV), are on par with biomass power and lower than wind.

Bagasse cogeneration has the potential to boost employment for neighboring populations and to increase farmers' income. It will also allow operational personnel to develop skills to use local equipment and technologies, thus improving the local socio-economy. As sugar mills tend to be located in rural areas near sugarcane plantations, bagasse cogeneration will prove beneficial to local populations by contributing to the expanding access to electricity supplies in areas otherwise distant from the grid. The advent of links to the network will facilitate the collection of electricity payments by electricity boards in rural areas whilst electricity boards will be able to serve rural consumers better through the upgrading of local and rural networks. The simultaneous increase in reliability and quality of power in the area will enhance the quality of life whilst reducing voltage and frequency variation and the associated damages that these may cause to network equipment. As it is a locally sourced fuel, bagasse will increase the reliability of the electrical supply by diversifying sources and reducing fossil fuel dependence.

3.4.2 Reduction in Greenhouse Gas (GHG) Emissions

As a biomass fuel, bagasse supplies a raw material for the production of natural, clean, and renewable energy, enabling its use to further government targets for renewable energy use. Bagasse combustion is environmentally friendly because it boasts low emission of particulates, SO_2, NO_x, and CO_2 compared to coal and other fossil fuels. For instance, in India and China, bagasse could displace coal, which, amongst other problems, has very high levels of ash. In terms of CO_2 and other GHGs, bagasse cogeneration would add no net emissions. Bagasse is generally viewed as a waste product that needs to be disposed of either by decomposition (composting) or combustion, both of which would release CO_2. Besides, if the bagasse was to be composted, it would also release methane, a kind of GHG that is 27 times more potent than CO_2. These benefits enable bagasse cogeneration to be a potentially significant player in international carbon (C) credit markets in the future, with sugar industries reaping the social and financial benefits of the added revenues. The harvesting of green sugarcane provides an emission reduction of approximately 30 kg of CH_4 and 0.80 kg of N_2O ha^{-1}. The methane emission lies in the range of 35–38 kg CH_4 ha^{-1}. Nitrous oxide emissions vary from 0.5 to 3 kg ha^{-1}. Net savings in CO_2 (equivalent) emissions, due to ethanol and bagasse substitution for fossil fuels correspond to 46.7× 10^6 t CO_2 (equivalent) yr^{-1}, nearly 20% of all CO_2 emissions from fuels in Brazil. Ethanol alone is responsible for 64% of the net avoided emissions. The emissions not resulting from the use of fossil fuels will be reduced from 19.5 kg of CO_2 (in 2005/2006) to 11.6 kg CO_2 Mg^{-1} of sugarcane in 2020 [8].

3.4.3 Bagasse-based Byproducts and Future Energy Assessment

Bagasse-based byproducts also address the challenge of the unavailability of fuel out of season. Surplus bagasse can be converted into bagasse logs of about 2–3 inches in diameter and 2–3 feet long to be used in place of the conventional fuel wood for domestic cooking. The production of charcoal from bagasse involves carbonization, mixing of molasses, and final carbonization of briquettes. Bagasse is a source of industrially important chemicals (Fig. 3.1) such as furfural, furfuryl alcohol, α-cellulose, xylitol, sucrolin, ethanol, activated carbon, hydrolyzed pith, and microbial protein from bagasse pith [9–10]. Considering the fast increase in green cane harvesting and the availability of straw as a supplemental fuel to bagasse, the potential of surplus power tends to increase. The following alternative technologies will be used in the comparisons of the possible different uses of sugarcane biomass for the future assessment of bagasse potential (Tab. 3.5).

Tab. 3.5 Avoided CO_2 emissions with bagasse utilization as fuel (measured as C).

	50% moisture (10^6 t yr^{-1})	Fuel oil replaced (10^6 t yr^{-1})[a]	Avoided C release (10^6 t yr^{-1})[b]
Bagasse production	76.0	–	
Bagasse utilization			
Sugar production	28.0	4.9	4.2
Energy sector (ethanol)	37.0	6.5	(5.5)[c]
Fuel, other sectors	7.0	1.2	1.0
Losses, other uses	4.0	–	–
Total			5.2[c]

a. Wet bagasse: 7.74 MJ kg^{-1}, LHV, boiler efficiency 74% (bagasse) and 82% (fuel oil), related to LHV.
b. Fuel oil: 0.86 kg C per kg fuel oil.
c. Bagasse as fuel for ethanol production is not considered as avoiding carbon release; it is treated here as an "internal transformation".

Present technology: Surplus power generation with a pure cogeneration system with a backpressure (BP) steam turbine generator and steam conditions at 22 bar/300°C, using only bagasse as fuel.

Advanced technology I: Surplus power generation with a condensing/extraction steam turbine (CEST) generator, steam conditions at 90 bar/520°C, using only bagasse as fuel.

Advanced technology II: Surplus power generation with a CEST generator, steam conditions at 90 bar/520°C, using all bagasse and 40% of the straw.

Advanced technology III: Surplus power generation using Biomass Integrated Gasification/Combined Cycle (BIG/CC), using all bagasse and 40% of the straw.

Advanced technology IV: Additional bioethanol production by the hydrolysis of the available biomass and generation of surplus power in a cogeneration system supplying the power required by both the conventional and hydrolysis bioethanol plants, with steam conditions at 65 bar/480°C, using all bagasse and 40% of the straw.

The process steam consumption for the conventional bioethanol plant was assumed to be 500 kg/tonne of cane for the present technology and 340 kg tonne of cane for all advanced technology cases. The output of useful forms of energy (bioethanol and electricity) for the five cases above is summarized in Tables 3.6 and 3.7 [11].

Tab. 3.6   Useful energy production for the different alternatives.

	Present	Adv. I	Adv. II	Adv. III	Adv. IV
Steam conditions					
Pressure (bar)	22	90	90	90	65
Temperature (°C)	300	520	520	520	480
Process steam (kg tc^{-1})	500	340	340	340	340
Power generation technology	BP	CEST	CEST	BIG/CC	BP/Hydrol
Bagasse (% total)	95	100	100	100	100
Straw (% total)	0	0	40	40	40
Bioethanol yield (L tc^{-1})	82	82	82	82	119
Surplus electricity (kWh tc^{-1})	5	81	145	194	44

tc = tonnes of cane processed.

Tab. 3.7   Energy and GHG emission balances for the alternative uses of sugarcane biomass.

	Present	Adv. I	Adv. II	Adv. III	Adv. IV
Bioethanol yields (L tc^{-1})	82	82	82	82	119
Surplus power (kWh tc^{-1})	5	81	145	194	44
Final products energy (HHV)					
Bioethanol (MJ tc^{-1})	1,919	1,919	1,919	1,919	2,785
Surplus electricity (MJ tc^{-1})	18	292	522	698	158
Surplus bagasse (MJ tc^{-1})	124	0	0	0	0
Total (MJ tc^{-1})	2,061	2,211	2,441	2,617	2,943
Primary energy recovery (%)	28.6	30.7	33.9	36.3	40.9
Avoided GHG					
Bioethanol (kg CO$_2$ e tc^{-1})	173	173	173	173	251
Electricity (Brazil average) (kg CO$_2$ e tc^{-1})	1	22	38	52	12
Bagasse (kg CO$_2$ e tc^{-1})	12	0	0	0	0
Total (kg CO$_2$ e tc^{-1})	186	195	211	225	263

Source: [11].

3.5   Sugarcane Trash: A Potential Biomass for Sustainable Energy

A sugarcane crop producing 115 tonnes of millable canes per hectare normally produces 7–9 tonnes of dry leaves (trash), 8–9 tonnes of stubble, and 4–6 tonnes of root. In India, 30–38 million tonnes of trash is available annually, which is either burnt in the field or is mulched due to its beneficial effect on the moisture conservation and weed control in ratoon raising. This waste chemically represents typical lignocellulosic material composed of approximately 40% cellulose, 25% hemicelluose, and 18%–20% lignin. One major application of trash is as crop mulch because it adds plant nutrients, conserves moisture, improves soil's fertility status, and sustains it with minimum use of energy intensive inputs like irrigation and chemical fertilizers. Sugarcane trash contains approximately 5.0 kg N, 1.5 kg P, and 5.7 kg K per tonne. Besides these macronutrients, the quantity of micronutrients and other substances in the trash is of considerable importance and the concentration of these is given in Table 1. The proportion of water-soluble substances ranges between 14%–18% of trash on dry weight basis. The water extract of the trash is acidic (pH 5.5) due to the presence of phenolic acids. The amount of phenolic acids in the trash is 33.0 kg Mg^{-1}. Nutrients are leached out easily through soaking the trash in water for 24 hours. Carbon and ash contents decrease in the trash eight-fold and 53% indicates that half the nutrients present in the trash are released without microbial decomposition. The residual biomass decomposes relatively faster in tropical and subtropical areas as compared to temperate regions. As the cost of energy required for irrigation and chemical fertilizers is increasing day by day and greater emphasis is being laid on sustainable agriculture, efficient methods are being explored to use crop residues. Sugarcane trash has a high $C:N$ ratio and contains water-soluble and insoluble substances.

The sugarcane trash, about 400–460 kg Mg^{-1}, contains lignin ranging from 230–260 kg Mg^{-1} of polysaccharides and can be freed from lignin so as to make celluloses and hemicelluloses available for enzymatic treatment reactions by either chemical or mechanical disruption. The disruption of the lignin seal is termed as a pretreatment process. Various strategies involving internal, chemical, and enzymatic pretreatments have been used to disrupt and crystallize structures of the celluloses and hemicelluloses [12–13].

3.6 Sugarcane Biomass for Biofuel Production

3.6.1 Chemical Composition of Sugarcane Biomass

Sugarcane biomass, which is a focus of 2nd generation ethanol production, is a lignocellulosic material composed of cellulose, hemicelluloses, and lignin. Cellulose is a linear polymer of glucose units linked by β $(1 \rightarrow 4)$—glycosidic bonds, forming cellobiose that is repeated several times in its chain. This cellulosic fraction can be converted into glucose by enzymatic hydrolysis, using cellulases, or by chemical means. Hemicellulose is a heteropolysaccharide composed by hexoses (D-glucose, D-galactose, and D-mannose), pentoses (D-xylose, L-arabinose), acetic acid, D-glucuronic acid, and 4-O-methyl - D-glucuronic acid units. The hemicelluloses are classified basically according to the sugars that are present in the main chain of polymer such as xylan, glucomannan, and galactan [14]. The hemicellulose differs substantially from cellulose to be amorphous, a character which makes it easier to be hydrolyzed than cellulose [15]. The hemicellulosic fraction can be removed from lignocellulosic materials by some type of pretreatment, like acid or hydrothermal hydrolysis, and liberating sugars, mainly xylose, that subsequently can also be fermented to ethanol [16–17].

Lignin is a complex aromatic macromolecule formed by the radical polymerization of three phenyl-propane alcohols, namely p-coumarilic, coniferilic, and synapilic. In the plant cell wall, lignin and hemicelluloses involve the cellulose elementary fibrils, providing protection against chemical and/or biological degradation [18]. The content of lignin and its distribution are the responsible factors for the recalcitrance of lignocellulosic materials to enzymatic hydrolysis, limiting the accessibility of enzymes. Therefore, the process of delignification can improve the conversion rates of enzymatic hydrolysis [19]. The lignin is primarily used as a fuel, but it can be chemically modified to be used as chelating agent [20], for removal of heavy metals from wastewater [21] or as precursor material for the production of value-added products as activated carbon [22], surfactants [23], and adhesives [24]. Sugarcane biomass is quantitatively composed of 38.4%–45.5% cellulose, 22.7%–27.0% hemicellulose, and 19.1%–32.4% lignin. Non-structural components of biomass, namely, ashes (1.0%–2.8%) and extractives (4.6%–9.1%), are the other substances that make the chemical composition of biomass. The ash content of sugarcane biomass is lower than the other crop residues like rice straw and wheat straw (with approximately 17.5% and 11.0%). Sugarcane biomass is also considered a rich solar energy reservoir due to its high yields and annual regeneration capacity [25].

3.6.2 Conversion of Sugarcane Biomass into Ethanol

Sugarcane bagasse (SB) and sugarcane straw (SS) are ideal feedstocks for 2nd generation (2G) ethanol production. These raw materials are rich in carbohydrates and renewable and do not compete with food/feed demands. However, the efficient bioconversion of SB/SS (efficient pretreatment technology, depolymerization of cellulose, and fermentation of released sugars) remains challenging to commercialize the cellulosic ethanol. Among the technological challenges, robust pretreatment and development of an efficient bioconversion process (implicating suitable ethanol producing strains converting pentose and hexose sugars) play key roles.

Cellulose and hemicellulose fractions are composed of mixtures of carbohydrate polymers. A number of different strategies have been envisioned to convert the polysaccharides into fermentable sugars. In general, the biological process of converting the lignocellulose biomass to fuel ethanol involves: (i) pretreatment either to remove lignin or hemicellulose so as to liberate cellulose (Figs. 3.4a and 3.4b), (ii) depolymerization of carbohydrate polymers to produce free sugars by cellulase-mediated action, (iii) fermentation of hexose and/or pentose sugars to produce ethanol, and (iv) distillation of ethanol (Fig. 3.5).

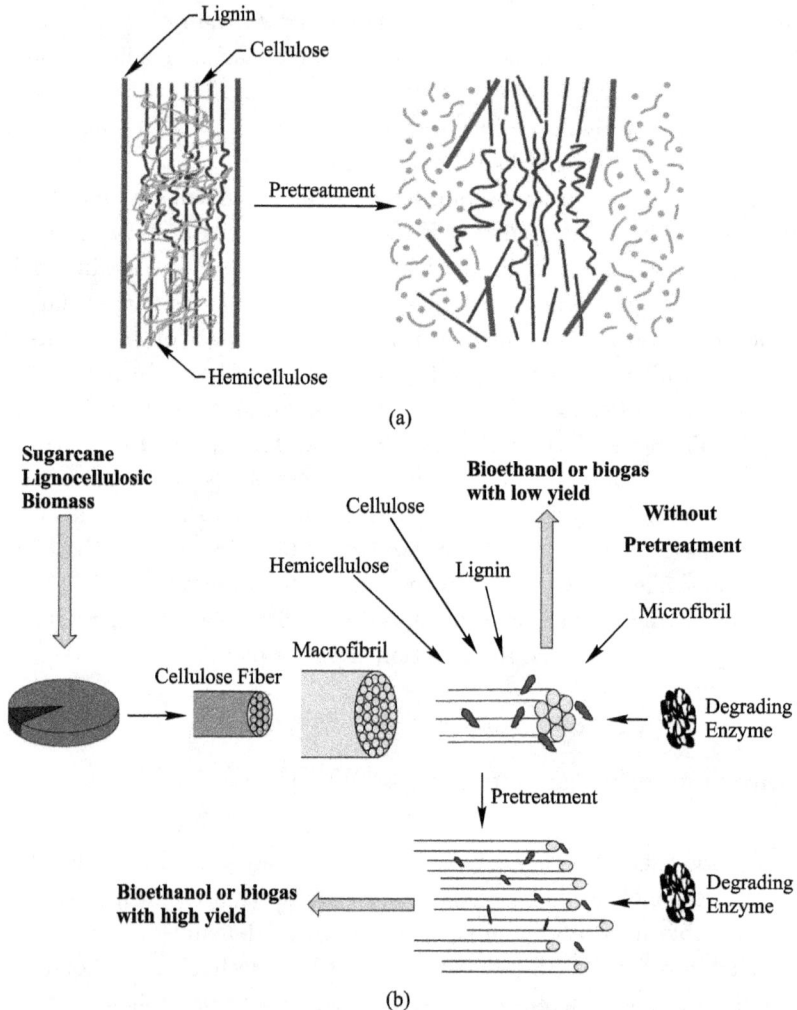

(a)

(b)

Fig. 3.4 (a) Lignocellulose model showing lignin, cellulose, and hemicellulose [26]; (b) Pretreatment scheme of sugarcane biomass [26].

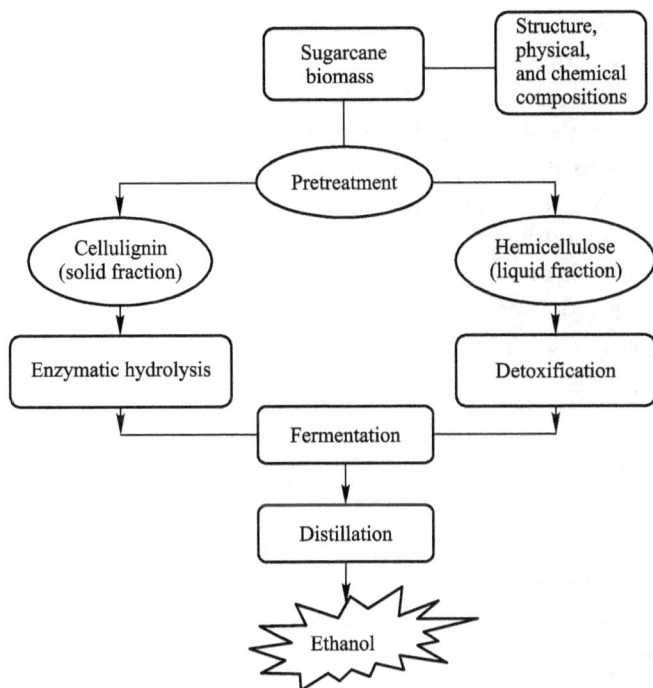

Fig. 3.5 Procedural flow diagram for the bioconversion of cane biomass into ethanol.

3.6.3 Pretreatment of Sugarcane Biomass

Ideally, the pretreatment of lignocellulosic biomass should: (i) increase the accessible surface area and decrystallize cellulose, (ii) partially depolymerize cellulose, (iii) solubilize hemicellulose and/or lignin, (iv) modify the lignin structure, (v) maximize the enzymatic digestibility of the pretreated material, (vi) minimize the loss of sugars, and (vii) minimize capital and operating costs [27–28]. Figure 3.5 presents scanning electronic microscopy (SEM) of SB before diluted sulfuric acid pretreatment and of cellulignin obtained after pretreatment. A rupture of the cellulose-hemicellulose-lignin strong matrix occurred after the pretreatment. In Figure 3.4b, an ordered structure of matrix can be seen, while Figure 3.4a presents a disordered structure of the cellulose-lignin complex. It is also possible to find empty spaces between the fibers as a consequence of the removal of hemicelluloses and low-crystallinity cellulose flocks [29]. In general, hydrolysate originating after diluted acid pretreatment is rich in the hemicellulose fraction. Various pretreatment technologies (alone or in combination) have been proposed in the literature. Broadly, pretreatment technologies can be categorized into four types: physical (mechanical), physicochemical, chemical, and biological pretreatments. Mechanical pretreatment increases the surface area by reducing the size of the SB or SS [30]. A high control of operation conditions is required in the physicochemical methods because these reactions occur at high temperatures and pressure [19]. Chemical methods degrade hemicellulose or remove lignin and thus loosen the structure of the lignin-hemicellulose network. Biological pretreatment methods are used for the delignification of lignocellulosic biomass [31]. However, the longer pretreatment times and loss of a considerable amount of carbohydrates can occur during this pretreatment [30]. Each method has its own specificity in terms of mechanistic application on cell wall components with the applied conditions [31]. Some types of pretreatments used for ethanol production are seen in Figure 3.5.

3.6.3.1 Physico-chemical pretreatments

Milling is a mechanical pretreatment that breaks down the structure of lignocellulosic materials and decreases the cellulose crystallinity [32]. Disadvantages of milling include the high power required by the machines and the consequent high energy costs. For sugarcane, bagasse pretreatment is necessary for a lot of cycles and many passes through the miller and the cycles usually have a long period of operation [33]. The pyrolysis process is carried out at high temperatures (more than 300°C). This process degrades cellulose rapidly into H_2, CO, and residual char [30]. After the separation of char, the recovered solution is primarily composed of glucose, which can be eventually fermented for ethanol production [34]. This process starts with the heating of the biomass. Primary pyrolysis reactions initiate at high temperatures to release volatiles, followed by condensation of hot volatiles and proceeded with autocatalytic secondary pyrolysis reactions [35]. The yield and quality of products after pyrolysis depends on several parameters, which can be categorized as process parameters, namely, temperature, heating rate, residence time, reaction time, reactor type, type and amount of catalyst, type of sweeping gas, and flow rate [35] and feedstock properties (particle size, porosity, cellulose, hemicellulose, and lignin content) [36]. Microwave pretreatment is considered as an alternative process for conventional heating. The main advantages of this process are the short reaction times and homogeneous heating of the reaction mixture [37]. Microwave-assisted pretreatment of SB/SS could be a useful process to save time and energy and minimum generation of inhibitors. It can be considered as one of the most promising pretreatment methods for changing the native structure of cellulose with the occurrence of the lignin and hemicellulose degradation and thus increasing the enzymatic susceptibility. Microwaves can be combined with the chemicals to further improve the sugar yield from the substrate.

Hydrothermal: Steam explosion (or hydrothermal) is one of the most common pretreatment methods. This pretreatment requires minimum or no chemical additions [38]. In this process, a mix of biomass and steam is maintained at a high temperature in a reactor, promoting the hemicellulose hydrolysis followed by a quick decompression ending the reaction [39].

Ammonia fiber explosion (AFEX): The AFEX process consists of liquid ammonia and steam explosion. It is an alkaline thermal treatment that exposes the lignocellulosic material to high temperatures and pressure followed by fast pressure release resulting in an increase of water-holding capacity and of digestibility of substrates (hemicellulose and cellulose) by enzymes, thus obtaining high sugar recovery [40–41].

CO_2 explosion: The CO_2 explosion is based on the hypothesis that CO_2 would form carbonic acid, increasing the hydrolysis rate of the pretreated material and thus aiding the penetration of CO_2 molecules into the crystalline structure of lignocellulosics [32, 42–43].

Hot water: The hot water is maintained in contact with the biomass for about 15 minutes at a temperature of 200–230°C. During this process, about 40%–60% of the total biomass is dissolved and all hemicellulose is removed.

Acid pretreatment: The most commonly used acid is H_2SO_4, where its contact with biomass promotes the hemicellulose breakdown of xylose and other sugars [26]. Acids lead to solubilization of hemicellulose at high temperatures or at high concentrated acid, releasing pentose sugars [26, 44] and facilitating the enzymatic hydrolysis of the remaining substrate (cellulignin) [26, 44–46].

Alkaline pretreatment: Alkaline pretreatment is a delignification process where a significant amount of hemicellulose is also solubilized by employing bases like sodium hydroxide, calcium hydroxide (lime), potassium hydroxide, ammonia hydroxide, and sodium hydroxide in combination with hydrogen peroxide or others [47]. The action mechanism of alkaline hydrolysis is believed to be the saponification of intermolecular ester bonds cross linking xylan hemicelluloses and other components [48].

Oxidative delignification: The oxidative delignification process causes the delignification and chemical swelling of the cellulose, improving enzymatic saccharification significantly. In this process, the lignin degradation is catalyzed by the peroxidase enzyme with the presence of H_2O_2 [48].

Organosolv: This process involves a strong inorganic acid as a catalyst, promoting the breakdown of lignin-lignin and carbohydrate-lignin bonds from the biomass [27].

Wet oxidation: The wet oxidation process occurs in the presence of oxygen or catalyzed air where the most-used catalyst is the sodium carbonate [49–50].

3.6.3.2 Biological pretreatment

Biological pretreatment is the alternative to chemical pretreatment to alter the structure of lignocellulosic materials. As the cost component makes the chemical operation a bit expensive and often leads to sugar losses due to the severity of the operational conditions [51], several microbial pretreatments have been attempted. Some bacteria, namely, *Cellulomonas carte, Cellulomonas uda, and Bacillus macerans*, and fungal species like *Trichoderma reesi, Trichoderma viridae, Aspergllis terreaus*, and *Aspergillus awamori*, have successfully delignified the sugarcane trash [52]. The pretreated trash has been converted into fermentable sugars by enzymatic saccharification. These sugars have been processed for the production of ethanol by fermentation at Bench scale. The most effective microorganism for biological pretreatment of lignocellulosic materials is white rot fungi [53]. These microorganisms degrade lignin through the action of lignin-degrading enzymes such as peroxidases and laccase [30]. This pretreatment is environmentally friendly because of its low energy use and mild environmental conditions [42]. However, the main disadvantages such as low efficiency, considerable loss of carbohydrates, long residence time, requirement of careful control of growth conditions, and space restrain its application [47].

3.6.4   Enzymatic Hydrolysis/Saccharification of the Cellulosic Fraction

The conversion of the cellulosic fraction into fermentable sugars involves the pretreatment of the raw material followed by its enzymatic hydrolysis. Enzymatic hydrolysis is an ideal approach for degrading cellulose into reducing sugars because of mild reaction conditions (pH between 4.8–5.0 and temperature between 45–50°C) and negligible byproduct formation with high sugar yields. However, enzymatic hydrolysis depends on optimized conditions for maximal efficiency (hydrolysis temperature, time, pH, enzyme loading, and substrate concentration) and suffers from end-product inhibition and biomass structural restraints [54–55]. For overcoming the end-product inhibition and reducing the time, hydrolysis and fermentation can be combined into so-called simultaneous saccharification and fermentation (SSF) or simultaneous saccharification and co-fermentation (SSCF). The enzymatic hydrolysis of the cellulosic fraction requires three classes of cellulolytic enzymes (cellulases): (i) endo-β-1, 4-glucanases (EG, E.C.3.2.1.4), which attacks regions of low crystallinity in the cellulose fiber, creating free chain ends, (ii) cellobiohydrolase or exoglucanase (CBH, E.C. 3.2.1.91), which degrades the molecule further by removing cellobiose units from the free chain-ends, and (iii) β-glucosidases (E.C. 3.2.1.21), which hydrolyses cellobiose to produce glucose [48].

Breakdown of hemicellulose requires several enzymes such as xylanase, b-xylosidase, glucuronidase, acetylesterase, galactomannanase, and glucomannanase [54]. Cellulase enzymes, when acting together with xylanases on delignified SB/ SS, exhibit a better yield due to the synergistic actions of the enzymes [56]. The enzyme source also has a major effect on the hydrolysis efficiency. Therefore, understanding the interaction between cellulases and pretreated biomass is vital to effectively develop low-cost pretreatment and enzyme properties that can lead to competitive ethanol costs [55].

3.6.5 Detoxification of Cellulosic and Hemicellulosic Hydrolysates

The main preoccupations in the pretreatment of lignocellulosic materials are to minimize the sugar degradation and subsequently minimize the formation of inhibitory compounds for microbial metabolism, on the one hand, and to limit the consumption of chemicals, energy, and water and the production of wastes on the other hand [57]. These compounds individually as well as synergistically affect the physiology of fermenting microorganisms. Therefore, it is essential to eliminate these inhibitory compounds or reduce their concentrations to obtain the satisfactory product yields during microbial fermentation of lignocellulose hydrolysates [58]. A number of methods like evaporation, neutralization, use of membranes, ion exchange resins, activated charcoal, and enzymatic detoxification using laccases and peroxidases have been attempted to detoxify the hydrolysates aiming at ethanol production. Considering that different lignocellulosic hydrolysates have different degrees of inhibition and that microorganisms have different inhibitor tolerances, the changes in the methods of detoxification depend on the source of the lignocellulosic hydrolysate and the microorganism being used [59].

3.6.6 Fermentation of Sugars from Sugarcane Biomass into Ethanol

Ethanol fermentation is a biological process in which sugars are converted by microorganisms to producing ethanol and CO_2. Even though there exist many methods and processes to use lignocellulosic materials for ethanol production, it is still difficult to obtain economic ethanol from lignocellulosics [56]. The availability of a robust genetic transformation system of *S. cerevisiae* along with a long history of this microorganism in industrial fermentation processes makes it most desired microorganisms for ethanol production. *S. cerevisiae* has high resistance to ethanol, consumes significant amounts of substrate in adverse conditions, and shows high resistance to inhibitors present in the medium [60]. Unfortunately, xylose metabolism presents a unique challenge for *S. cerevisiae* to assimilate pentose sugars due to the absence of genes required for the assimilation of these molecules. The maximum utilization of all sugar fractions is essential to obtain an economic and viable conversion technology for bioethanol production from SB and SS. To obtain the desired ethanol yields from SB/SS hydrolysates, it is essential that the hemicellulose fraction be fermented with same conversion rates as the cellulose fraction [60].

Hemicellulose hydrolysate contains primarily pentose sugars (xylose and arabinose) and some amounts of hexose sugars like mannose, glucose, and galactose [61]. A variety of yeast, fungi, and bacteria are capable of assimilating pentose, but only a few are promising candidates for efficient xylose fermentation into ethanol. In yeasts, the assimilation of D-xylose follows the pathway where the sugar passes through a pool of enzymatic processes to enter in the phosphopentose pathway [62]. There are several microorganisms capable of assimilating pentose sugars, but only few species are capable of assimilating sugars to produce ethanol at an industrial scale. The ethanol purification occurs in three steps: distillation, rectification, and dehydration. A highly concentrated ethanol solution is obtained in the first two steps (about 92.4 wt %). The mixture is then dehydrated in order to obtain ethanol anhydrous by a dehydration method. The dehydration can be realized by azeotropic distillation and extractive distillation.

3.6.7 Pyrolysis of Sugarcane Biomass

As previously discussed, sugarcane lignocellulosic biomass can be converted to energy either through a biochemical pathway or through a thermochemical pathway. The biochemical conversion requires several unit operations and faces the challenges of high pretreatment and enzyme costs, low fermentability of mixed sugar stream, especially 5-carbon sugar, the generation of inhibitory soluble compounds (acetic acid, furfural, 5-hydroxymethyl furfural and phenolic compounds, etc.), and degradation of sugars. However, the thermochemical pathway involves the conversion of lignocellulose biomass to bio-oil and syngas (mixture of CO and H_2) followed by conversion of syngas to ethanol using either chemical catalysts or microbial agents. In sugar industries, bagasse is used as feedstock for cogeneration during milling season. The sugar factories do not cogenerate during the off season because of the lack of alternative biomass supply. Because the factories do not cogenerate electricity throughout the year, there occurs significant loss in C credits and ethanol production. In such a scenario, bagasse and trash storage as a feedstock seems to be the only solution, at first glance, but their storage and handling on a large scale is an extremely expensive, difficult, and risky operation because of low density and self-combustion properties. This creates not only a lack of an alternative energy carrier for electricity with storage capability for use during off-season but also affects ethanol production. The conversion of SS and SB into bio-oil and syngas thus helps in overcoming the barriers for the cheap and sustainable production of a sustainable alternative energy carrier (electricity and syngas for the production of ethanol).

Bagasse and trash transformation into high-density renewable fuels, like charcoal and bio-oil, can significantly increase the profitability of sugarcane plantations. Thus, energy recovery from sugarcane bagasse by pyrolysis technology may be worthwhile. On average, a hectare of sugarcane generates about 10 tonnes of trash. Because it has no value as cattle fodder and because it is also fairly resistant to decomposition, the trash is burnt *in situ* in order to clear the field for the next crop. In Maharashtra, India, more than 4 million tonnes of trash are destroyed in this way. Pyrolysing the trash and converting it into fuel briquettes can thus be a profitable, small-scale, rural business.

3.7 Conclusions

The last three decades of vigorous developments in pretreatment technologies, microbial biotechnology, and downstream processing have made it a reality to harness the sugarcane residues for the production of many products of commercial significance on a large scale without jeopardizing the food/feed requirements. Sugarcane bagasse and sugarcane straw are the attractive renewable feedstocks for energy production. The judicious use of these feedstocks shall provide a sustainable supply of drop-in ethanol, industrial enzymes, organic acids, single-cell proteins, and steam and electricity generation. Thus, a long-term sustainable 2nd generation ethanol production process from sugarcane residues can be established by a planned and complete utilization of SB and SS, proper pretreatment and detoxification strategies, in-house cellulase production, development of cellulolytic strains and ethanol-producing strains from pentose and hexose sugars showing inhibitor resistance, ethanol tolerance, and faster sugar conversion rates, saccharification, and fermentation along with cheaper and faster distillation. Further integration of bioethanol-/bio-oil-producing units with sugars/distilleries for the co-utilization of machinery, reactors, and other equipment along with maximum byproduct utilization (lignin, furans, and yeast cell mass) shall lead sugarcane crops into the largest energy industries with production estimates of about 450 GJ ha^{-1} yr^{-1}.

References

[1] Fulton L. Biomass agricultural sustainability, markets and policies. International energy agency (IEA) biofuels study interim report, IEA France, OECD 2004; 105–112.

[2] Karimi M, Rajabi Pour A, Tabatabaeefar A, Borghei A. Energy analysis of sugarcane production in plant farms, American-Eurasian J. Agric. & Environ. Sci. 2008; 4 (2): 165–171.

[3] FAOSTAT (2008) http://faostat.fao.org/default.aspx.

[4] Valdes JL, Puig J, Torres A, Rodrigues ME, Prado R. Revista ICIDCA Sobre los Derivados de la Cana Azucar 1989; 23(3): 38.

[5] Paturau JM. By-products of the Cane Sugar Industry. Elsevier Scientific Publishing Co., Amsterdam 1982.

[6] Paturau JM. Electricity export from cane sugar factories. F.O. Lichts. Guide to the Sugar Factory Machine Industry 1984; A75–A88.

[7] Paturau JM. Alternative uses of sugarcane and its byproducts in agroindustries (English). In: Sugarcane as feed. Proceedings of FAO Animal Production and Health Paper no. 72; Expert Consultation on Sugarcane as Feed, Santo Domingo (Dominican R.), 7–11 July 1986 Sansoucy, R. (Ed.) Aarts, G. (Ed.) Preston, T.R. (Ed.) / Rome (Italy), FAO, 1988: 24–45.

[8] Cerri CEP, Maia SMF, Galdos MV, Feigel BJ, Bernoux M. Brazilian greenhouse gas emissions: the importance of agriculture and livestock. Sci. Agric. 2009; 66: 831–843.

[9] Tannenbaum SR, Wang DIC. Single-Cell Protein II, MIT Press, Cambridge, MA 1975.

[10] Avgerinos G, Wang DIC. Direct microbiological conversion of cellulose to ethanol. Annual Reports on Fermentation Processes 1980; 4: 165–192.

[11] Marcelo KP. Sustainability of sugarcane bioenergy. Centro de Gestão e Estudos Estratégicos (CGEE), 2012: 359.

[12] Parisi F. Advances in lignocellulosic hydrolysis and in the utilization of hydrolyzates. Adv Biochem Eng Biotechnol 1989; 38: 53–87.

[13] Playne MJ. Increased digestibility of bagasse by pretreatment with alkali and steam explosion. Biotechnol Bioeng 1984; 26: 420.

[14] Fengel D, Wegener G. Wood Chemistry, Ultrastructure, Reactions. Walter de Gruyter, Berlin, Germany 1989.

[15] Singh A, Mishra P. Microbial pentose utilization. Current Applications in Biotechnology, vol. 33 of Progress in Industrial Microbiology 1995.

[16] Boussarsar H, Rogé B, Mathlouthi, M. Optimization of sugarcane bagasse conversion by hydrothermal treatment for the recovery of xylose. Bioresource Technology, 2009; 100(24): 6537–6542.

[17] Canilha L, Santos VTO, Rocha GJM. A study on the pretreatment of a sugarcane bagasse sample with dilute sulfuric acid. Journal of Industrial Microbiology and Biotechnology, 2011; 38: 1467–1475.

[18] Kuhad RC, Singh A, Eriksson KE. Microorganisms and enzymes involved in the degradation of plant fiber cell walls. Advances in Biochemical Engineering and Biotechnology, 1997; 57: 45–125.

[19] Taherzsadeh MJ, Karimi K. Enzymatic-based hydrolysis processes for ethanol from lignocellulosic materials: a review. Bioresources 2007; 2: 707–738.

[20] Goncalves AR, Soto-Oviedo MA. Production of chelating agents through the enzymatic oxidation of Acetosolv sugarcane bagasse lignin. Applied Biochemistry and Biotechnology A 2002; 98–100: 365–371.

[21] Stewart D. Lignin as a base material for materials applications:chemistry, application and economics. Industrial Crops and Products 2008; 27(2): 202–207.

[22] Fierro V, Torné-Fernández V, Montané D, Celzard A. Adsorption of phenol onto activated carbons having different textural and surface properties. Microporous and Mesoporous Materials, 2008; 111 (1–3): 276–284.

[23] Chum HL, Parker SK, Feinberg DA. The Economic Contribution of Lignin to Ethanol Production from Biomass, Solar Energy Research Institute, Golden, Colo, USA. 1985.

[24] Benar PP. ao acetosolv de bagac, o de cana e madeira de eucalipto (Acetosolv pulping of bagasse and Eucalyptus wood) [M.S. thesis], Chemical Institute, Campinas University, Sao Paulo, Brazil 1992.

[25] Pandey A, Soccol CR, Nigam P, Soccol VT. Biotechnological potential of agro-industrial residues: sugarcane bagasse. Bioresource Technology, 2000; 74(1): 69–80.

[26] Mosier N, Wyman C, Dale B. Features of promising technologies for pretreatment of lignocellulosic biomass. Bioresource Technology 2005; 96(6): 673–686.

[27] Holtzapple MT, Humphrey AE. The effect of organosolvent pretreatment on the enzymatic hydrolysis of poplar. Biotechnology and Bioengineering, 1984; 26(7): 670–676.

[28] Margeot A, Hahn-Hagerdal B, Edlund M, Slade R, Monot F. New improvements for lignocellulosic ethanol. Current Opinion in Biotechnology, 2009; 20(3): 372–380.

[29] Rocha GJM, Martin C, Soares IB, Souto Maior AM, Baudel HM, de Abreu CAM. Dilute mixed-acid pretreatment of sugarcane bagasse for ethanol production. Biomass and Bioenergy 2011; 35 (1): 663–670.

[30] Kumar P, Barrett DM, Delwiche MJ, Stroeve P. Methods for pretreatment of lignocellulosic biomass for efficient hydrolysis and biofuel production. Industrial and Engineering Chemistry Research 2009; 48 (8): 3713–3729.

[31] Chandel AK, Chan EC, Rudravaram R, Narasu Ml, Rao LV, Ravinda P. Economics and environmental impact of bioethanol production technologies: an appraisal. Biotechnology and Molecular Biology Reviews 2007; 2: 14–32.

[32] Sun Y, Cheng J. Hydrolysis of lignocellulosic materials for ethanol production: a review. Bioresource Technology 2002; 83(1): 1–11.

[33] Hideno AH, Tsukahara Inoue K. Wet disk milling pretreatment without sulfuric acid for enzymatic hydrolysis of rice straw. Bioresource Technology 2009; 100(10): 2706–2711.

[34] Sarkar P, Bosneaga E, Auer M. Plant cell walls throughout evolution: towards a molecular understanding of their design principles. Journal of Experimental Biology 2009; 60: 3615–3635.

[35] Acikalin, Karaca F, Bolat E. Pyrolysis of pistachio shell: effects of pyrolysis conditions and analysis of products. Fuel 2012; 95: 169–177.

[36] Agblevor FA, Besler S, Wiselogel AE. Fast pyrolysis of stored biomass feedstocks. Energy & Fuels 1995; 9(4): 635–640.

[37] Balcu I, Macarie CA, Segneanu AE, Oana R. Combined microwave-acid pretreatment of the biomass. In: Progress in Biomass and Bioenergy Production, SS, Shaukai (Ed.). In Tech, Rijeka, Croatia 2011; 223–238.

[38] Kaar WE, Gutierrez CV, M.Kinoshita C. Steam explosion of sugarcane bagasse as a pretreatment for conversion to ethanol. Biomass and Bioenergy 1998; 14(3): 277–287.

[39] Agbor VB, Cicek N, Sparling R, Berlin A, Levin DB. Biomass pretreatment: fundamentals toward application., Biotechnology Advances 2011; 29: 675–685.

[40] Sarkar N, Ghosh, SK, Bannerjee, S, Aikat, K. Bioethanol production from agricultural waste: an overview. Renewable Energy 2012; 37: 19–27.

[41] Bals B, Wedding C, Balan V, Sendich E, Dale B. Evaluating the impact of ammonia fiber expansion (AFEX) pretreatment conditions on the cost of ethanol production. Bioresource Technology 2011; 102(2): 1277–1283.

[42] Hamelinck CN, Hooijdonk van G, Faaij APC. Ethanol from lignocellulosic biomass: techno-economic performance in short-, middle- and long-term. Biomass and Bioenergy, 2005; 28(4): 384–410.

[43] Cardoso WS, Santos FA, Mota AM, Tardin FD, Resende ST, Queiroz JH. Pré-Tratamentos de Biomassa para Produc, ão de Etanol de Segunda Gerac, ão. Revista Analytica 2012; 56: 64–76.

[44] Alvira PE. Tomás-Pejó EM. Ballesteros M, Negro MJ. Pretreatment technologies for an efficient bioethanol production process based on enzymatic hydrolysis: a review. Bioresource Technology 2010; 101(13): 4851–4861.

[45] Taherzsadeh MJ, Karimi K. Pretreatment of lignocellulosic wastes to improve ethanol and biogas production: a review. International Journal of Molecular Sciences 2008; 9 (9): 1621–1651.

[46] Palmqvist E, Hahn-Hägerdal B. Fermentation of lingo cellulosic hydrolysates. II: inhibitors and mechanisms of inhibition. Bioresource Technology 2000; 74(1): 25–33.

[47] Zheng Y, Pan Z, Zhang R. Overview of biomass pre-treatment for cellulosic production. International Journal of Agricultural and Biological Engineering, 2009; 2: 51–68.

[48] Sun Y, Cheng, J. Hydrolysis of lignocellulosic material for ethanol production: a review. Bioresource Technology 2002; 96: 673–686.

[49] Carvalheiro F, Duarte LC, Gırio FM. Hemicellulose biorefineries: a review on biomass pretreatments. Journal of Scientific and Industrial Research 2008; 67(11): 849–864.

[50] Bjerre AB, Olesen AB, Fernqvist T. Pretreatment of wheat straw using combined wet oxidation and alkaline hydrolysis resulting in convertible cellulose and hemicellulose. Biotechnology and Bioengineering 1996; 49: 568–577.

[51] Kellar F, Hamilton J, Nguyen Q. Microbial pretreatment of biomass. Applied Biochemistry and Biotechnology 2003; 105–108: 27–41.

[52] Singh P, Suman A, Tiwari P, Arya N, Gaur A, Shrivastava AK. Biological pretreatment of sugarcane trash for its conversion to fermentable sugars. World Journal of Microbiology and Biotechnology 2008; 24: 667–673.

[53] Sarkar N, Ghosh SK, Bannerjee S, Aikat K. Bioethanol production from agricultural wastes: an overview. Renewable Energy 2012; 37(1): 19–27.

[54] Duff SJB, Murray WD. Bioconversion of forest products industry waste cellulosics to fuel ethanol: a review. Bioresource Technology 1996; 55(1): 1–33.

[55] Milagres AMF, Carvalho W, Ferraz AL. Topochemistry, porosity and chemical composition affecting enzymatic hydrolysis of lignocellulosic materials. In Routes to Cellulosic Ethanol, M. S. Buckeridge and G. H. Goldman, Eds., Springer, Berlin, Germany 2011.

[56] Giese EC, Chandel AK, Oliveira IS, Silva SS. Prospects for the bioethanol production from sugarcane feedstock: focus on Brazil. In: Sugarcane: Production, Cultivation and Uses, J. F. Goncalves and K. D. Correa, Eds., Nova Science Publishers, New York, NY, USA 2011.

[57] Luz, SM, Goncalves AR, Del' Arco Jr AP, Leão AL, Ferrão PMC, Rocha GJM. Thermal properties of polypropylene composites reinforced with different vegetable fibers. Advanced Materials Research 2010; 123–125: 1199–1202.

[58] Chandel AK , Silva SS, Singh OV. Detoxification of lignocellulosic hydrolysates for improved bioethanol production. In: Biofuel Production-Recent Developments and Prospects, M. A. S. Bernardes, Ed., Tech, Rijeka, Croatia 2011; 225–246.

[59] Anish R, Rao M. Bioethanol from lignocellulosic biomass part III hydrolysis and fermentation. In: Handbook of Plant-Based Biofuels, A. Pandey (Ed.). CRC Press, Portland, Ore, USA. 2009; 159–173.

[60] Lin Y, Tanaka S. Ethanol fermentation from biomass resources: current state and prospects. Applied Microbiology and Biotechnology 2006; 69(6): 627–642.

[61] Shen J, Wyman CE. A novel mechanism and kinetic model to explain enhanced xylose yields from dilute sulfuric acid compared to hydrothermal pretreatment of corn stover. Bioresource Technology 2011; 102: 9111–9120.

[62] Bettiga, M, Hahn-Hägerdal B, Gorwa-Grauslund MF. Comparing the xylose reductase/xylitol dehydrogenase and xylose isomerase pathways in arabinose and xylose fermenting Saccharomyces cerevisiae strains. Biotechnology for Biofuels 2008; 1: 16.

Chapter 4

Jatropha (*Jatropha curcas* L.) as a New Biofuel Feedstock for Semi-arid and Arid Regions and Its Agro-ecological Sustainability Issues

Yiftach Vaknin, Uri Yermiyahu, and Asher Bar-Tal

4.1   Introduction

Biofuels are generally considered as a solution to sustainable development. They have the potential to provide certain but limited levels of energy security and their development and utilization significantly reduces emissions of greenhouse gas (GHG) and harmful pollutants. In developed countries, biofuels are seen as a new market opportunity due to their ability to absorb surplus agricultural production while maintaining productive capacity in the rural sector. However, in developing countries, biofuels can contribute to rural development in three main areas that include (i) employment creation, (ii) income generation, and (iii) replacement of traditional biomass, which is an inefficient and unsustainable energy resource, with modern and sustainable forms of bioenergy [1]. The clean development mechanism (CDM), established by the Kyoto Protocol, promotes industrialized nations to provide resources to developing countries in order to support their sustainable developments. At the same time, it promotes the reduction of global (GHG) emissions, since it is becoming practically impossible to reduce emissions in these developed countries [2].

   Jatropha curcas L. (JCL), as a biofuel feedstock of choice primarily in developed countries, has been reviewed numerous times over the past 15 years [1, 3–17]. During this time, researchers worldwide have attempted to explore its broad traits as a feedstock for biodiesel as well as other biofuels and to utilize its valuable byproducts.

Then, why do it again? Especially when only recently, various sustainability issues of JCL as a biodiesel feedstock were thoroughly reviewed [18–19]. Several details come to mind:

1. Data has been accumulating regarding its cultivation and, by some, it is no longer considered a crop of primarily wild attributes.

2. Ever since the bursting of what we call the "Jatropha bubble" around 2008 [20], proper research and development have been producing validated protocols for its propagation and cultivation.

3. Being tropical in nature, it is cultivated in similar climatic regions around the world while arid and semi-arid regions are being carefully explored.

4. Quite a lot of information regarding the utilization of its byproduct has accumulated, thus making it more sustainable and more economical.

5. Reading all of these reviews reveals that a lot of data regarding JCL is passed around, often without proper validation, and misinformation is still quite prevalent.

6. Sustainable cultivation and utilization of JCL is a primary goal that can be achieved in vast regions of the developing world given that proper research and development are done.

7. Over the past seven years, the authors of this chapter have been exploring the potential of this plant under arid and semi-arid conditions in the Israeli Negev desert and coastal plains, respectively [Personal information].

In the following sections, we shall review various aspects of JCL as a feedstock for biodiesel as well as other biofuels while debating its level of sustainability in the present and future cultivation on a global scale, particularly in the arid and semi-arid regions of the world. A case study of JCL cultivation under conditions of the arid Negev desert will be briefly described and discussed.

4.2 Systematics and Global Distribution

JCL is a small tree or large shrub of the Euphorbiaceae family and there are about 170 known species of the genus *Jatropha*, mostly native to the New World; although, 66 species have been identified as originating in the Old World [3]. JCL is a diploid species with a 2n chromosome number of 22 [21]. It is native to Mexico and continental Central America [3, 22], but is now cultivated widely in tropical and subtropical countries worldwide [1, 3, 5–6, 23–25]. JCL is believed to have been spread by the Portuguese from its center of origin in Central America via Cape Verde and Guinea Bissau to tropical and sub-tropical countries in Africa and Asia, where it is now widespread [3]. JCL plants are believed to have a lifespan of 30 to 50 years or more [1]. In 2008, JCL was globally grown on an estimated area of 900,000 ha with more than 85% being in Asia, chiefly Myanmar, India, China, and Indonesia. Africa accounts for around 12%, mostly in Madagascar and Zambia, but it is also in Tanzania and Mozambique. Latin America accounts for around 3%, mostly in Brazil. The planted area of JCL was projected to grow to approximately 13 million ha by 2015 [26]. Governments have been the main drivers for JCL cultivation and developed specific programs for this crop with most production aimed for the local markets [26].

4.3   Vegetative Growth and Sexual Reproduction

JCL grows to 3–5 m high under favorable conditions. It has succulent spreading branches and the leaves, varying in size and shape, are arranged alternately on the stem. The branches and stems contain semi-transparent latex and because they are soft and hollow, they are of poor economic value. Approximately five roots are formed from seedlings, having one central tap root and the rest are peripheral [3]. A tap root is not usually formed by vegetatively propagated plants (*e.g.*, stem cuttings or tissue culture).

JCL is a monoecious plant with separate male and female flowers residing on the same bloom clusters (Figs. 4.1a–b), at various male to female ratios of 4–31:1 [27–31], which decrease as the age of the plant increases [32]. In permanently humid and hot regions, JCL will grow and bloom throughout the year [3]. In more seasonal regions, JCL will bloom with two distinct peaks—during summer and autumn, depending on water availability either by precipitation [27] or by irrigation [Vaknin *et al.*, personal information]. Plant productivity starts after the first year and typically becomes stable when trees are 2–4 years old. The economic production of the JCL plants extends from the first year after planting to 40 years. However, the tree life and fruit production span over ~100 years [10].

The plant is self-compatible, with a small percentage (12%) of fruit set by apomixes [29]. Cross-pollination results in a higher fruit set percentage than self-pollination [27, 29, 31, 33]. Optimal pollination of JCL was suggested to be by cross-pollinating insects [29, 34] belonging to at least seven systematic orders including bees, ants, thrips, flies, butterflies, bugs, and beetles [27, 31, 35]. Among the various insect visitors of JCL, *Apis* spp. was the most frequent and the most effective pollinator [36–38]. Insecticides are generally avoided on JCL plants because they can harm pollinators, which are essential to fruit productivity [39]. Currently, the basic requirements for JCL pollination services are largely neglected and should be addressed when it is cultivated on a commercial scale [40].

Fig. 4.1 Flowers, fruits, and seeds of *Jatropha curcas*: (a) Bloom cluster with male (M) and female (F) flowers; (b) Male (M) and female (F) flowers; (c) Fruit cluster with all fruit developmental stages from green (G) to yellow (Y), yellowish brown (YB), and mature brown (B); (d) Dry fruits with seeds; (e) Mature seeds. The scale bar is 1 cm.

The fruits of JCL are capsule-like and about 2.5–4 cm in diameter. The immature fruits are green and fleshy, becoming yellow and dark brown as they ripen (Fig. 4.1c). Mature fruits split to release up to three black seeds, each about 2 cm long and 1 cm wide (Figs. 4.1d–e). Fruits are mature and ready to harvest around 50–120 days after flowering, depending on the weather (personal information). Freshly harvested seeds show that dormancy and post-ripening drying is necessary before they can germinate [15, 41].

JCL is considered as having an invasive potential through seed dispersal in some parts of the world [17, 42–44]. Due to these environmental concerns, Australia's Northern Territory and Western Australia, South Africa, Brazil, Fiji, Honduras, India, Jamaica, Panama, Puerto Rico, and El Salvador have declared it a noxious weed [1]. However, some field observers have stated that the plant is not invasive [4, 12]. There is little evidence on it actually becoming a threat as an invasive weed. In Australia, the occurrence of JCL invasiveness is along rivers and other sources of flowing water. The size of the seed and its toxic content do not promote natural dispersion by wind or animals and flowing water is probably one of the only ways to further disperse it. When grown under arid or semi-arid conditions, with limited water availability, the chances of it becoming invasive are probably nonexistent [Vaknin *et al.* personal information].

4.4 Optimal and Sub-optimal Climate and Growth Conditions

JCL is widely distributed in the wild and cultivated tropical areas of Central America, South America, Africa, India, south-eastern Asia, and Australia, where it typically grows between 15°C and 40°C with rainfalls between 250 and 3000 mm [7, 45]. Optimum ecological conditions for JCL are in the warm sub-humid tropics and subtropics with cultivation limits at 30°N and 35°S, annual precipitation of 1000 to 1500 mm [7], temperatures of 20°C to 28°C with no frost, average minimum temperatures of the coldest months above 8–9°C [46], and where the soils are free-draining sands and loams with no risk of water-logging [1]. Outside the tropics, suitable growing opportunities for JCL are found in warm temperate climates with no frost risk, characterized by either having dry seasons or being fully humid [46–47]. Tropical climates with no dry seasons and subtropical deserts have moderate yield potentials. A major constraint for the extended use of JCL seems to be the lack of knowledge on its potential yield under sub-optimal and marginal conditions and the growing and management practices are poorly developed or documented [8, 25].

Drylands cover approximately 50% of the global area including arid, semi-arid, and dry sub-humid regions, where the average precipitation is less than the potential water loss due to evaporation and transpiration ($ET_p < 0.65$) [48]. JCL is well-adapted to arid and semi-arid conditions down to 200–300 mm and is, therefore, considered to be drought-tolerant [14, 17, 20, 26, 49–51].

A study comparing the climatic conditions in the area of JCL's natural distribution with climatic conditions in JCL plantations worldwide [52] has revealed that JCL is not naturally common in regions with annual precipitation of less than 950 mm. Furthermore, approximately 85% of the native specimens were located in regions with tropical climates, while only 2.5% were located in semi-arid regions and none were found in arid regions. Globally, however, plantations were less situated in tropical climates (~50%) and relatively more situated in temperate semi-arid and arid climates (~30% and 20%, respectively). Respective yields show that in arid regions, yield is cut by half without supplementary irrigation. However, very little is known about its water use or water use efficiency as a crop [8].

JCL can grow on moderately sodic, saline, degraded, and eroded soils [53–55]. Fruit production, however, declines dramatically under dry conditions as it sheds its leaves under stressful conditions [5, 39]. Rainfall induces flowering; however, heavy rains at the time of flowering could lead to the complete loss of flowers [14].

Very high temperatures can depress yields of JCL [56]. Vegetative growth can be excessive at the expense of seed production if too much water is applied; for example, with continuous-drip irrigation [Vaknin *et al.*, personal information]. High precipitation (> 700 mm) is likely to cause fungal attack and restrict root growth in all but the most free-draining soils [10]. JCL is often described as having a low nutrient requirement because it has adapted to growing in poor soils. However, growing a productive crop requires correct fertilization and adequate rainfall or irrigation.

4.5 Propagation

JCL is easily propagated by both generative (direct seeding or indirect by production of seedlings) and vegetative (direct planting of cuttings or plants from tissue culture) methods [5, 57]. The advantages of vegetative propagation include: true-to-type seedlings, faster crop growth, higher survival rate, early flowering, and enhanced yield. Disadvantages include susceptibility to drought, adventitious root system, and shorter lifetime [58–60]. The advantages of generative propagation include a tap-root system and drought tolerance, whereas its disadvantages include genetic variability, slow growth and development, and late flowering [61]. For quick establishment of hedges and for erosion control, direct planting of cuttings is considered easier, whereas for long-life plantations for seed oil production, plants propagated from seeds are better suited [3]. Based on the authors' vast experience in clonal propagation, it is suggested that the establishment of commercial plantations through clonal propagation is now a plausible reality [Vaknin *et al.*, personal information].

4.6 Uses and Abuses of JCL

4.6.1 Traditional Non-fuel Uses

JCL was originally used as a live fence to contain or exclude farm animals, to control erosion, to reclaim land, for medicinal purposes, for various pests and disease control, and as an oil source for soap production, lamp fuel, as a lubricant, for cooking oil, and as a direct substitute for diesel fuel [3–5, 14, 17]. All parts of the JCL plant contain toxins such as phorbol esters, curcins, and trypsin inhibitors. Phorbol esters could form very potent bio-compounds as pesticides against disease vectors in animals [14]. Some varieties found in Mexico and Central America are known to be non-toxic [8]. Detoxification of the seed cake to render it usable as a livestock feed is possible, but it is unlikely to be economically feasible on a small scale [1]. JCL flowers attract bees, providing the opportunity for honey production in conjunction with other uses of the JCL tree [5].

4.6.2 Feedstock for Biofuels

Biomass conversion for energy purposes produces solid, liquid, or gaseous biofuels that offer better adaptation for clean and efficient utilization [62]. The most promising biomass energy conversion options that are currently pursued in most parts of the world today can be divided into three main platforms: thermo-chemical (charcoal production, gasification, and pyrolysis), physical-chemical (pressing and/or extraction and an optional esterification), and bio-chemical (alcohol fermentation and anaerobic digestion) conversions. Utilization of JCL for energy purposes may cover all three options.

4.6.2.1 Non-refined seed oil

The non-refined seed oil of JCL can be used as fuel by simple diesel engines of rural water pumps, electricity generators, tractors, and trucks, even without being refined [3, 49, 51, 62]. During the Second World War, it was used as a diesel substitute in Madagascar, Benin, and Cape Verde, while its glycerine byproduct was used for the manufacture of nitroglycerine [1]. The direct use of plant oils and/or blends with fossil fuels was generally considered unsatisfactory and impractical for both direct and indirect diesel engines due to poor fuel atomization, piston ring sticking, fuel injector choking, fuel pump failure, and the dilution of lubricating oil [63]. However, Pramanik [64] established that 40%–50% of JCL oil can be substituted for diesel without any engine modification or preheating of the blends.

4.6.2.2 Biodiesel

The potential significance of JCL as a feedstock for biodiesel production is most remarkable. Azan *et al.* [65] examined 75 non-edible oilseed plants and found that JCL outmatches all of the others as a source of biodiesel. Recently, the growing interest in JCL has initiated research and breeding programs worldwide, all concerned with its potential as an oil crop for biodiesel production [3, 8, 13, 16]. At the same time, international and national investors were rushing to establish JCL cultivation in large areas of Belize, Brazil, China, Egypt, Ethiopia, Gambia, Honduras, India, Indonesia, Mozambique, Myanmar, the Philippines, Senegal, and the United Republic of Tanzania [66].

As a commercial crop, it is now planted for its seed, which contains 20%–40% non-edible oil [9, 12, 67–68]. Unlike the major domesticated oilseed crops, rapeseed, soybean, and oil palm, JCL is exhibiting traits largely attributed to wild plants [68–69] and there are currently very few agronomically improved varieties of JCL available [1, 10, 13, 26].

Estimates of seed oil yields per hectare, under the current conditions of crop establishments, are highly variable, ranging from 0.1 up to 5 tons [17, 39, 49, 70]. Trabucco *et al.* [46] mapped JCL seed yield worldwide in response to climate for past, present, and future climate conditions. They found the yields to be significantly affected by annual average temperatures, minimum temperatures, annual precipitation, and precipitation seasonality. Poor knowledge on JCL's oil yield under sub-optimal and marginal conditions makes it difficult to predict yields from future plantations under these conditions [8]. Genetic improvement of JCL is currently being done by selection of "elite" genotypes from a wide range of germplasm [71], by induced mutation [72], by introgression of desirable traits using inter-specific hybridization [21, 73] and by insertion of specific genes using biotechnological interventions [16, 74–75].

JCL accessions available in India showed modest levels of genetic variation, while a wide variation has been found between the Indian and Mexican genotypes [76]. This was expected since variability is higher at the center of origin, *i.e.*, the intertropical Americas. However, the precise center of origin within the intertropical Americas has not yet been established [74].

The fuel properties of JCL biodiesel are close to those of fossil diesel and match the American and European standards [24]. Because of its lower pour point, JCL oil can be used to complement palm oil and give a mixed product compatible with consumption in Asian countries [77].

4.6.2.3 Biogas

The defatted seed cake's high organic matter content makes it suitable for biogas generation [70, 78]. Experiments have shown that some 60% more biogas was produced from JCL seed cakes in anaerobic digesters than from cattle dung and that it had a higher calorific value [11, 79]. After fermentation, most nutrients remain and can be further used as fertilizer to maintain JCL production at a sustainable level [8]. The fruit husks and seed shells are not optimally suitable as substrates in biogas digesters because of their very low digestibility and, thus, degradability [14]. Nevertheless, a feasibility test of JCL seed shell as an open-core gasifier feedstock revealed that using this technology may reach maximum gasification efficiency of nearly 70% [80].

4.6.2.4   Jet fuel

While diesel and gasoline fuels could be replaced by natural gas, ethanol, methanol, fuel cells (hydrogen), or batteries, aircraft engines will be using kerosene or kerosene-like fuels for many years to come. Vegetable oils are a proven feedstock for the production of jet fuels. Over the past ten years, the potential for the use of biofuels in aviation on a global scale has been seriously developed, with the first commercial flights using bio-jet fuel commencing in autumn, 2011. Recently, the need for a renewable feedstock for jet fuels became urgent in view of the new EU regulations stating that all flights in and out of EU airports are to be included in the EU Emissions Trading Scheme for 2012 [81], making the use of renewable jet fuel mandatory. If cultivated properly, JCL could deliver strong environmental and socioeconomic benefits while reducing GHG emissions by up to 60% compared to petroleum jet fuel [82].

4.6.2.5   Pyrolized bio-oil

The fruit and seed cakes of JCL can be converted to pyrolysis oil using a low-temperature conversion process [83]. The conversion rate to oil was 23% and 19% for fruit and seed cakes, respectively. When the pyrolysis oil was added to final concentrations of 2%, 5%, 10%, and 20% (w/w) to commercial diesel fuel, the density, viscosity, sulfur content, and flash point of the mixtures were found to be within the Brazilian standards of the diesel directive of the National Petroleum Agency (ANP No. 15, of 19/7/2006).

4.6.2.6   Burnt biomass

The seed shell and the fruit husk both have high energy contents that make them potentially important generators of energy through burning [3, 14]. The seed shell of JCL has 45%–47% lignin and the fruit husk also has a high energy content and, hence, both of these materials could be used for generating energy through burning [84]. Furthermore, the ash that remains after shell combustion is high in sodium and potassium, making it suitable for soil enrichment [70].

4.6.3 Utilization of JCL byproducts

There are various possibilities for utilizing the byproducts of JCL while minimizing waste, adding value for the producers, and reducing the C cost of the oil as a biofuel thus improving the sustainability and environmental impact of utilizing JCL [1]. The JCL de-fatted seed cake can be used for mulching [85] or can be returned to the soil as a substitute for chemical fertilizer with 1.0 kg of seed cake replacing 0.15 kg of $N : P : K$ ($40 : 20 : 10$) chemical fertilizer [39]. It can be used for biogas formation with plans for producing biogas from the seed cake more prevalent in Asia and Latin America than in Africa [26]. The JCL seed cake can also be converted to briquettes for domestic or industrial combustion, with one kilogram of briquettes combusting completely in 35 min at 525–780°C [70]. The JCL seed cake was also investigated as a substrate for the industrial production of enzymes such as proteases and lipases [86].

According to Singh *et al.* [70], a holistic approach to the utilization of JCL fruit will give three times the energy of biodiesel alone. JCL fruit husks can be used for direct combustion as they make up around 35%–40% of the whole fruit by weight and have a calorific value approaching that of fuel wood [1]. Further uses of the fruit husks include the production of compost, for example, by incorporating effective lignocellulolytic fungal consortium, which can reduce the phytotoxicity of the degraded material, thus producing better-quality compost [85]. Generally neglected, JCL byproducts include pruned branches that can be returned to the soil or used as fuel either directly or after transformation to either biogas or liquid fuels through pyrolysis [87].

4.7 JCL as a Sustainable Alternative to Fossil Fuels

A sustainable system is termed as the co-existence of the human species along with other species while maintaining the productivity and elasticity of economic systems in a regenerative and stable environment [18]. In a more specific context, agriculture is termed sustainable when current and future food and biofuel demands can be met without unnecessarily compromising economic, ecological, social, or political needs [88]. It is necessary to define the fraction of farmland, waste land, or barren land that could be used for the production of biofuels in a sustainable manner without conflicting with food security and environmental issues [2].

A global attempt to define socially and environmentally acceptable modes of biofuel production, through the application of sustainability standards, was reported by the ESG project [89]. Utilizing a holistic approach, super-governmental organizations like the EC and various UN agencies, national and sub-national governments, and corporations and civil society organizations are all making an effort to define such standards and codes of conduct with various levels of success.

In the interest of clarity and simplicity, we suggest that biofuels could be termed "sustainable" when several basic requirements are met: (i) they are produced from renewable feedstocks; (ii) their environmental impact is significantly lower than fossil fuels [25, 89–90]; (iii) they do not compete with food production [26]; and (iv) they enhance socioeconomic development [91–92].

JCL is globally claimed to be a sustainable feedstock for biofuel production reclaiming marginal and degraded lands in semi-arid and arid regions while enhancing socioeconomic development without competing with food production or depleting natural carbon stocks and ecosystem services [26, 91–92]. In the following sections, we shall test various aspects of these claims and attempt to weed-out misconceptions and non-validated conclusions.

4.7.1 Environmental Impacts

Environmental implications of biofuels are commonly assessed using a life cycle analysis—a complete life cycle comparison of a fossil fuel with a biofuel [93]. Biofuel production requires non-renewable resources including fuels consumed by farm machinery in land preparation, planting, tending, irrigation, harvesting, storage, and transport and fuels used to produce herbicides, pesticides, and fertilizers as well as energy required for feedstock transformation into biofuels [94]. Energy requirements are generally lower for perennial crops, such as JCL, than for annual crops, which involve greater use of machinery and a higher level of chemical inputs. Several attempts to conduct JCL life cycle assessments were made [1, 39, 95–98], showing a positive energy balance and impact on global warming potential, thus fitting into the context of sustainable development.

Soil degradation and, particularly, the desertification of dry lands is a global problem. Enhancing carbon sequestration in degraded agricultural lands could have direct environmental, economic, and social benefits for local people. Therefore, initiatives that sequester C are welcomed for the improvement in degraded soils, plant productivity, and the consequent food safety and alleviation of poverty in dry land regions [99]. Ogunwole *et al.* [100] presented a series of advantages of JCL cultivation, in a degraded Indian entisol, to soil structural stability and C and N content, thus increasing the potential for the carbon sequestration rate. This soil structural recovery under JCL cultivation was further implied as a sustainable improvement in the surface integrity of these soils, ensuring more water infiltration rather than runoff and erosion.

Conversion of rainforests, peatlands, savannas, or grasslands to produce biofuel crops releases 17 to 420 times more CO_2 than the annual GHG reductions that these biofuels would provide by displacing fossil fuels [101]. Globally, there are huge areas of degraded former croplands available in the developing world and, in many tropical regions, land degradation and soil erosion are major threats to existing land-use patterns [14, 25]. Furthermore, the over-reliance on biomass for energy needs in developing countries results in its over-exploitation, which may lead to desertification, especially when coupled with adverse environmental factors such as drought [62]. Therefore, to become sustainable, JCL should not be grown on formerly well-balanced ecosystems but rather on disturbed and deteriorating lands. Even then, growing JCL on marginal lands may result in loss of biodiversity [102].

Under arid and semi-arid conditions, JCL can reclaim marginal soils by exploring the soil with its roots, recycling nutrients from deeper soil layers, providing shadow to the soil, and reducing risks of erosion and desertification [8, 96, 103–104]. The root structural mechanism of JCL seedlings supports this claim by suggesting that the lateral roots decrease soil erodibility through additional soil cohesion, whereas the taproots and sinkers increase resistance against shallow land sliding, enable exploitation of subsurface soil moisture, and thus enhance vegetative cover, even in very dry environments [105]. It was further suggested that even sand dunes could be stabilized greatly by the ecosystem reconstruction of degraded land, particularly in arid/dry regions [19].

It is expected that in a relatively short period of time, JCL cultivation will help to improve water retention and soil conditions, thus reclaiming the land and making it again suitable for staple crop production [1, 14]. Sanderson [20], however, was less adamant and concluded that although JCL may not be a savior plant, as previously mentioned as transforming vast quantities of desert lands into biofuel-producing "moneymakers", it is likely to find its niche as a local alternative in certain developing countries.

As far as yield is concerned, growing JCL under semi-arid conditions is not very favorable [91]. JCL trees can survive drought conditions; however, under these conditions, the tree cannot produce maximum seed yields [39, 45]. To ensure the most sustainable exploitation of JCL, the efforts should be concentrated on the alleviation of the constraints that limit its cultivation on marginal lands and in traditional farming systems [102]. Provided that the JCL crop is handled with environmental sensitivity, especially in terms of water use, its diverse uses could make it a useful, sustainable bio-resource [106].

JCL has also been implicated as having high potential as a soil phytoremediator, especially in removing pollutants such as heavy metals [107–111], industrial hydrocarbons [112], and pesticides. In another case, incorporating fly-ash waste from coal-fired power plants, in an attempt to amend the soil, has improved JCL's photosynthetic rate, proving that waste could be utilized while improving plant performance [113]. However, these efforts and others are still in their infancy and further experience and validation should be accumulated before large-scale projects are practiced for industrial or domestic use [114].

4.7.2   Socioeconomic Impacts

Some 200 million people are believed to be directly affected by desertification and more than one billion people are at risk. The future sustainability of dry land ecosystems and the livelihoods of people living in them depend directly on the actions taken for land-use management. These activities should include soil and water conservation for improved land-use management practices and farming systems, taking into account health, social, and economic issues when developing strategies and policies to improve land management [99].

The global hype of JCL could be harnessed to increase rural development by considering small-scale, small-holder, community-based JCL initiatives for local use, like small JCL plantations, agroforestry systems with JCL intercropping, and agro-silvo-pastoral systems [92]. The primary characteristics of small-holder agriculture in semi-arid developing countries are its diversity in space, its variability through time, and its multidimensionality in terms of the ways it operates and survives [115], all resulting from the desire to be highly responsive to a varied, changeable, and hazardous environment [99]. A recent study analyzing the economic feasibility of sustainable small-holder bioenergy production under semi-arid conditions in Tanzania revealed that JCL oil was too expensive to be used as a substitute for fuel wood but was economically viable as a feedstock for biodiesel [116]. Recent reports on the socioeconomic sustainability of small-scale JCL cultivation in the eastern provinces of Zambia [117] and Mali [118], from the farmers' perspectives, revealed that it had a positive effect on the socioeconomic sustainability by providing them with better chances to earn money. The extra income may lead to new investments in both the farm and the family, such as sending children to school, which is an investment in the future.

JCL biofuel production could be especially beneficial to poor producers, particularly in semi-arid, remote areas that have little opportunity for alternative farming strategies, few alternative livelihood options, and increasing environmental degradation [1–2, 19, 62]. For the present, the main pro-poor potential of JCL is within a strategy for the reclamation of degraded farmland along with local processing and utilization of the oil and byproducts [2].

Ethical concerns about the social conditions of biofuel production (*e.g.*, labor rights, women rights, child labor, land tenure security, and more) should not be neglected [89]. They should support the notion that biofuel production avoids any coercive measures that leave people worse off than they would be in the absence of biofuel production. JCL cultivation has also led to concerns that it might displace food crops in food-insecure regions, particularly, Africa. When developed nations such as the European Union countries and the United States use foods such as corn, canola, and beets as feedstocks for biofuel, it may have a secondary impact on food price hikes. However, the main factors causing this phenomenon are the larger global demand for food caused by higher incomes in countries such as China and India, high agricultural input prices, and the surges in oil prices. Most JCL varieties, unlike some other biofuel feedstocks, are inedible and therefore do not create a direct conflict with food production [63]. Furthermore, there is no loss of land for food production or other purposes as only degraded lands where profitable food production would not be possible are foreseen to be used [5, 13–14]. An analysis of JCL cultivation on a global scale [26] has revealed that only 1.2% of areas planted with JCL had been used for food production in the five years prior to the start of the project. An additional advantage of JCL cultivation is that it can even promote food production when inter-planted in alleys with staple crops because the food crops will profit from the nutrition and the shelter effects of the perennials. Furthermore, under the umbrella of the JCL plant, maize, sorghum, millet, and other staple crops will profit from the advanced management practices of this energy plant [5, 14, 19].

In developing countries, yield improvement of food crops could promote a shift from agricultural lands that are traditionally used for food production to energy crop cultivation. Therefore, more efficient agro-techniques as well as better allocation of areas for food crops could increase land availability for the production of biofuels without threatening food security [119].

4.8 Significance of Irrigation and Fertilization for JCL Cultivation

Until now, the prevalent misconceptions regarding JCL cultivation were that it grows in all types of soils [2], it requires very little irrigation [120], it is drought-tolerant [121], and it uses little water compared to other biofuel crops [1], making it a more sustainable choice. However, there is no scientific support to these claims [8]. Regretfully, this has led several JCL projects to failure, not reaching the targeted yields expected from this crop.

Information on JCL water use and water use efficiency as a crop is quite meager. For the species *Jatropha pandurifolia* L. and *Jatropha gossypifolia* L., a water use efficiency of 3.62 and 2.52 mmol CO_2 mmol^{-1} H_2O was reported [122]. These values are in the range of other oil-seed species like soybean (3.9 mmol CO_2 mmol^{-1} H_2O) and oil palm (3.95–4.42 mmol CO_2 mmol^{-1} H_2O). The growth of JCL plants is dependent on rainfall or on irrigation and under a long dry season without irrigation, the leaves are wilting and fall off.

In the semi-arid regions of western India, where environmental conditions for agriculture are often influenced by low and erratic rainfall, frequent droughts, poor soil conditions, and unreliable irrigation water supply, the locals utilize a community lift irrigation system to grow food crops in sufficient amounts [123]. It was suggested that the same irrigation system can be adopted for the expansion of biofuel crops.

A study conducted on the feasibility of JCL in Tamil Nadu, India [124] revealed that the initial misconception that JCL needs water mainly during the first year, for initial survival only [125], was probably incorrect. Continuous irrigation was found to significantly increase the number of fruiting periods per year, from one up to three, depending on the level and frequency of irrigation [126].

Relatively recently, no quantitative data on water need, water productivity, or water use efficiency of JCL was available [45]. Concurrently, it was suggested that in arid and semi-arid areas, JCL productivity was at risk of being very low without supplementary irrigation [8, 47]. It was further supported by Jingura [121] claiming that while JCL production in Zimbabwe will be mainly centered in the drier parts of the country, irrigation will become essential to reach the potential of the plant.

The claim that JCL has low nutrient requirements [8] does not fit well with the fact that its leaves, fruits, and seeds are rich in N, P, and K and are widely used as organic fertilizers [4]. Since JCL is not an N-fixing species, to maintain its productivity, fertilizers will have to be added to the soil [5]. Balota *et al.* [127] stated that although it has adapted to low fertility soils, JCL requires soil acidity corrections and the addition of a considerable amount of fertilizer for high productivity.

The contradiction between the previous assumptions of drought resistance and low nutrients demand and the findings of yield response to irrigation and the high nutrients content of plant organs raises the question: what do we actually know about JCL water and fertilizer requirements, especially in arid and semi-arid regions? And how can we implement this knowledge on a commercial scale? In recent years, scattered data regarding JCL's response to various levels of irrigation and fertilization has been trickling down, as described in the next section.

4.8.1   Effects of Irrigation on Pot-grown JCL Plants

Young pot-grown seedlings of various JCL accessions were exposed to different
levels of drought stress [47, 128]. Drought was found to significantly reduce
leaf area, biomass, and relative growth rate, but had no effect on specific leaf
area, daily range in leaf water potential, leaf water content, transpiration efficiency, or aboveground biomass water productivity [47]. Seedlings under extreme drought stress (no irrigation) stopped growing, started shedding leaves,
and showed shrinking stem diameters [128]. When drought as well as heat stress
were imposed on young pot-grown JCL seedlings, major changes in key physiological processes of the plants were revealed [129]. Drought was more damaging
in terms of oxidative stress and photosynthetic damage than heat stress; however, the combination of both exhibited a negative interactive response.

The effects of different soil media on growth and chemical constituents of
JCL pot-grown seedlings were tested under several water regimes [130]. With
increasing water supply, growth parameters including plant height, number of
leaves, and fresh and dry weight of leaves and stem as well as the chemical
content of chlorophyll, carotenoids, and N, P, and K levels in the leaves were
significantly increased. At the same time, root length and fresh and dry weight
of roots as well as N, P, K, and proline content in the roots decreased. These
growth parameters and chemical constituents tended to increase by using clay
media as compared to sand media, suggesting that clay media probably enhanced
water availability. However, clay may be problematic in cases of flooding.

Drought tolerance of JCL was evident only under moderate water stress. Under severe water stress, the plants experienced decreased foliar metabolism and
relative leaf water content [131], decreased root space structure, and lower water
use efficiency [132].

4.8.2   Effects of Irrigation on Field-grown JCL Plants

An attempt to assess the optimum irrigation intervals and planting density for
better growth and yield of JCL, under dryland conditions, was made in Tamil
Nadu, India [133]. Planting JCL at 2 m × 2 m spacing and irrigating the plants
once every 15 days increased the number of branches and plant girth, which
ensured better productivity.

When rain-fed plants were compared to irrigated ones in Haryana, India [25],
the irrigated plants performed vegetatively significantly better. Irrigation at
different time intervals did not have any effect on plant performance. Sexual
reproduction, however, was not tested and expected yield was suggested to increase with irrigation, but no supportive evidence was provided.

The effects of various water stress levels on the oil yield of JCL in the arid region of Enshas, Egypt [50] is one of the few studies conducted on field-grown plants, emphasizing yield production as well as oil quality. Plants exposed to four levels of water stress (125%, 100%, 75%, and 50% of potential evapotranspiration (ET_p)) revealed that the highest characteristics of JCL seed oil were recorded for the treatment of 100% of ET_p. Water stress, however, had no significant effect on the fatty acid composition of JCL seed oil.

A similar study was conducted by the authors of this chapter in a test plot (Fig. 4.2) in the arid Negev desert of Israel (mean annual precipitation ranges at 100–200 mm; Israel Meteorological Service). Plants exposed to three levels of irrigation (100%, 40%, and 10% of ET_p) revealed that the highest vegetative as well as reproductive characteristics of JCL were recorded with irrigation of 100% ET_p (Tab. 4.1; Fig. 4.3). The plants shed their leaves during winter (December–February) and new vegetative and reproductive buds emerged in the following late spring (May) and early summer (June). The plants then proceeded to grow vegetatively and bloomed twice—summer bloom period (June–July) and fall bloom period (October–November). Having started with a relatively similar number of pruned branches, the plants developed a similar number of branches and inflorescences at all irrigation levels. However, the "reproductive potential", described here as number of female flowers per plant, was significantly reduced with decreasing irrigation during both bloom periods, as more female flowers were produced per inflorescence (Tab. 4.1). This alone would explain significant differences in oil yield per plants. However, we found that the realization of the "reproductive potential" was also affected with a reduced percentage of fruit and seed sets under the most extreme level of water (*i.e.*, 10% ET_p). Additionally, seed size as well as seed oil concentrations were significantly reduced with decreasing levels of irrigated water (Tab. 4.2). Oil yield, which is a function of

Fig. 4.2 A *Jatropha curcas* irrigation test plot at Hazerim in the Israeli Negev desert.

Tab. 4.1 Effects of three levels of irrigation—high (0.1), mid (0.4), and low (1.0) ET_p—on the vegetative growth and reproductive success of JCL during the first (I) and second (II) bloom periods in 2010.

Bloom period		Irrigation level (ET_p)		
		10%	40%	100%
I	Initial branches	24.25 ± 1.34^a	26.19 ± 0.85^a	25.25 ± 1.47^a
	Developing branches	55.63 ± 2.96^a	62.19 ± 1.97^a	60.38 ± 3.04^a
	Inflorescences	1.06 ± 3.82^a	31.81 ± 2.84^a	31.75 ± 4.00^a
	Female flowers	177.71 ± 26.33^b	192.43 ± 21.62^b	278.49 ± 39.74^a
	Fruit set (%)	24.40 ± 3.30^b	53.26 ± 4.28^a	58.51 ± 5.00^a
	Seed set (%)	22.31 ± 3.43^b	50.65 ± 4.25^a	54.91 ± 4.39^a
	Fruits	46.31 ± 10.70^b	104.56 ± 13.54^a	143.81 ± 19.31^a
	Seeds	136.77 ± 32.23^b	297.16 ± 38.59^b	406.71 ± 53.98^a
II	Inflorescences	20.56 ± 3.46^b	39.81 ± 2.42^a	41.88 ± 3.32^a
	Female flowers	17.53 ± 7.63^c	215.58 ± 33.22^b	619.39 ± 76.02^a

Note: All data are shown as means $\pm$ SE. Means of each trait, separately, that are not significantly different are marked with the same letter.

Fig. 4.3 Vegetative response of *Jatropha curcas* to three levels of irrigation—high (a), mid (b), and low (c)—at Hazerim in the Israeli Negev desert. The picture was taken mid-winter when the plants had already shed their leaves.

reproductive success and seed oil content, is therefore detrimentally affected at almost all aspects when the plants are exposed to decreasing levels of irrigation under arid conditions.

Tab. 4.2 The effects of three levels irrigation—high (0.1), mid (0.4), and low (1.0) ET_p—on seed and seed oil traits of JCL plants during the first bloom period in 2010.

	Irrigation level (ET_p)		
	10%	40%	100%
Seed weight (g)	0.53 ± 0.02^c	0.58 ± 0.02^b	0.72 ± 0.02^a
Kernel weight (g)	0.32 ± 0.01^b	0.35 ± 0.02^b	0.45 ± 0.01^a
Seed oil (%)	39.66 ± 0.81^b	40.56 ± 1.16^b	46.95 ± 1.43^a
Seed protein (%)	37.04 ± 0.73^a	35.11 ± 0.80^a	31.74 ± 1.00^b
Seed oil (g)	0.13 ± 0.01^b	0.15 ± 0.01^b	0.21 ± 0.01^a
Seed protein (g)	0.13 ± 0.00^a	0.12 ± 0.00^a	0.11 ± 0.00^b
C18 : 1 (%)	41.79 ± 1.18^a	41.72 ± 0.97^a	39.04 ± 0.85^a
C18 : 2 (%)	31.38 ± 1.16^b	31.49 ± 0.98^b	34.79 ± 1.00^a
C16 : 0 (%)	14.52 ± 0.23^a	14.57 ± 0.22^a	14.06 ± 0.24^a
C18 : 0 (%)	6.92 ± 0.21^a	6.77 ± 0.17^a	6.40 ± 0.20^a
Unsaturated (%)	74.56 ± 0.29^b	74.68 ± 0.27^b	75.86 ± 0.31^a
Saturated (%)	22.64 ± 0.25^a	22.36 ± 0.23^a	21.40 ± 0.30^b

Note: All data are shown as means $\pm$ SE. Means of each trait, separately, that are not significantly different are marked with the same letter.

4.8.3 Effects of Fertilization on JCL Plants

In all of the reported experiments, the fertilizers were applied at various amounts as basal solid fertilizers. A strong response of JCL seedlings to P dose has been reported [127, 134–135]. De Souza *et al.* [135] reported a biomass positive response of JCL seedlings to K application in a 120-day growth period. The recommended rates of P and K in their study were 25 and 67 mg dm^{-3}, respectively. The critical levels, corresponding to the recommended P rates, were 13 and 74 mg dm^{-3} for K in soil (Mehlich-1) and the N, P, and K levels in the shoot dry matter of JCL were 37.4, 2.1, and 35.7 g kg^{-1}, respectively.

Maia *et al.* [136] used the "missing element technique", with omission of liming and each one of the macro and micronutrients, and reported that plants without the nutrients P, K, and liming showed limited growth. The nutrients N, P, K, Ca, Mg, and liming affected the shoot biomass, while the roots were more affected by the absence of N, P, Mg, and Ca. These results suggest that the macronutrients were more limiting to the growth of the plant.

Fertigation—the simultaneous application of water and fertilizers via drip irrigation—is the most efficient fertilization and irrigation practice; therefore, it is the recommended method in semi-arid and arid regions. The authors of this chapter have studied the responses of JCL plants to a range of N (6–200 mg L^{-1}), P (0.2–19.3 mg L^{-1}), and K (5–209 mg L^{-1}) concentrations in the irrigating solutions in short pot experiments. The concentrations of P and K in all N treatments were 4.4 and 80 mg L^{-1}, respectively, the concentrations of N and K in all P treatments were 75 and 80 mg L^{-1}, respectively, and the concentrations of N and P in all K treatments were 75 and 4.4 mg L^{-1}, respectively. The growth medium was perlite to minimize interaction of the nutrients with the medium. Daily irrigation in excess was applied for maintaining constant concentrations in the root zone.

The plants showed positive non-linear responses in plant size to elevated concentrations of N, P, and K in the irrigating solutions (Fig. 4.4). Similar responses

Fig. 4.4 *Jatropha curcas* plant shoot and root size as a function of N (upper), P (middle), and K (lower) concentrations in the irrigation water.
The values are in mg L^{-1}.

of the growth parameters (height, stem diameter, and the number of leaves) to N, P, and K concentrations were obtained (data not presented). The optimal concentrations of N, P, and K were 47, 1.2, and 105 ppm [1] , respectively.

Very little is known about the nutrient requirements of JCL as a field-grown crop. An effort to evaluate the effects of N and P on the vegetative growth and productivity of JCL was made in the semi-arid region of Gujarat, India [54]. Both plant height and canopy width were significantly increased with the applications of N and P. Seed yield was also increased by more than 60% compared to control levels of zero fertilization.

The effects of increasing levels of N, P, and K under rain-fed conditions (average annual rainfall of 900 mm) were tested in a field trial in Khon Kaen, Thailand [137]. Application of fertilizer significantly increased branch number, branch length, and fruit yield, especially under mid-levels of fertilization (312.5 kg ha^{-1}). Application of higher rates depressed growth as well as yield.

4.9   Conclusions

The future of energy resources is going to be based on versatility and ingenuity. At the moment, biofuel feedstocks are being developed and tested in many countries across the globe and JCL is just one of a long list of potential biofuel sources. Even after more than 15 years of research and development, JCL has yet to prove its viability on a commercial scale. Accumulated data on this unique oil and biomass crop suggest that it is going to become a major feedstock for biodiesel as well as other biofuels, primarily in developed countries, encountering socioeconomic as well as ecological adversities. Many of these countries are looking for a sustainable solution for their problems and JCL is claimed, by some, to be the solution, although major concerns are voiced [11, 102]. However, suitable land available for energy production is scarce and marginal land encountering problems of desertification and erosion are plentiful, yet avoided, for lack of knowledge on how to use them for the benefit of the local population.

Here, we suggest an unlikely habitat such as a dry and hot desert as having near-optimal conditions to grow JCL on a commercial scale, assuming that water, fertilizer, and insect pollinators are sufficiently provided. The availability of appropriate land for JCL cultivation in tropical countries is quite low and expansion of land requires the destruction of natural habitats filled with flora and wildlife. A survey conducted by the authors of this chapter of JCL cultivation in tropical countries across the globe revealed a common problem with soil-borne diseases as well as flooding damages due to poor drainage. Both problems and the potential of it becoming an invasive species are usually absent when JCL is

---

1   1 ppm = 10^{-6}.

grown in an arid environment with low precipitation and low relative humidity. In the deserts, for example, virgin land is amply available with no history of prior crop cultivation accompanied by residual traces of soil-borne diseases.

It is commonly agreed that the production of biofuel feedstocks from predominantly rain-fed agriculture faces increasing risks from drought and other elements of weather [138]. Water stress can be detrimental to seed oil production when optimal water requirements are met. When irrigation levels are too low, the plants conserve energy by restraining vegetative and reproductive growth and when irrigation levels are too high, vegetative growth is enhanced and seed production is reduced, probably due to a significant amount of photosynthetic production trans-located to the vegetative parts [19].

The recent exponential increase in cultivation and research of JCL, on a global scale, provides us with a flood of information involving a multitude of aspects. Yet, when it comes to growing JCL on a commercial scale, especially in a new environment, it usually involves a lot of guesswork and many mistakes are made, producing far below the targeted yields. All aspects of JCL utilization should be properly researched, including selection of elite genetic material, propagation through seeds or clones, provision of agronomic support, hand or mechanical harvest, seed oil extraction, byproduct utilization, biofuel production, and more. Taking care of one aspect, while all others are left severely underdeveloped, will subsequently end with the demise of the entire project. Indeed, a vast amount of information has been gathered by top research groups around the world; however, only a fraction has been released to the public as research papers or scientific reports.

It is our firm belief that JCL is to become a prime feedstock for biodiesel as well as other biofuels in the near future. The enormous need for jet fuel produced from seed oil is but one example of how this crop is going to be utilized for years to come. The current need for biofuel feedstock, which is very high, is only destined to increase and almost any amount of biofuel produced today is going to be immediately consumed and a lot more is needed. Commercial companies as well as research groups around the world already have a significant portion of the information needed for utilizing this plant in a sustainable manner. However, only a fraction of this information is publicly available as a result of what we coin as "false competition". Increasing oil yield of JCL in a given country probably will not interfere with the revenue acquired from growing JCL in a distant country or even in a neighboring one, as all feedstock will be locally consumed. Indeed, increasing the level of the energetic independence of some countries may be problematic for others but this is an important matter to be discussed elsewhere. It is our firm belief that countries embracing this crop, as well as other biomass crops, will walk a step closer to energy independency; jobs will be provided for people in impoverished regions, land quality will improve, and the outcome will be a significant improvement of the quality of life.

References

[1] FAO (Food and Agricultural Organization of the United Nations). *Jatropha*: a smallholder bioenergy crop—The potential for pro-poor development. Integrated Crop Management. Vol. 8. By Brittaine R, Lutaladio N. Rome, 2010.

[2] Jain SK, Kumar S, Chaube A. *Jatropha* biodiesel: key to attainment of sustainable rural bioenergy regime in India. Arch Appl Sci Res 2011; 3: 425–435.

[3] Heller J. Physic nut. *Jatropha curcas* L. promoting the conservation and use of underutilized and neglected crops. Institute of Plant Genetics and Crop Plant Research, Gatersleben/International Plant Genetic Resources Institute, Rome, 1996. (Accessed November 21, 2012, at http://www.bio-nica.info/biblioteca/Heller1996Jatropha.pdf)

[4] Gübitz GM, Mittelbach M, Trabi M. Exploitation of the tropical oil seed plant *Jatropha curcas* L. Bioresource Technol 1999; 67: 73–82.

[5] Openshaw K. A review of *Jatropha curcas*: an oil plant of unfulfilled promise. Biomass and Bioenergy 2000; 19: 1–15.

[6] Dde Jongh, J. General data on *Jatropha*. In: Rijssenbeek W, Ed. Handbook on *Jatropha curcas*. Eindhoven: FACT foundation, 2006: 4–9. (Accessed November 21, 2012, at http://www.globalbioenergy.org/uploads/media/0603_FACT_Foundation_-_Jatropha_Handbook.pdf)

[7] Foidl N, Foidl G, Sanchez M, Mittelbach M, Hackel S. *Jatropha curcas* L. as a source for the production of biofuel in Nicaragua. Bioresource Technol 1996; 58: 77–82.

[8] Jongschaap REE, Corre WJ, Bindraban PS, Brandenburg WA. Claims and facts on *Jatropha curcas* L.: global Jatropha curcas evaluation, breeding and propagation programme. Report 158. Wageningen, the Netherlands: Plant Research International, 2007.

[9] Kumar A, Sharma S. An evaluation of multipurpose oil seed crop for industrial uses (*Jatropha curcas* L.): a review. Ind Crop Prod 2008; 28: 1–10.

[10] Carels N. Jatropha curcas: a Review. Ad Bot Res 2009; 50: 39–86.

[11] Cooke E. Jatropha Biofuels, Miracle Plant or Economic Hazard? E555 Energy & the Environment. School of Public & Environmental Affairs Indiana University-Bloomington, 2009: 1–26. (Accessed November 21, 2012, at http://www.imaginehaitian.org/wp-content/uploads/2010/06/cooke_final_report_jatropha_biofuels.pdf).

[12] Henning RK. The *Jatropha* System, An Integrated Approach of Rural Development. baganí Rothkreuz 11, D-88138 Weissensberg, Germany, 2009: 1–105. (Accessed November 21, 2012, at http://betuco.be/agroforestry/Jatropha%20-%20integrated%20approach%20of%20rural%20development%205x.pdf).

[13] King AJ, He W, Cuevas JA, Freudenberger M, Ramiaramanana, D, and Graham I. Potential of *Jatropha curcas* as a source of renewable oil and animal feed. J. Exp Bot 2009; 60: 2897–905.

[14] Makkar HPS, Becker K. *Jatropha curcas*, a promising crop for the generation of biodiesel and value-added coproducts. Eur J Lipid Sci Technol 2009; 111: 773–787.

[15] Verma KC, Gaur AK. *Jatropha curcas* L.: substitute for conventional energy. World J Agric Sci 2009; 5: 552–556.

[16] Divakara BN, Upadhyaya HD, Wani SP, Laxmipathi GCL. Biology and genetic improvement of *Jatropha curcas* L.: a review. Appl Energ 2010; 87: 732–742.

[17] Parawira W. Biodiesel production from *Jatropha curcas*: a review. Sci Res Essays 2010; 5: 1796–1808.

[18] Kumar S, Chaube A, Jain SK. Sustainability issues for promotion of *Jatropha* biodiesel in Indian scenario: a review. Renew Sust Energ Rev 2012; 16: 1089–1098.

[19] Pandey VC, Singh K, Singh JS, Kumar A, Singh B, Singh RP. *Jatropha curcas*: a potential biofuel plant for sustainable environmental development. Renew Sust Energ Rev 2012; 16: 2870–2883.

[20] Sanderson K. Wonder weed fails to flourish. Nature 2009; 461: 328–329.

[21] Dehgan B. Phylogenetic significance of interspecific hybridization in Jatropha (Euphorbiaceae). Syst Bot 1984; 9: 467–78.

[22] ICRAF (International Centre for Research in AgroForestry). A tree species reference and selection guide—*Jatropha curcas*, 2008. (Accessed November 21, 2012, at http://www.worldagroforestrycentre.org/sea/products/afdbases/af/asp/ SpeciesInfo.asp?SpID=1013#Ecology).

[23] Augustus GDPS, Jayabalan M, Seiler GJ. Evaluation and bioinduction of energy components of *Jatropha curcas*. Biomass and Bioenergy 2002; 23: 161–164.

[24] Tiwari AK, Kumar A, Raheman H. Biodiesel production from *Jatropha curcas* with high free fatty acids: an optimized process. Biomass and Bioenergy 2007; 31: 569–75.

[25] Behera SK, Srivastava P, Tripathi R, Singh JP, Singh N. Evaluation of plant performance of *Jatropha curcas* L. under different agro-practices for optimizing biomass—A case study. Biomass and Bioenergy 2010; 34: 30–41.

[26] GEXSI. New feedstocks for biofuels—global market study on *Jatropha*—Final Report. pp 187. Berlin: GEXSI LLP, 2008. (Accessed November 21, 2012, at http://www.bioenergie.de/kraftstoffe-der-zukunft/Vortraege/zelt.pdf).

[27] Raju AJS, Ezradanam V. Pollination ecology and fruiting behaviour in a monoecious species, *Jatropha curcas* L. (Euphorbiaceae). Curr Sci India 2002; 83: 1395–1398.

[28] Bhattacharya A, Datta K, Datta SK. Floral biology, floral resource constraints and pollination limitation in *Jatropha curcas* L. Pak J Biol Sci 2005; 8: 456–460.

[29] Chang-wei L, Kun L, You C, Yong-yu S. Floral display and breeding system of *Jatropha curcas* L. For Stud China 2007; 9: 114–119.

[30] Tewari DN. *Jatropha* and Biodiesel. 1st ed. New Delhi: Ocean Books Ltd, 2007.

[31] Kaur K, Dhillon GPS, Gill RIS. Floral biology and breeding system of *Jatropha curcas* in North-Western India. J Tropic For Sci 2011; 23: 4–9.

[32] Prakash AR, Patolia JS, Chikara J, Boricha GN. Floral biology and flowering behaviour of Jatropha curcas. In: FACT Seminar on *Jatropha curcas* L. Agronomy and Genetics. Wageningen, the Netherlands: March 26–28. Wageningen: FACT Foundation; Article no. 2, 2007.

[33] Dhillon RS, Hooda MS, Handa AK, Ahlawat KS, Kumar YS, Singh N. Clonal propagation and reproductive biology of *Jatropha curcas* L. Indian J Agroforest 2006; 8: 18–27.

[34] Kun L, Wei-Lun Y, Chang-Wei L, Breeding system and pollination ecology in *Jatropha curcas*. For Res 2007; 20: 775–781.

[35] Rianti P, Suryobroto B, Atmowidi T. Diversity and effectiveness of insect pollinators of *Jatropha curcas* L. (Euphorbiaceae). Hayati J Bios 2010; 17: 38–42.

[36] Qing Y, Ping PD, Biao DZ, Liang WZ, Xiang SQ. Study on pollination biology of *Jatropha curcas* (Euphorbiaceae). J S China Agric Univ 2007; 28: 62–66.

[37] Abdelgadir HA, Johnson SD, Van Staden J. Approaches to improve seed production of *Jatropha curcas* L. Drakensville resort, South Africa. Proceedings of the 34th Annual Conference of the South African Association of Botanists (SAAB), 2008: 359.

[38] Chang-Wei L, Kun L, Xiao-Ming C, You C, Yong-yu S. Foraging and main pollinators of *Jatropha curcas* in dry-hot valley. Kunchong Zhishi, 2008; 45: 121–127.

[39] Prueksakorn K, Gheewala SH. Full chain energy analysis of biodiesel from *Jatropha curcas* L. in Thailand. Environ Sci Technol 2008; 42: 3388–3393.

[40] Vaknin Y. The significance of pollination services for biodiesel feedstocks, with special reference to *Jatropha curcas* L.: a Review. Bioenergy Res 2012; 5: 32–40.

[41] Jøker D, Jepsen J. *Jatropha curcas* L. Seed leaflet No. 83. Danida Forest Seed Centre, Humlebaek, Denmark, 2003. (Accessed November 21, 2012, at http://curis. ku.dk/portal-life/files/20648145/jatropha_curcas_83.pdf).

[42] Low T, C Booth. The Weedy Truth about Biofuels. Invasive Species Council, Melbourne, 2007: 46. (Accessed November 21, 2012, at http://www.invasives. org.au/documents/file/reports/isc_biofuels_revised_march08.pdf).

[43] Barney JN, DiTomaso JM. Non-native species and bioenergy: are we cultivating the next invader? BioScience 2008; 58: 64–70.

[44] GISP (Global Invasive Species Programme). Biofuel crops and the use of non-native species: mitigating the risks of invasion. Nairobi, Kenya, 2008.

[45] Achten WMJ, Verchot L, Franken YJ, *et al. Jatropha* bio-diesel production and use. Biomass and Bioenergy 2008; 32: 1063–1084.

[46] Trabucco A, Achten WMJ, Bowe C. Global mapping of Jatropha curcas yield based on response of fitness to present and future climate. GCBCB Bioenergy 2010; 2: 139–151.

[47] Maes WH, Achten WMJ, Reubens B, Raes D, Samson R, Muys B. Plant–water relationships and growth strategies of *Jatropha curcas* L. seedlings under different levels of drought stress. J Arid Environ 2009a; 73: 877–884.

[48] UNEP (United Nations Environment Programme). World Atlas of Desertification. Nairobi, 1992.

[49] Rijssenbeek W. Jatropha in developing countries—a sustainable bio-energy production. For the 2nd EPOBIO Workshop, Athens, Greece, 2007. (Accessed November 21, 2012, at http://epobio.net/workshop0705/presentations/ WinifriedRijssenbeek.pdf).

[50] Abou Kheira AA, Atta NMM. Response of *Jatropha curcas* L. to water deficits: yield, water use efficiency and oilseed characteristics. Biomass and Bioenergy 2009; 33: 1343–1350.

[51] Siddharth J, Sharma MP. Prospects of biodiesel from *Jatropha* in India: a review. Renew Sust Energ Rev 2010; 14: 763–771.

[52] Maes WH, Trabucco A, Achten WMJ, Muys B. Climatic growing conditions of *Jatropha curcas* L. Biomass and Bioenergy 2009b; 33: 1481–1485.

[53] Dagar J, Tomar O, Kumar Y, Bhagwan H, Yadav R, Tyagi N. Performance of some under-explored crops under saline irrigation in a semi-arid climate in North-West India. Land Degrad Dev 2006; 17: 285–299.

[54] Patolia JS, Ghosh A, Chikara J, Chaudhary DR, Parmar DR, Bhuva HM. Response of *Jatropha curcas* grown on wasteland to N and P fertilization. In: Expert Seminar on *Jatropha curcas* L. Agronomy and Genetics. Wageningen, the Netherlands. FACT Foundation, 2007. (Accessed November 21, 2012, at http:// www.jatropha-alliance.org/fileadmin/documents/knowledgepool/PatoliaGhosh_ Jatropha_Wasteland_N_and_P_fertilization.pdf).

[55] Shekhawat NS, Rathore JS, Phulwaria M, *et al.* Cultivation of *Jatropha curcas* on wastelands of drought prone regions: projections and realities. In: Expert Seminar on *Jatropha curcas* L. Agronomy and Genetics. Wageningen, the Netherlands. FACT Foundation, 2007.

[56] Gour VK. Production practices including post-harvest management of Jatropha curcas. In: Singh B, Swaminathan R, Ponraj V, Eds. Proceedings of the biodiesel conference toward energy independence—focus of *Jatropha*, Hyderabad, India, June 9–10. New Delhi, 2006: 223–251.

[57] Attaya AS, Geelen D, Belal AH. Progress in *Jatropha curcas* tissue culture. Am.-Eurasian J Sustain Agric 2012; 6: 6–13.

[58] Kochhar S, Kochhar VK, Singh SP, Katiyar RS, Pushpangadan P. Differential rooting and sprouting behaviour of two *Jatropha* species and associated physiological and biochemical changes. Curr Sci India 2005; 89: 936–939.

[59] Kochhar S, Singh SP, Kochhar VK. Effect of auxins and associated biochemical changes during clonal propagation of the biofuel plant—*Jatropha curcas*. Biomass and Bioenergy 2008; 32: 1136–1143.

[60] Noor Camellia NA, Thohirah LA, Abdullah NAP, Mohd Khidir O. Improvement on rooting quality of *Jatropha curcas* using indole butyric acid (IBA). Res J Agric Biol Sci 2009; 5: 338–343.

[61] Jimu L, Nyakudya IW, Katsvanga CAT. Establishment and early field performance of *Jatropha curcas* L. at Bindura University farm, Zimbabwe. J Sustain Dev Africa 2009; 10: 445–469.

[62] Ackom EK, Ertel J. An alternative energy approach to combating desertification and promotion of sustainable development in drought regions. Forum der Forschung 2005; 18: 74–78.

[63] Agarwal D, Agarwal AK. Performance and emissions characteristics of *Jatropha* oil (preheated and blends) in a direct injection compression ignition engine. Appl Therm Eng 2007; 27: 2314–2323.

[64] Pramanik K. Properties and use of *Jatropha curcas* oil and diesel fuel blends in compression ignition engine. Renew Energ 2008; 28: 239–248.

[65] Azan MM, Waris A, Nahar NM. Prospects and potential of fatty acid methyl esters of some non-traditional seed oils for use as biodiesel in India. Biomass and Bioenergy 2005; 29: 293–302.

[66] FAO (Food and Agricultural Organization of the United Nations). The state of food and agriculture. Biofuels: prospects, risks and opportunities. Rome, 2008. (Accessed November 21, 2012, at http://www.fao.org/docrep/011/i0100e/i0100e00.htm).

[67] Dde Oliveira JS, Leite PM, de Souza LB, *et al.* Characteristics and composition of *Jatropha gossypiifolia* and *Jatropha curcas* L. oils and application for biodiesel production. Biomass and Bioenergy 2009; 33: 449–453.

[68] Pant KS, Khosla V, Kumar D, Gairola S. Seed oil content variation in *Jatropha curcas* Linn. in different altitudinal ranges and site conditions in H.P. India. Lyonia 2006; 11: 31–34.

[69] Kaushik N, Kumar K, Kumar S, Kaushik N, Roy S. Genetic variability and divergence studies in seed traits and oil content of *Jatropha (Jatropha curcas* L.) accessions. Biomass and Bioenergy 2007; 31: 497–502.

[70] Singh RN, Vyas DK, Srivastava NSL, Narra M. SPRERI experience on holistic approach to utilize all parts of *Jatropha curcas* fruit for energy. Renew Energ 2008; 33: 1868–1873.

[71] Sunil N, Varaprasad KS, Sivaraj N, Kumar TS, Abraham B, Prasad RBN. Assessing *Jatropha curcas* L. germplasm in-situ—a case study. Biomass and Bioenergy 2008; 32: 198–202.

[72] Dwemahyani I, Ishak. Induced mutation on *Jatropha (Jatropha curcas* L.) for improvement of agronomic characters variability. Atom-Indonesia 2004; 30: 53–60. (Accessed November 21, 2012, at http://digilib.batan.go.id/atom-indonesia/fulltex/v30-n2-7-2004/Ita-Dwimahyani-Ishak.pdf).

[73] Parthiban KT, Kumar RS, Thiyagarajan P, Subbulakshmi V, Vennila S, Rao MG. Hybrid progenies in *Jatropha*—a new development. Curr Sci India 2009; 96: 815–23.

[74] Sujatha M, Reddy TP, Mahasi MJ. Role of biotechnological interventions in the improvement of castor (*Ricinus communis* L.) and *Jatropha curcas* L. Biotechnol Adv 2008; 26: 424–35.

[75] Mukherjee P, Varshney A, Johnson TS, Jha TB. Jatropha curcas: a review on biotechnological status and challenges. Plant Biotechnol Rep 2011; 5: 197–215.

[76] Basha SD, Sujatha M. Inter and intra-population variability of *Jatropha curcas* (L.) characterized by RAPD and ISSR markers and development of population-specific SCAR markers. Euphytica 2007; 156: 375–386.

[77] Shah S, Gupta MN. Lipase catalyzed preparation of biodiesel from Jatropha oil in a solvent free system. Process Biochem 2007; 42: 409–414.

[78] Staubmann R, Foidl G, Foidl N, *et al.* Biogas production from *Jatropha curcas* Press-Cake. Appl Biochem Biotech 1997; 63–65: 457–467.

[79] Abreu F. Alternative by-products from *Jatropha*. In: International Consultation on Pro-poor Jatropha Development. 10–11 April 2008, Rome: IFAD, 2008. (Accessed November 18, 2012, at http://www.ifad.org/events/jatropha/).

[80] Singh RN, Vyas DK. Feasibility study of *Jatropha* seed husk as an open core gasifier feedstock. Renew Energ 2007; 32: 512–517.

[81] EU directive 2008/101/EC of the European parliament and of the council of 19 November 2008, amending directive 2003/87/EC so as to include aviation activities in the scheme for greenhouse gas emission allowance trading within the Community. Official Journal of the European Union, 2009; 8: 3–21. (Accessed November 21, 2012, at http://eur-lex.europa.eu/LexUriServ/LexUriServ. do?uri=OJ:L:2009:008:0003:0021:EN:PDF).

[82] Sims B.Yale University conduct *Jatropha* sustainability study. Biodiesel Magazine, 2011. (Accessed November 21, 2012, at http://www.biodieselmagazine. com/articles/7696/boeing-yale-university-conduct-jatropha-sustainability-study).

[83] Figueiredo MK, Romeiro GA, Silva RVS, *et al.* Pyrolysis oil from the fruit and cake of *Jatropha curcas* produced using a low temperature conversion (LTC) process: analysis of a pyrolysis oil-diesel blend. Energ Power Eng 2011; 3: 332–338.

[84] Makkar HPS, Aderibigbe AO, Becker K. Comparative evaluation of a non-toxic and toxic variety of *Jatropha curcas* for chemical composition, digestibility, protein degradability and toxic factors. Food Chem 1998; 62: 207–215.

[85] Sharma DK, Pandey AK, Lata. Use of *Jatropha curcas* hull biomass for bioactive compost production. Biomass and Bioenergy 2009; 33: 159–162.

[86] Mahanta N, Gupta A, Khare SK. Production of protease and lipase by solvent tolerant Pseudomonas aeruginosa PseA in solid-state fermentation using Jatropha curcas seed cake as substrate. Bioresource Technol 2008; 99: 1729–1735.

[87] Manurung R, Wever DAZ, Wildschut J, *et al.* Valorisation of *Jatropha curcas* L. plant parts: nut shell conversion to fast pyrolysis oil. Food Bioprod Process 2009; 87: 187–196.

[88] DFID. Agricultural sustainability. A working paper 2004; 23. (Accessed November 21, 2012, at http://dfid-agriculture-consultation.nri.org/summaries/wp12. pdf).

[89] ESG (Earth System Governace project). Constructing sustainable biofuels: governance of the emerging biofuel economy, 2009. (Accessed November 21, 2012, at http://www.earthsystemgovernance.org/ac2009/papers/AC2009-0423.pdf).

[90] Janulis P. Reduction of energy consumption in biodiesel fuel life cycle. Renew Energ 2004; 29: 861–71.

[91] Francis G, Edinger R, Becker K. A concept for simultaneous wasteland reclamation, fuel production, and socio-economic development in degraded areas in India: need, potential and perspectives of Jatropha plantations. Nat Resour Forum 2005; 29: 12–24.

[92] Achten WMJ, Maes WH, Aerts R, *et al. Jatropha*: from global hype to local opportunity. J Arid Environ 2009; 74: 164–165.

[93] German Institute for Standardization. ISO 14040:2006(E) & ISO 14044:2006(E). Environmental Management—Life Cycle Assessment—Requirements and Guidelines. Berlin: Beuth Verlag, 2006.

[94] Sagar AD, Kartha S. Bioenergy and Sustainable Development? recommend. A newsletter of the community for energy, environment and development, 2007; 4: 1–5. (Accessed November 21, 2012, at http://www.energycommunity.org/ documents/BiofuelsSustainabileDevelopment.pdf).

[95] Tobin J, Fulford DJ. Life Cycle Assessment of the production of biodiesel from *Jatropha*. MSc Dissertation. The University of Reading, UK, 2005.

[96] Ndong R, Montrejaud-Vignoles M, Girons OS, *et al.* Life cycle assessment of biofuels from *Jatropha curcas* in West Africa: a field study. GCB lob Change Biol Bioenerg 2009; 1: 197–210.

[97] Whitaker M, Heath G. Life cycle assessment of the use of *Jatropha* biodiesel in Indian locomotives. Technical Report NREL/TP-6A2-44428, 2009: 102. (Accessed November 21, 2012, at http://www.nrel.gov/biomass/pdfs/44428.pdf).

[98] Achten WMJ, Almeida J, Fobelets V, *et al.* Life cycle assessment of *Jatropha* biodiesel as transportation fuel in rural India. Appl Energ 2010a; 87: 3652–3660.

[99] FAO (Food and Agricultural Organization of the United Nations). Carbon sequestration in dryland soils. Rome, 2004. (Accessed November 21, 2012, at ftp:// ftp.fao.org/agl/agll/docs/wsrr102.pdf).

[100] Ogunwole JO, Chaudhary DR, Ghosh A, Daudu CK, Chikara J, Patolia JS. Contribution of *Jatropha curcas* to soil quality improvement in a degraded Indian entisol. Acta Agr Scand B-S P 2008; 58: 245–251.

[101] Fargione J, Hill J, Tilman D, Polasky S, Hawthorne P. Land clearing and the biofuel carbon debt. Science 2008; 319: 1235–1238.

[102] Mergeai G. *Jatropha curcas*: what sustainability? Tropicultura 2008; 26: 1.

[103] Fairless D. Biofuel: the little shrub that could–maybe. Nature 2007; 449: 652–655.

[104] Wiesenhütter J. CCD (Convention Project to Combat Desertification) project. Use of the physic nut (*Jatropha curcas* L.) to combat desertification and reduce poverty. Possibilities and limitations of technical solutions in a particular socio-economic environment, the case of Cape Verde. Deutsche Gesellschaft für Technische Zusammenarbeit (GTZ), 2003: 18. (Accessed November 21, 2012, at http://www.underutilized-species.org/documents/Publications/use_of_jatropha_curcas_en.pdf).

[105] Reubens B, Achten WMJ, Maes WH, *et al.* More than biofuel? *Jatropha curcas* root system symmetry and potential for soil erosion control. J Arid Environ 2011; 75: 201–205.

[106] Askew MF. Natural plant oils and fats as renewable resources: a sustainable option? Lipid Technol 2007; 19: 254–257.

[107] Agamuthu P, Abioye OP, Aziz AA. Phytoremediation of soil contaminated with used lubricating oil using *Jatropha curcas*. J Hazard Mater 2010; 179: 891–4.

[108] Juwarkar AA, Yadav SK, Kumar P, Singh SK. Effect of biosludge and biofertilizer amendment on growth of *Jatropha curcas* in heavy metal contaminated soils. Environ Monit Assess 2008; 145: 7–15.

[109] Kumar GP, Yadav SK, Thawale PR, Singh SK, Juwarkar AA. Growth of *Jatropha curcas* on heavy metal contaminated soil amended with industrial wastes and Azotobacter—a greenhouse study. Bioresource Technol 2008; 99: 2078–2082.

[110] Mangkoedihardjo SS, Surahmaida. *Jatropha curcas* L. for phytoremediation of lead and cadmium polluted soil. World Appl Sci J 2008; 4: 519–522.

[111] Jamil S, Abhilash PC, Singh N, Sharma PN. *Jatropha curcas*: a potential crop for phytoremediation of coal fly ash. J Hazard Mater 2009; 172: 269–75.

[112] Yadav SK, Juwarkar AA, Kumar GP, Thawale PR, Singh SK, Chakrabarti T. Bioaccumulation and phyto-translocation of arsenic, chromium and zinc by *Jatropha curcas* L.: impact of dairy sludge and biofertilizer. Bioresour Technol 2009; 100: 4616–4622.

[113] Agamuthu P, Abioye OP, Aziz AA, Phytoremediation of soil contaminated with used lubricating oil using *Jatropha curcas*. J Hazard Mater 2010; 179: 891–894.

[114] Mohan S. Growth of biodiesel plant in flyash: A sustainable approach. Response of *Jatropha curcus*, a biodiesel plant in fly ash amended soil with respect to pigment content and photosynthetic rate. Procedia Environ Sci 2011; 8: 421–425.

[115] Abhilash PC, Yunus M. Can we use biomass produced from phytoremediation. Biomass and Bioenergy 2011; 35: 1371–1372.

[116] Mortimore M, Adams WM. Working the Sahel: Environment and Society in Northern Nigeria. NY, USA: Routledge, 1999.

[117] Wiskerke WT, Dornburg V, Rubanza CDK, Malimbwi RE, Faaij APC. Cost/benefit analysis of biomass energy supply options for rural smallholders in the semi-arid eastern part of Shinyanga Region in Tanzania. Renew Sust Energ Rev 2010; 14: 148–165.

[118] Andreasson K, Richard MA. Sustainable miracle? Determining the socio-economic sustainability of small scale Jatropha cultivation in the Eastern Province of Zambia. ARBETSRAPPORTER Kulturgeografiska institutionen. Nr. 783. Uppsala Universitet 2011: 1–39. (Accessed November 21, 2012, at http://uu.diva-portal.org/smash/record.jsf?pid=diva2:488242).

[119] Favretto N, Stringer LC, Dougill AJ. Cultivating clean energy in Mali: policy analysis and livelihood impacts of *Jatropha curcas*. Paper prepared for the conference "Energy and People: Futures, complexity and challenges" organized by the Environmental Change Institute. Oxford, 2011. (Accessed November 21, 2012, at http://www.ukerc.ac.uk/support/tiki-download_file.php?fileId=1942&display).

[120] IEA (International Energy Agency). Sustainable production of second-generation biofuels. potential and perspectives in major economies and developing countries. Information paper. Paris, 2010: 218.

[121] NOVOD. *Jatropha*, an alternate source for biodiesel. 2007. (Accessed November 21, 2012, at http://www.novodboard.com/Jatropha-english.pdf).

[122] Jingura RM. Technical options for optimization of production of *Jatropha* as a biofuel feedstock in arid and semi-arid areas of Zimbabwe. Biomass and Bioenergy 2011; 35: 2127–2132.

[123] Li Guo T. The photosynthesis and water use efficiency of eight garden tree species. Forest Research 2002; 15: 291–296.

[124] Agoramoorthy G, Hsu MJ, Chaudhary S, Shieh P. Can biofuel crops alleviate tribal poverty in India's drylands? Appl Energ 2009; 86: 118–124.

[125] Ariza-Montobbio P, Lele S. Jatropha. Plantations for biodiesel in Tamil Nadu, India: viability, livelihood trade-offs and latent conflict. Ecol Econ 2010; 70: 189–195.

[126] Paramathma M, Venkatachalam P, Sampathrajan A, *et al.* Cultivation of Jatropha and biodiesel production. Professor and Nodal Officer, Center of Excellence in Biofuels. Agricultural Engineering College & Resarch Institute, Tamil Nadu Agricultural University, Coimbatore, 2007: 180.

[127] Tomomatsu Y, Swallow B. Jatropha curcas Biodiesel production in Kenya economics and potential value chain development for smallholder farmers. Working Paper 54. Nairobi: World Agroforestry Centre, 2007.

[128] Balota EL, Machineski O, Scherer A. Mycorrhizal effectiveness on physic nut as influenced by phosphate fertilization levels. Revista Brasileira de Ciencia do Solo 2012; 36: 23–32.

[129] Achten WMJ, Maes WH, Reubens B, *et al.* Biomass production and allocation in *Jatropha curcas* L. seedlings under different levels of drought stress. Biomass Bioenergy 2010b; 34: 667–676.

[130] Silva EN, Ferreira-Silva SL, Fontenele AV, Ribeiro RV, Viégas RA, Silveira JAG. Photosynthetic changes and protective mechanisms against oxidative damage subjected to isolated and combined drought and heat stresses in *Jatropha curcas* plants. J. Plant Physiol 2010; 167: 1157–1164.

[131] Mazhar AAM, Abd El Aziz NG, El. Habba E. Impact of different soil media on growth and chemical constituents of *Jatropha curcas* L. seedlings grown under water regime. J Am Sci 2010; 6: 549–556.

[132] Arcoverde GB, Rodrigues BM, Pompelli MF, Santos MG. Water relations and some aspects of leaf metabolism of *Jatropha curcas* young plants under two water deficit levels and recovery. Braz. J. Plant Physiol 2011; 23: 123–130.

[133] Kenan L, Qiliang Y, Zhenyang GE, Xiaogang L. Simulation of *Jatropha curcas* L. root in response to water stress based on 3D visualization. Procedia Engineering 2012; 28: 403–408.

[134] Swaminathan C, Vivekanandan P, Kennedy RR. Growth response of *Jatropha curcas* to spacing and irrigation in dry lands. J Maharashtra Agric Universities 2010; 35: 460–461.

[135] De Lima RDS, Severino LS, Gheyi HR, Sofiatti V, Arriel NHC. Phosphorus fertilization on growth and contents of macronutrients in *Jatropha curcas* seedlings. Revista Ciencia Agronomica 2011; 42: 950–956.

[136] De Souza PT, de Barros Silva E, Grazziotti HP, Fernandes LA. NPK fertilization on initial growth of physic nut seedlings in quartzarenic neossol. Revista Brasileira de Ciencia do Solo 2011; 35: 559–566.

[137] Maia JTL, Guilherme, DD, Paulino, MAD, Silveira, HRD, Fernandes, LA. Effect of omission of macro and micronutrients on growth of physic nutrient. Revista Caatinga 2011; 24: 174–179.

[138] Suriharn B, Sanitchon J, Songsri P. Kesmaka T. Effects of pruning levels and fertilizer rates on yield of physic nut (*Jatropha curcas* L.). Asian J Plant Sci 2011; 10: 52–59.

[139] Cotula L, Dyer N, Vermeulen S. Fuelling exclusion? The biofuels boom and poor people's access to land. London: IIED and FAO, 2008: 72. (Accessed November 21, 2012, at http://pubs.iied.org/pdfs/12551IIED.pdf).

Chapter 5

Environmental Aspects of Willow Cultivation for Bioenergy

Achim Grelle

5.1   Introduction

Forests play important roles in the global carbon (C) cycle, firstly because of their vast C stocks in biomass and soil and secondly, because of the large continuous fluxes of C into and out of forest ecosystems. The C stored in forest biomass is renewable at a time scale of decades and comprises a potential source of bioenergy. The soil C stock, on the other hand, has built up over thousands of years, especially in boreal regions where the largest soil C stocks are located and stored in a relatively stable state as long as soil temperatures are kept low. However, if changes such as rising temperatures or physical disturbances happen to the soil, soil organic matter may be increasingly decomposed and the soil C stock may be diminished, accompanied by emissions of CO_2.

On the long term, the C balances of natural forest ecosystems are close to neutral, meaning that C uptake by photosynthesis is more or less balanced by C emissions due to respiration. CO_2 is respired from living biomass by plant metabolism (autotrophic respiration) and from soil organisms during decomposition of organic matter (heterotrophic respiration). A newly established forest builds up a soil C stock by accumulation of aboveground (leaves, dead wood) and belowground (roots) litter. Since all of that carbon is taken from the atmosphere through photosynthesis, such young forests are significant C sinks. The more C is accumulated in the soil, the more CO_2 will be emitted by heterotrophic respiration. Eventually, C emissions will become large enough to balance uptake and the old-growth, unmanaged forests will thereby become C-neutral with respect to the atmosphere. In boreal regions, this happens at a time scale of millennia, *i.e.*, unmanaged forests that have established after the last ice age are now generally close to equilibrium with respect to the C balance.

Forest management affects the C balance in several ways. Firstly, biomass is removed and used for things like biofuel, for example. This implies that CO_2 is emitted instantaneously by combustion rather than successively by decomposition and respiration. Secondly, harvesting implies removal of photosynthetically active foliage and, thereby, elimination of a C sink. Third, management generally implies physical disturbance of the forest soil and clear-cutting exposes the soil to changes in light, temperature, and hydrological conditions. These factors affect the decomposition of organic matter and heterotrophic respiration.

Consequently, a forest clear-cut loses large amounts of C for about a decade after harvesting. After that, it turns into a C sink again, but it takes another decade to compensate for the losses during the clear-cut phase by new C uptake. Therefore, for a forest managed by clear-cutting the average annual C uptake during the entire rotation period is only about half of the "potential" uptake, *i.e.*, the amount of C that the forest takes up each year during its mature phase. However, the total uptake can be optimized by harvesting the forest when the annual C uptake drops below the mean annual uptake, *i.e.*, the length of the rotation period can be chosen for maximum C uptake rather than for maximum yield, which is not necessarily the same.

5.2 Willow Plantations

The challenge of forestry in terms of sustainable production of bio-fuel is to maximize yield and thereby substitute fossil fuels and, at the same time, conserve or even increase the soil C stock. This might be achieved by "continuous cover forestry", *i.e.*, selective annual harvest of single trees that avoids clear-cutting, keeping the canopy closed and the C sink intact. Another feasible approach is intense cultivation of fast-growing tree species such as willow, which gives high yield and short rotation periods and, consequently, short, yet frequent, clear-cut phases. Naturally, the high rate of biomass increase is associated with large C uptake from the atmosphere.

This even has political and economic impacts because, in some countries, short-rotation forestry may be accounted for to fulfill commitments under the Kyoto Protocol. In Sweden, for example, willows (*i.e.*, different varieties of *Salix* sp.) are commercially grown on 15,000 ha of farmland for production of fuel for district heating plants, usually combined with power generation. Ashes from the combustion are usually recycled to the willow plantations. Besides the potential of fast biomass production and C uptake, willow plantations even have the capability to treat and recycle wastewater by enhanced denitrification, by plant uptake of nitrogen (N) and other nutrients from the wastewater, and by uptake and absorption of pollutants such as heavy metals. The uptake of nutrients, in turn, results in enhanced growth (*e.g.*, [2], [3]). By means of separation devices in a combustion power plant, pollutants such as heavy metals can be extracted and processed before recycling the ashes.

A beneficial system is the combination of a willow plantation with a nearby district heating plant and a wastewater treatment plant, such as that in Enköping, Sweden. There, a 75 ha willow plantation adjacent to a wastewater treatment plant is irrigated with a fraction of the wastewater during summertime. The sewage sludge that is produced in the tertiary water treatment step is dewatered by sedimentation and centrifugation. The supernatant water from this process corresponds to less than 1% of the total water flow in the wastewater treatment plant, but contains about 25% of the N entering the system. During summer, the supernatant water is mixed with treated wastewater and distributed to the willow plantation through drip pipes that are laid out in every double row of willow plants. Furthermore, the field is drained by drainage pipes at approximately 60 cm depth.

After establishing a willow plantation in spring (*e.g.*, May), mechanical weeding is conducted during the first summer. During the following winter, plants are cut back to promote sprouting and canopy closure. Usually, the plantation is harvested during the third or fourth winter after establishment. There, the shoots are cut and the root system is left in place for several rotation periods. Therefore, re-growth immediately commences during the season following harvest.

5.3 Carbon Sequestration and Greenhouse Gas Fluxes

Key variables in determining the environmental impacts of biofuel production include fluxes of greenhouse gas (GHG) within the soil-plant-atmosphere system. In particular, fluxes of C between the atmosphere and the plantation (photosynthesis and respiration) and between the vegetation and the soil (soil C storage) have to be quantified and related to C emissions by combustion of biomass. By biomass inventory, the beneficial effects of fossil fuel substitutions by willow plantations are relatively well-known. However, less is known about the corresponding C sequestration. To study the overall carbon and GHG balance of a willow plantation, budgets of different compartments can be estimated separately, as a complement to the determination of the total net ecosystem fluxes.

C storage in different plant sections can be quantified by biomass sampling and soil carbon storage can be estimated by subsequent sampling of soil probes or by process-based simulation models. To assess the total budget, atmospheric fluxes of total ecosystem GHG exchange can be measured quite accurately by micrometeorological methods.

5.3.1 Estimates of Growth and Carbon Sequestration

The C sequestration of a willow plantation can be estimated by means of a conceptual model, such as the one presented by Aronsson [1]. In the model, the C sequestration is a result of (i) C incorporation into harvestable shoots, (ii) C incorporation into non-harvestable coarse roots and stumps, and (iii) incorporation of leaf and fine-root litter-C into the humus pool, *i.e.*, the pool of slow-degradable soil organic matter.

1. The annual aboveground growth of willow plants, *i.e.*, the growth of harvestable shoots, can be estimated by a combination of destructive and non-destructive measurements [11]. There, the diameters of a large number of shoots are measured at a certain height (*e.g.*, 95 cm). Several shoots of all diameter classes are then harvested and dried to obtain the dry weights and, thereby, the biomass content. This gives a mathematical relationship between shoot diameter and biomass content, which can be used to estimate the total amount of biomass in the plantation on the basis of diameter measurements. Furthermore, maps can be produced showing the local distribution of biomass, which reflects variations in growing conditions within the plantation.

2. The C incorporation into non-harvestable stumps can be calculated using templates of biomass partitioning, such as the ones presented by Rytter [10]. However, little is known about the build-up of coarse, structural roots in short-rotation coppice systems. Therefore, this component of the C sink is often omitted.

3. The annual growth of fine roots is a major component of the total biomass increment in a willow plantation. However, after the first few years of plant establishment, the annual die-off of fine roots is of similar magnitude. Consequently, the living fine-root biomass is fairly constant. Thus, literature data from, *e.g.*, amount of leaf- and fine-root litter formation can be used to calculate the annual net C buildup in humus.

However, the biomass allocation pattern varies considerably with growth conditions such as N availability and is likely to be unevenly distributed in somewhat heterogeneous willow plantations. Nonetheless, to estimate average carbon sequestration, it is sufficient to treat the annual growth of leaves, stump wood, and fine roots as constant fractions of shoot growth. Assuming that shoots, stumps, and humus fractions have C proportions of 50%, the following equations can be used to calculate the plants' annual C-sequestration:

$$\text{Shoot C} = \text{Shoot Growth (SG)} \times 0.50 \tag{5.1}$$

$$\text{Stump Growth} = 0.05 \times \text{SG} \tag{5.2}$$

$$\text{Stump C} = \text{Stump Growth} \times 0.50 \tag{5.3}$$

$$\text{Fine Root Litter} = 0.69 \times \text{SG} \tag{5.4}$$

$$\text{Fine Root Litter to Humus} = 0.50 \times \text{Fine Root Litter} \tag{5.5}$$

$$\text{Fine Root Litter C to Humus} = \text{Fine Root Litter to Humus} \times 0.50 \tag{5.6}$$

$$\text{Leaf Litter} = 0.31 \times \text{SG} \tag{5.7}$$

$$\text{Leaf Litter to Humus} = 0.42 \times \text{Leaf Litter} \tag{5.8}$$

$$\text{Leaf Litter C to Humus} = \text{Leaf Litter to Humus} \times 0.50 \tag{5.9}$$

By combining Equations (5.1)–(5.9), the total C-sequestration (C_{SEQ}) becomes:

$$C_{\text{SEQ}} = \text{Shoot Growth} \times 0.76 \tag{5.10}$$

By relating biomass increment to shoot growth, measurements of shoot diameters at many locations all over the plantation (e.g., 5 sample points per hectare) can be used to produce maps of biomass increment such as Fig. 5.1 (which corresponds to Fig. 5.2 in [5]). Here, the annual biomass increments for two differently aged parts of the Enköping willow plantation are shown for two consecutive years. In 2003, the left part of the plantation (ca. 4.5 ha) had been harvested and in 2004, the right part (ca. 12 ha) had been harvested, both four years after establishment. During 2003, the average growth in the field was 10.2 t dry matter (DM) ha^{-1} (8.9 t DM ha^{-1} in the 4.1 ha large one-year-old harvested part and 10.8 t DM ha^{-1} in the 9.8 ha large three-year-old part). During 2004, the average growth was 9.4 t DM ha^{-1} (17.0 t DM ha^{-1} in the two-year-old part and 6.3 t DM ha^{-1} in the one-year-old harvested part). During autumn, 2004, litterfall of 945 kg DM ha^{-1} was measured. A common problem of regrowing willow plantations in Sweden is damages by night frosts during late spring and early summer. This may limit shoot growth throughout the season. However, a shoot growth of 2–3 m during the first year can be achieved (Fig. 5.2).

5.3.2 Eddy Flux Measurements

Micrometeorological measurements by the eddy flux methodology (also called eddy-covariance) are excellent tools to continuously study net fluxes of GHG or other air constituents *in situ* with high time resolution.

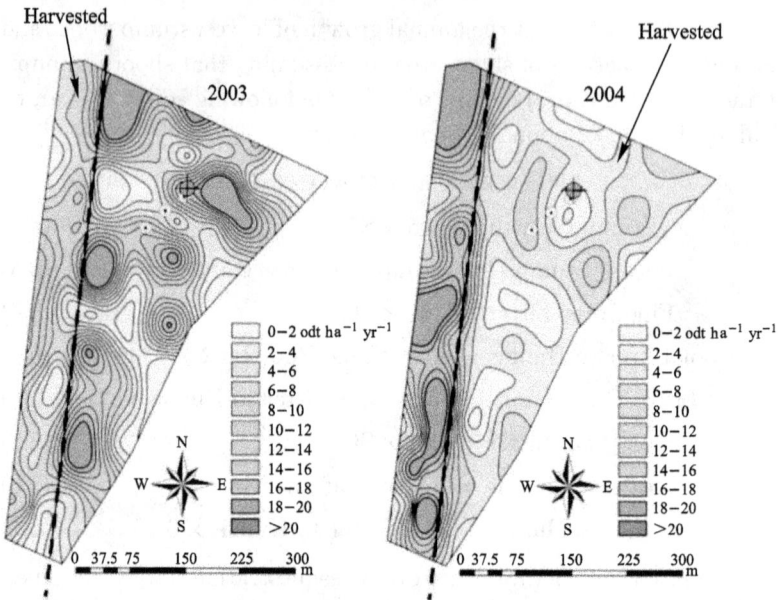

Fig. 5.1 Variations of biomass increments in the Enköping willow plantation during 2003 and 2004. The dotted line indicates the boundary between the areas that were harvested during 2002/2003 and 2003/2004, respectively. The cross indicates the position of the flux system.

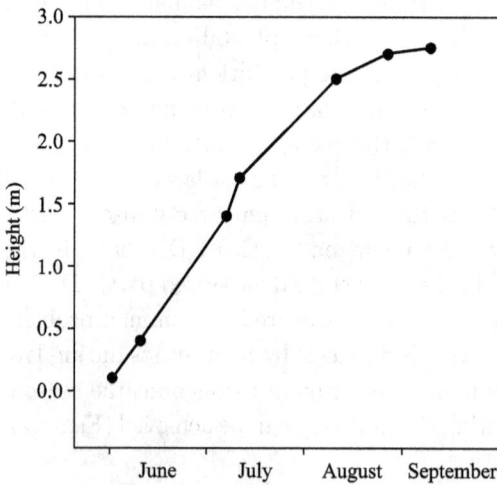

Fig. 5.2 Shoot growth in the Enköping willow plantation during 2004.

If a vegetated surface absorbs (or emits, respectively) a GHG, the GHG concentration close to the surface will decrease (or increase, respectively). Successively, a vertical gradient of GHG concentration will arise in the atmospheric boundary layer close to the surface, created through the consumption (or emission, respectively) by the vegetation, and spread by molecular diffusion.

Wind in the planetary boundary layer is virtually always turbulent, *i.e.*, it contains irregular vortices (eddies) of different sizes and structures. The eddies are three-dimensional, which means that they add irregular vertical components to the air movement. These vertical components mix up the concentration gradient in the boundary layer: each time an air parcel is forced upward by an eddy, it contains less (resp. more) GHG than its surroundings; each time a parcel is forced downward, it contains more (resp. less) GHG than its surroundings. The vegetation continuously reinforces the gradient, while the turbulence acts towards leveling it. In particular, this means that there is a well-defined correlation between local GHG concentration and momentary vertical air movement: each time an air parcel is forced to move vertically, it carries some GHG from the source (higher concentration) to the sink (lower concentration).

With accurate and fast sensors, we can measure local air movements and GHG concentrations: we observe in which direction air parcels move and, at the same time, in the same place, we measure how much GHG they carry with them. In other words, we continuously follow the ecosystem's breathing. To capture contributions of all relevant small eddies, we have to do short and quick measurements (typically, 20 times per second); to capture the even larger ones, we have to integrate our measurements over a longer time period (typically, 30 minutes).

Eddy flux sensors are normally installed some meters above the ecosystem to be studied. This way, it is assured that the air movement and GHG concentration measurements are done within the gradient created by the underlying soil-vegetation system. Consequently, the determined vertical fluxes of GHG are representative for the interaction of the studied ecosystem with the atmosphere.

However, there are certain restrictions for the method's applicability. To obtain a representative gradient in the boundary layer, the spatial extensions of the ecosystem must be sufficiently large and to obtain representative measurements from all wind directions, the ecosystem must be sufficiently homogeneous, flat, and horizontal. However, commercial willow plantations of several hectares in size often meet these criteria. To validate the representativeness of measured ecosystem fluxes, the concept of source area analysis (also called footprint analysis) is a useful methodology. Here, turbulence statistics are evaluated to model the three-dimensional transport of particles from the ecosystem surface to the heights of the sensors. The horizontal travel distance together with the wind direction indicates the location of the source areas of measured fluxes. Distance and extent of the source area depend on the surface roughness and the height of the sensor above the surface and varies with wind speed, atmospheric stability, and turbulence intensity.

The total net flux between an ecosystem and the atmosphere is usually the tiny balance between two large gross fluxes, *e.g.*, photosynthesis and respiration, and is quite sensitive to small changes in any of the gross components. By eddy flux measurements, the net flux is determined directly, which yields much higher accuracy than calculating the differences between estimated gross fluxes. On the other hand, net flux measurements cannot be readily used to determine uptake and emissions separately. Net CO_2 fluxes from willow plantations reflect the growing conditions. The stand development as well as the beginning and the end of the growing season can be clearly seen in the CO_2 balance. Furthermore, varying shoot growth due to frost damage or weed competition, for example, results in variations of the net CO_2 uptake.

In Figure 5.3, mean diurnal courses of CO_2 fluxes from the Enköping plantation are shown for different seasons. During winter, there is a small but consistent loss of CO_2 by respiration throughout the day. Respiration increases with temperature and is therefore larger during the rest of the season. However, during the daytime in the growing season, even photosynthesis takes place and this more than compensates for the CO_2 losses by respiration. It can be seen as large negative (downward) fluxes of CO_2 during daytime.

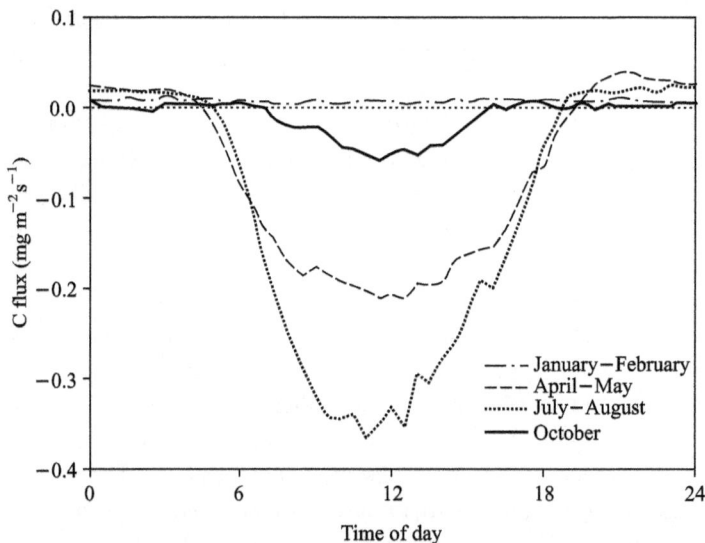

Fig. 5.3 Mean diurnal courses of CO_2 fluxes from the Enköping willow plantation in mass units of CO_2-carbon. Positive values mean emission, negative values mean uptake.

To assess seasonal C budgets, the measured CO_2 fluxes are integrated over time. This way, the total CO_2 exchange between the ecosystem and the atmosphere can be determined for different time periods. In Figure 5.4, such inte-

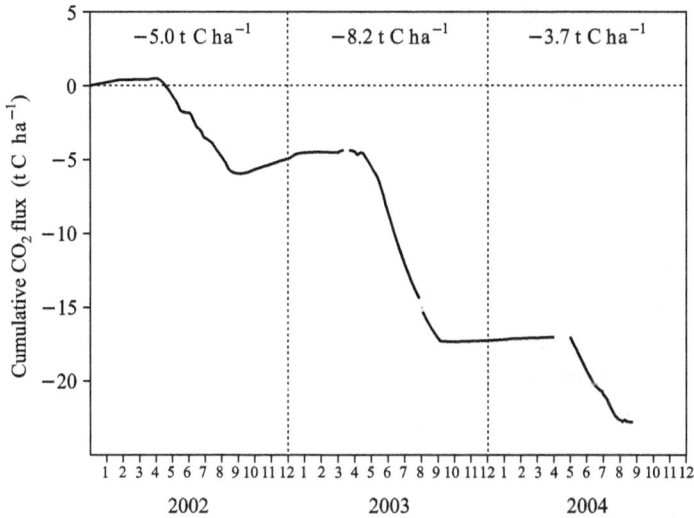

Fig. 5.4 Cumulative CO_2 fluxes from the Enköping willow plantation, expressed in mass units of CO_2-C. Negative values mean uptake. The numbers denote annual sums of C uptake, except for 2004, where the measurements ceased in September. Breaks in the line indicate periods where gaps in the data have been filled by linear interpolation of the cumulative fluxes.

grated (cumulative) fluxes are shown for three consecutive, contrasting years from the Enköping willow plantation. For easier comparison with biomass estimates, the CO_2 fluxes here are converted into fluxes of C. Again, negative values mean downward fluxes (uptake) and positive values mean upward fluxes (emission).

During the first year, there was significant growth of weeds that suppressed willow development by competition. Consequently, the total carbon uptake was relatively small (5 t C ha^{-1} yr^{-1}). During the second year, the plantation grew very well and the largest net uptake was observed (8.2 t C ha^{-1} yr^{-1}). During the third year, the uptake was smaller and the measurements stopped at the end of September, so a full annual budget could not be determined.

In forests, the leaf area index (LAI), *i.e.*, the ratio between leaf area and ground area, is the main factor controlling C uptake. It is closely related to stand age and the degree of canopy closure determines the time when a regrowing forest turns from a C source into a C sink. In a fast-growing willow plantation, however, during the first growing season after harvest, it takes only about one month for the canopy to close more or less and during the following years, the LAI around mid-summer is fairly constant, with only slight increases during the establishment phase, which can be one to three years.

5.3.3 Closing the Carbon Budget

There are considerable differences between the amounts of C taken up by the whole plantation, as measured by eddy flux, and the amounts allocated to the willow shoots, as determined by diameter measurements. This is expected because the eddy flux system measures the CO_2 uptake of the total ecosystem, while shoot growth only represents one of several C sinks in the system. Other components of C uptake are stump growth, weed growth, and leaf- and fine-root growth, which eventually produces litter that enters the soil humus pool.

There are not many reports about C allocation to willow roots but with the results from a study by Rytter [10] and a simple model by Aronsson [1], we can calculate the C allocation by stump and root growth and leaf and fine-root litter production as a function of shoot growth. Adding these modeled amounts of belowground C allocation to the shoot and weed growth gives a sum that agrees very well with the measured ecosystem fluxes (Fig. 5.5).

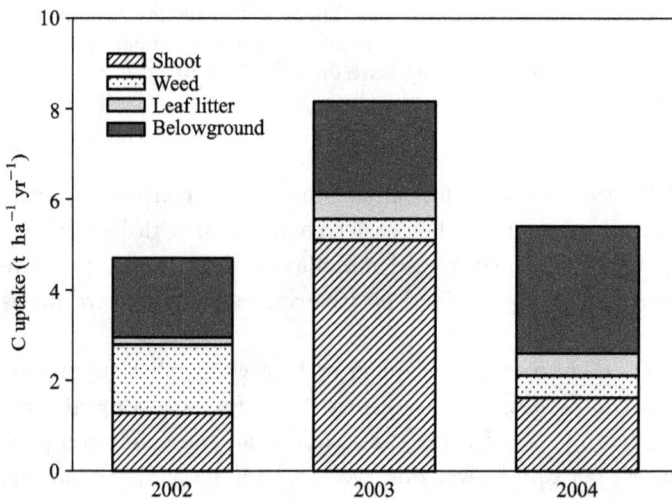

Fig. 5.5 C allocation to different compartments of the Enköping willow plantation, determined by a combination of measurements and modeling.

An interesting feature of this budget is the C allocation to the soil. It means that not only the C that is allocated in the shoots—and is later re-emitted to the atmosphere by combustion—is taken up from the atmosphere, but even an additional amount that is used to increase the soil C pool. At low temperatures, such as in Scandinavia, the soil C pool can grow large and remain stable for a very long time, as can be seen in the boreal forest soils that contain massive amounts of C. That means that willow plantation soils, unless they are disturbed by tillage, for example, have the potential to contribute to long-term C sequestration that counteracts rising atmospheric CO_2 concentrations.

5.3.4   The Fertilization Effect

The effect of N fertilization on willow shoot growth may vary between 10% and 73% (*e.g.*, [7]). Yield effects due to fertilization by wastewater are yet to be explored; however, from practical experience, it is reasonable to assume a growth increase of 50% as an effect of wastewater fertilization. That means that in Enköping wastewater irrigation led to an extra C sequestration between 2 and 4 t C ha^{-1} yr^{-1}.

At the same time, even nitrous oxide (N_2O) is usually emitted from fertilized, cultivated soils. N_2O is another GHG that is about 300 times more efficient than CO_2 in terms of global warming potential (GWP). To assess the total GHG budget of the plantation, even those emissions have to be taken into account and related to the C balance. Normally, fluxes of non-CO_2 GHG are expressed as CO_2-equivalents, which means that their GWP is taken into account and related to the GWP of CO_2. In other words, the GHG fluxes are expressed in corresponding amounts of CO_2 that would give the same contribution to global warming. At the Enköping plantation, fluxes of N_2O were also measured by eddy flux and the annual emissions corresponded to 4 t CO_2 ha^{-1} yr^{-1}. We assume that all N_2O emissions were due to fertilization by wastewater irrigation. If, on the other hand, the wastewater had been treated in a conventional way instead of using it for irrigation, this treatment would also have caused certain N_2O emissions, which corresponds to 25% of the observed emissions from the willow plantation. To estimate the effect of wastewater fertilization on the plantation's GHG balance, we relate the C uptake due to the growth increment by fertilization to the N_2O emissions caused by fertilization (Fig. 5.6). The result is a reduction of the beneficial fertilization effect by ca. 20%. Thus, despite this reduction, wastewater irrigation is still favorable from an environmental point of view.

5.3.5   What Are the Limits?

In 1984, an experimental willow plantation was established in Uppsala, Sweden. This "model forest" was managed with optimum care, which means rigorous weeding and exact controlled irrigation and fertilization according to plant requirements. CO_2 fluxes were measured by means of another micrometeorological

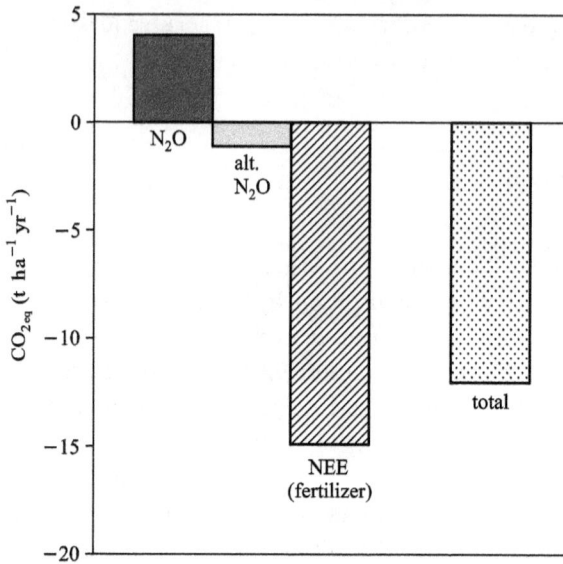

Fig. 5.6 The N-related GHG budget of the Enköping willow plantation, expressed in CO_2 equivalents.

"N_2O" means direct N_2O emissions from the salix plantation, "alt. N_2O" means emissions that would have occurred if the wastewater had been treated differently than by salix irrigation, and "NEE (fertilized)" means fertilization effect of wastewater-N on net ecosystem exchange of CO_2. "Total" is the overall effect of N-fertilization on the GHG fluxes (sum of the three columns).

technique based on vertical gradients of wind speed and gas concentrations [9]. The model forest took up more than twice as much C as the Enköping plantation, which illustrates the potential of short-rotation willow plantations to act as terrestrial C sinks and sources for biofuel. Even though such optimum treatment is currently not commercially feasible, it defines some upper limit of C uptake that can be achieved in mid-Sweden. In the perspective of conventional forestry, the C uptake of the model forest corresponds to 12 times the uptake of an average managed spruce forest in the same region.

5.3.6 Substitution Efficiency and Climate Effect

When biomass is combusted as a substitute for fossil fuel, C is oxidized that has previously been absorbed from the atmosphere during the growth of the corresponding biomass. During the next rotation ("generation") of the vegetation, a similar amount of C will again be absorbed and so on. This way, the term "renewable energy" is justified. The degree of substitution efficiency, also referred to as C neutrality, describes the ability of measures such as short rotation forestry to substitute fossil fuel by renewable energy, $i.e.$, to reduce net emissions of CO_2 in the long term [8]. If the whole chain of production and consumption of biofuel does not lead to any rise of CO_2 concentration in the atmosphere in the long term, it is 100% substitution-efficient. If it, on the other hand, leads to the same concentration increase as would fossil fuel, the substitution efficiency is zero. Here, the alternative fate of the biomass, which is what would happen to the biomass if it was not used as biofuel, is of central importance.

If, for example, biofuel was supplied by timber that could have otherwise been used to build houses that stand for centuries, this biofuel would not be C-neutral: its substitution efficiency would be close to zero. In practice, however, harvest residuals such as branches and tops, or even stumps, are used as biofuel. In those cases, the biomass would stay in the forest and decay slowly, if it was not used as biofuel. That means that its C content would be emitted as CO_2 in any case, but more slowly: at a time scale of decades instead of immediately. However, as long as no additional changes such as soil disturbance occur to the ecosystem, this form of biofuel can be considered C-neutral: it has high substitution efficiency.

For willow plantations such as the one in Enköping, the substitution efficiency may even be above 100%. This is because the plantation actually takes up more C from the atmosphere than what is emitted by combustion, since a part of it stays in the ground. Provided that the soil C storage is stable in the long term, this means that the whole chain of production and consumption of biofuel from willows leads to C sequestration rather than to C emission, on top of the effect of fossil fuel substitution that is achieved anyway.

Based on the C budget of the Enköping plantation during 2003, approximately 8 t C ha^{-1} were taken up from the atmosphere, of which ca. 5 tonnes were allocated to the shoots and consequently re-emitted by combustion later. That means that almost 3 tonnes of C entered the humus pool per hectare, which, if it can be considered stable, means that the substitution efficiency is 160%.

However, the time lag between emissions by instant burning and slow decomposition cannot be neglected when climate effects are considered. Since CO_2 has a very long atmospheric lifetime as GHG, we can assume that the major part of emitted combustion CO_2 is still present in the atmosphere after a century (for simplicity, we neglect uptake by oceanic or terrestrial sinks as they do not affect the principle process). During that century, the whole amount of CO_2 is active as GHG and participates in radiative forcing, which can be illustrated by the time integral of the emission (Fig. 5.7). If, on the other hand, the biomass was to slowly decay in the forest, the corresponding CO_2 would be emitted successively during the century, slowly increasing its contribution to radiative forcing.

Because only a fraction of the total CO_2 is present in the atmosphere for most of the time period, the corresponding time integral of emissions is, at all times,

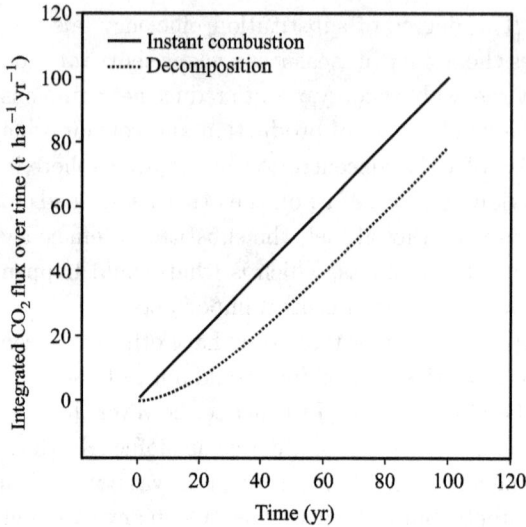

Fig. 5.7 Time integral of CO_2 emissions from woody biomass. The solid line indicates emission from instant burning, the dotted line indicates emissions from slow decomposition of biomass left in the ecosystems. The numbers in this example are typical for tree stumps.

smaller than the integral of instant burning. That means that at a time scale of centuries, instant burning of biofuel contributes more to climate change than slow decomposition, because total CO_2 spends more time in the atmosphere, even though the same total amount will be emitted in the end. In other words, even 100% C-neutral or substitution-efficient biofuel will not be "climate-neutral" because of the effect of instant burning. More details about that concept can be found in [6]. However, it is important to note that even incompletely climate-neutral biofuel is much more favorable to the climate system than fossil fuel.

5.4 Conclusions

The energy content of biomass from willow plantations is ca. 19 MJ (kg DM)$^{-1}$ [4]. That means that an average production of 7.5 t DM ha^{-1} yr^{-1} for the whole rotation period leads to a potential energy production of $142,500$ MJ ha^{-1} yr^{-1}. This corresponds to an energy flux density that can simply be expressed on a basis of smaller units in W m^{-2}. Converting hectares to square meters and years to seconds, it corresponds to a continuous energy flux density of ca. 0.5 W m^{-2}. If we relate that to the incoming solar radiation that is 340 W m^{-2} on a global average, it becomes obvious that alternative technology such as solar panels (if we assume an efficiency of 10%) can easily produce much higher energy fluxes. In particular, solar panels with a surface corresponding to 1.3% of the willow plantation area could potentially produce a similar amount of energy per year. This is a rough estimate disregarding latitude, cloudiness, orientation of solar panels, etc., but it illustrates the order of magnitude. Furthermore, once established, arrays of solar panels require much less maintenance and processing than willow plantations.

On the other hand, investment costs for solar panel arrays of that magnitude are relatively high—according to our rough estimate, ca. 1 ha of solar panels would be needed to replace the energy production of the 75 ha Enköping willow plantation. However, a comparison with the real world shows that this is an underestimation because operational solar power plants in the same area already produce an amount of energy corresponding to a 1 ha willow plantation on an effective area of 310 m^2, oriented at an angle of 40° from the horizontal plane [12]. This implies that more than 2.3 ha of such solar panels would be needed to annually produce the same amount of energy as the Enköping willow plantation. Furthermore, electricity cannot substitute solid fuel in every case—there is a societal need for solid, liquid, and gaseous fuels as well as electricity. Considering that, and taking into account the beneficial effects of willow plantations on soil chemistry, *i.e.*, filtering of pollutants and increases of the soil carbon stock, short rotation forestry is a favorable strategy for substitution of fossil fuel by bioenergy.

References

[1] Aronsson PG. Nitrogen Retention in Vegetation Filters of Short-Rotation Willow Coppice. Doctoral thesis. Silvestria 161. Swedish University of Agricultural Sciences, 2000.

[2] Aronsson PG, Bergström LF. Nitrate leaching from lysimeter-grown short-rotation willow coppice in relation to N-application, irrigation and soil type. Biomass and Bioenergy 21, 2001; 21: 155–164.

[3] Dimitriou I, Aronsson PG. Nitrogen leaching from short-rotation willow coppice after intensive irrigation with wastewater. Biomass and Bioenergy 26, 2004; 26: 433–441.

[4] Dzurenda L, Geffertova J, Hecl. V. Energy characteristics of wood-chips produced from *Salix viminalis*—Clone ULV. Drvna Industrija 61, 2010; 61: 27–31.

[5] Grelle A, Aronsson PG,Weslien P, Klemedtsson L, Lindroth A. Large carbon-sink potential by Kyoto forests in Sweden—a case study on willow plantations. Tellus 59B, 2007; 910–918.

[6] Grelle A, Strömgren M, Hyvönen R. Carbon balance of a forest ecosystem after stump harvest. Scandinavian Journal of Forest Research 27, 2012; 27: 762–773.

[7] Hofmann-Schielle C, Jug A, Makeschin F, Rehfuess KE. Short-rotation plantations of balsam poplars, aspen and willows on former arable land in the Federal
Republic of Germany. I. Site-growth relationships. Forest Ecology and Management 121, 1999; 121: 41–55.

[8] Lindholm EL. Energy use and environmental impact of roundwood and forest
fuel production in Sweden. Doctoral thesis no 2010:40. Faculty of Natural Resources and Agricultural Sciences, Swedish University of Agricultural Sciences,
2010; ISBN 978-91-576-7453-1.

[9] Lindroth A, Cienciala E. Water use efficiency of short-rotation *Salix viminalis* at
leaf, tree and stand scales. Tree Physiology 16, 1996; 16: 157–262.

[10] Rytter RM. Fine-root production and carbon and nitrogen allocation in basket
willows. Doctoral thesis. Silvestria 39. Swedish University of Agricultural Sciences,
1997; ISBN 91-576-5323-2.

[11] Verwijst T, Telenius B. Biomass estimation procedures in short rotation forestry.
Forest Ecology and Management 121, 1999; 121: 137–146.

[12] Hebyanläggningen, Teknisk beskrivning Hebyanläggningen, Solel i Sala & Heby
ekonomisk förening, 2013 (Accessed January 11, 2013, at http://solelisalaheby.se/
anlaggningarna).

Part III
Biofuels and Biogeochemical Impacts

Chapter 6

Short Rotation Forestry for Energy Production in Italy: Environmental Aspects and New Perspectives of Use in Biofuel Industry

Sara Bergante, Terenzio Zenone, and Gianni Facciotto

6.1 Introduction

In recent years, increasing requirements of energy, due to fast economic growth in Asia, the uncertainty of the political situation, and catastrophic events of war in the areas of production and refinement of the oil and natural gas, have caused rises in fossil fuel prices. Oil has increased from $25 per barrel in 2003 to over $130 per barrel in 2008, while in 2013 the price was assessed to be $90 per barrel. Renewable energy has become an important means to secure an European energy supply and to reduce greenhouse gas (GHG) emissions. According to the Biomass Action Plan by the European Union (EU) Commission [1], the use of bioenergy should cover about 20% of the energy demand by 2020. A considerable amount of this percentage can be covered by residual biomass (agricultural and forest wastes, wet and dry manure, municipal solid waste, industrial waste wood, etc.) and dedicated energy crops (oil and starch crops, perennial or annual grasses, and short rotation forestry (SRF)) characterized by high production per hectare and low environmental pressure [2]. Among the dedicated woody crops grown as SRF, poplar and willow represent the most common species cultivated in Italy. Since some biomass power plants have become operational, the energetic use of poplar rapidly increased, especially the by-products of traditional poplar cultivation. For this reason, the price of raw biomass had a considerable increase in a short period of time. To feed the energetic power plants and to avoid competition with the wood industries, several regions provide financial support to SRF crops within the programs of rural development. In Italy, SRF occupies about 7,000 ha, most of them located in the northern part of the country. SRF plantations consist of densely planted, high-yielding varieties with very short rotations of two to six years. When harvested, the crops are usually converted into wood chips, which can be used for energy production.

The dedicated woody crops are grown with three different cultural models, characterized by different lengths of rotation and densities of trees. These can be described as:

1. Annual rotation, generally with over 10,000 trees ha^{-1}.
2. Biennial or triennial rotation, with density among 4,000 and 10,000 trees ha^{-1} (Fig. 6.1).
3. Five to six year rotation, with density among 1,100 and 2,000 trees ha^{-1}.

The productivity of these plantations can range from 3–4 to 15–20 t ha^{-1} yr^{-1} of dry matter in the first rotation cycle or both poplar and willow clones [3]. High variability is reported for poplar and willow SRF crops in other European countries [4–5]. Even if research initiatives concerning clone/provenance selection for biomass production have been successfully implemented in Italy [6–7] and in Europe [8–9]: This wide variability suggests the importance of factors different from the genetic characteristics. Water availability appeared to be the principal factor affecting the establishment of poplar and willow energy plantations. Variations in the rainfall regime consequent to climate changes could seriously influence land suitability to SRF crops [10].

Besides biomass production, SRF crops can provide a series of ecological services that have become more and more important over the last few years. Restoration of quarry sites or polluted lands, stabilization of river banks, and creation of ecological corridors are often accomplished with SRF cultivation. Recently, SRF plantations were used in research projects to evaluate the possibility of applying wastewater and sludge with different origins. While this is a quite new approach in Europe, there is an increasing interest in such systems for treating and reusing waste residues and simultaneously producing biomass for energy. Poplar and willow (Salicaceae), however, are not the only species used in SRF. Black locust (*Robinia pseudoacacia* L.), Eucaly ptus (*Eucalyptus* spp.) and other species, such as Siberian elm (*Ulmus pumila* L.) and the Princess tree (*Paulownia tomentosa* L.), are under investigation in several experimental projects. Black locust is characterized by high growth rates and N fixation properties (Figs. 6.1 and 6.2). Therefore, it does not require N fertilizers and, up to now, has had no phytosanitary problems. The Princess tree and Siberian elm as well as poplar and willow show high growth rates and are suitable for frequent coppicing.

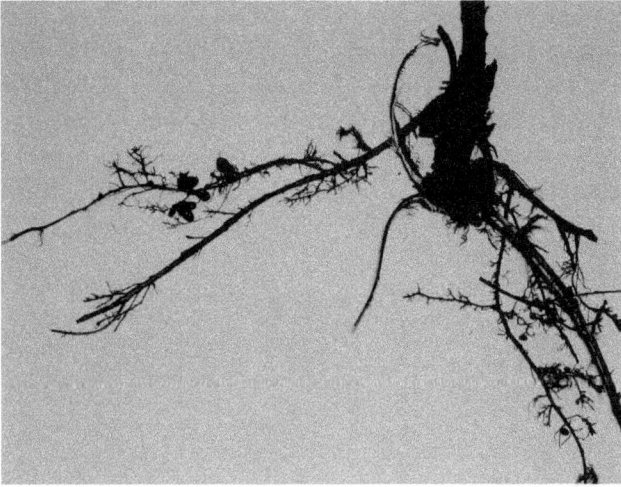

Fig. 6.1 *Robinia pseudoacacia* one-year-old roots with nodules. The *Rhizobium* is normally present in the soil but can generate nodules only on black locust roots. The combination of black locust-*Rhizobium* allows plants to grow on marginal soils, even in water-scarce conditions. The nitrogen in the leaves is released in the soil with the litter (up to 60 kg ha^{-1} yr^{-1}) during autumn/winter.

Fig. 6.2 Four-year-old black locust SRF plantation at Castellamonte, Italy. After five years of growth, with a survival of 95%, production reached 44.5 Mg ha^{-1}. In other Counties, black locust is used for honey production (Hungary), cattle shadow (Argentina), soil restoration (Bulgaria and Germany), poles for vineyards (France). In Italy it is used mainly for biomass and honey production.

The abovementioned species, when cultivated in areas with suitable soils and climatic conditions, are able to provide multiple environmental services, ensuring, at the same time, biomass production for commercial use (Fig. 6.3). In addition, recent and ongoing development of 1st and 2nd generation biofuels has resulted in an increase in the demand for biomass from the bioenergy market. This development, influenced not only by the energy market but also by the food market, is strongly affected by the competitiveness of the prices of biofuels compared to fossil fuels.

Fig. 6.3 Poplar SRF with a density of $8,300$ trees ha^{-1} in Lombriasco, Italy. At the end of the third year of growth, in a not-so-suitable soil (high clay content) and without irrigation, biomass productions was 18 Mg ha^{-1}.

Sustainability issues associated with the availability of farmland for dedicated biomass cultivations, however, remain the underlying factors affecting the development of SRF. Land is a limited resource and incorrect plans of no-food biomass production could trigger competitive mechanisms with food crops and generate social, economic, and environmental negative consequences.

6.2   Ecological Services Provided by SRF

6.2.1   Buffer Strips and Ecological Corridors

Despite the restrictions and safety measures suggested by the EU, the Nitrates Directive [11], and the Water Framework Directive [12], many water bodies, lagoons, and rivers present a high degree of eutrophication, while underground waters are at risk of contamination from nitrates [13–14]. The agri-environmental measures of the EU and the agricultural policy over the past decade have financed, through farmland owners, the cultivation of strips of trees called "buffer strips". These cultivations are linear structures of varying height and width that go along rivers and streams (Fig. 6.4). The buffer strips are cultivated in intensely farmed areas and are able to provide a filter for water run-off from fields into the collecting water bodies (rivers, lakes, etc.) Agricultural practices are responsible for 50%–80% of the total N flow into water bodies [15]. The filtered waters contain, in addition to pesticides and herbicides, considerable quantities of N compounds that are easily leached into the soil. Buffer strips of SRF and perennial grasses (switchgrass, miscanthus, and giant reed) are able to develop an extensive root system in the soil to ensure uniform water filtering. The action is carried out by the absorption of eutrophic waters and removal by means of the respiratory activity of soil microorganism communities. Initial evaluations, conducted in experimental plots using poplar clones have shown that N removal was approximately 120 kg ha^{-1} yr^{-1} [16] and 200 kg ha^{-1} yr^{-1} [17]. The buffer strips with high density (5,000–8,000 trees ha^{-1} with spacing from 2.50–3 m × 0.40–1.00 m) may be harvested after every three or four years. In order to maintain constant filtering activity, one or more rows can be harvested in alternate years. Studies conducted between 1999 and 2010 [18–20] have shown an increasing nitrate absorption capacity from the first to the third year of the plantation (with removal of up to 85% of nitrates present). The same studies also verified the important denitrifying action of the microorganism communities

Fig. 6.4 Poplar stand near a water canal. The high plantation density generates a dense root barrier near water a body. Poplars and willows are able to bear periods of submersion and are usually cultivated on marginal soils near rivers.

in the rhizosphere, which are able to remove between 70 and 170 kg ha^{-1} yr^{-1}, depending on the season, soil conditions, and measurement depth. Buffer strips can also contribute to a considerable accumulation of organic matter in the soil, through the litter and promotion of microorganism activity [21–22]. Haycock and Pinay [23] showed how the buffer strips can ensure the removal action also during the winter season, thanks to the microbial activities.

Buffer strips, when suitably designed, can also be considered as ecological corridors that link one area to another, allowing undisturbed movement of wild animals. For this purpose, buffer strips with different species (herbaceous, shrubs, and trees of different heights) are preferable because they are able to create an environment that develops at varying heights and provides separate ecological niches for different species [24]. Poplar buffer strips and other tree species are preferable to strips with a single herbaceous species. Studies conducted in northern Europe and North America have shown the presence of a considerable amount of bird species (between 24 and 41) in SRF plantations [25–27]. Moreover, SRF with willow and black locust does not require the use of pesticides and this can promote the colonization of a large number of insects compared to the intensive monoculture. Studies conducted in Great Britain on commercial SRF of willows and poplars identified a large number of bird species in the external rows of the crops compared to the surrounding areas with traditional crops. Also, several groups of insects present on top of the canopy were identified and considered an excellent food source for birds [28]. Due to the *salicaceae's* ability to tolerate flooding for long periods, poplars and willows can grow near rivers and, in many cases, they have reduced the speeds of the waters that cause catastrophic damages to cultivation and inhabited areas (Fig. 6.5) [29].

Fig. 6.5 Poplar stands for industry wood production in the Po river valley in Piedmont, Italy.

6.2.2   Fertirrigation: Disposal of Livestock, Urban and Industrial Wastewaters

The use of water originated from urban water treatment plants and livestock farms for the irrigation of non-food crops such as those of SRF represent a new approach that is currently under investigation. Several research projects have been conducted in countries where the establishment of SRF has had a long tradition (e.g., Sweden, Denmark) or the paucity of water resources has led to alternative uses of wastewater, such as Spain, Italy, and Portugal. However, irrigation of SRF with wastewater or the application of sludge is not a common practice outside Europe [30–31].

The possibility of disposing livestock wastewaters is of great interest to farmers: Many intensive livestock farms have to face the problem of disposing manure and slurry while still complying with the legal provisions of the "Nitrates Directive" that limit the N rate within a range of 340 and 170 kg N ha^{-1} yr^{-1}. Over the last decade, this has led to a dramatic increase of prices for leased land in areas with high-density of livestock farming. In many cases, the lands are leased for the sole purpose of spreading or burying waste resulting in a high risk of soil degradation and contamination of surface and deep waters. However, several studies [32–34] have shown that the use of wastewaters in crop farming, if implemented correctly, may offer an advantage not only for the environment but also for the crops. The use of livestock wastewaters in SRF cultivations has been studied by many research groups in Europe; however, it is known that high N concentration can cause damage to the plants when the manure comes into direct contact with the plant tissue. However, if used appropriately, incorporation into the soil or surface spreading in association with a good water supply, manure and manure slurry have beneficial effects on crop growth and maintenance of soil fertility [35].

The fertirrigation can also be carried out with other types of wastewaters. In Italy, for example, an experimental SRF plantation has recently become operational to collect waters resulting from olive processing and the production of pellets using poplar wood. In this site, delimited by a complex layer of geotextile sheeting and equipped with a water collection and pumping system, 10 poplar clones were irrigated using an underground system of waters containing wastewater from olive oil production. Soil samples have shown a progressive rebalancing of pH values (from 4.7 to 7.6) and a reduction in polyphenols achieved in just a few months. The plants, after an initial restricted growth period due to the unsuitable environment, achieved a survival rate of 99% and they were harvested at the end of the second year, generating a yield of 16.7 t ha^{-1} of dry matter [36].

6.2.3 Soil Erosion Control

One of the best definitions of soil quality for agricultural use was provided by Doran and Parkin in 1994: "the capacity of a reference soil to interact with the ecosystem for the purpose of maintaining the productivity of crops, the quality of the environment and promoting the health of animals and plants" [37]. In Italy, more than 77% of the land is threatened by the risk of erosion due to the geological characteristics of the country. The intensive farming practices lead to a depletion of more than 1.5% of soil organic C every year due to mineralization [38]. The phenomenon of soil erosion is also a source of growing concern for the international scientific community. In 2006, soil conservation was the subject of further attention from the European Commission, which adopted the final version of the Communication 179 COM (2002) entitled "Towards a Thematic Strategy for Soil Protection" [39]. Due to its near-permanent soil cover, SRF leads to a reduction in soil erosion risks compared with most agricultural crops and, in particular, arable crops characterized by frequent tillage. The decrease in erosion due to planting SRF will be greater if a cover crop is used to stabilize soils during the first two growing seasons [40]. Given the crop characteristics of most SRF and energy grasses (permanent soil cover, low input use, little use

Fig. 6.6 Poplar SRF with high density ($1,600$ trees ha^{-1}) after the fifth year in Pavia, Italy. Dense canopy cover protects the soil and produces a high litter matter (about 40% of the aboveground dry wood production).

of machinery, and outside planting and harvesting), the establishment of such permanent energy crops is likely to bring benefits for soil erosion, compared to annual agricultural crops (Fig. 6.6).

The soil erosion associated with SRF may involve only the first phases of the plantations because of the limited dimensions of the plants that leave the soil vulnerable to the atmospheric agents (rain, winds) and the frequent management during the initial stages of the plantation.

6.2.4 CO_2 Uptake and Carbon Sequestration

The agricultural sector is expected to significantly contribute to climate change mitigation. The European Climate Change Program (ECCP) has identified the technical measures to achieve this goal and the promotion of woody bioenergy crops is viewed as one among the most efficient means. On the basis of the final document of the ECCP entitled, "Working Group Sinks Related to Agricultural Soils", short-rotation coppice (SRC) and perennial vegetation are expected to have a sequestration potential in the soil of about 2–7 t CO_2 ha^{-1} yr^{-1} [41]. It should be pointed out that these estimates do not consider the benefits arising from fossil fuel replacement, which are far greater than the effects of C sequestration. However, as the production of woody crops for energy typically involves intensive cultural practices, especially in the case of SRC, and when we seek to fully assess the bioenergy benefits regarding climate change mitigation, the energetic and environmental costs associated with management treatments need to be carefully evaluated. To evaluate the contribution of SRF for producing renewable energy, two aspects should be taken into consideration: the quantity of fossil fuel energy required for generating each unit of renewable energy and the GHG emissions that are released directly and indirectly as a consequence of the cultivation [42]. SRF requires a limited supply of N and has been suggested as a potential tool for limiting the emissions of this molecule into the atmosphere [43].

In Italy, studies conducted to evaluate the C budget of poplar SRF have shown a net C uptake in different locations and environmental conditions. Within the Kyoto project [44–45], input-output analyses of the GHG involved in SRF of poplars and conventional poplar plantations (PP) was performed in order to evaluate the full GHG balance and the energy efficiency. The GHG balances of each plantation considered in the study included the estimates of the CO_2 uptake by the aboveground biomass and soil (absorptions) and the CO_2, CH_4, and N_2O emitted during the field operation carried out according to standard cultural schemes. For the PP case, two schemes were hypothesized according to the intensity of the cultural practices (popular plantations under high input,

PPH and popular plantations under low input, PPL, respectively). The CO_2 uptake by the aboveground biomass was set to 90–60 t dry matter ha^{-1} yr^{-1} (equal to 16.5–11 t CO_2 ha^{-1} yr^{-1} for PPH and PPL, respectively). In the case of SRF, a mean annual productivity of 13.5 t dm ha^{-1} yr^{-1} corresponding to 29.4 t CO_2 ha^{-1} yr^{-1} was considered. Soil CO_2 uptake was assumed in both PP and SRC to be equal to 2.27 t ha^{-1} yr^{-1}. CO_2, CH_4, and N_2O emissions from agriculture machinery use were calculated considering diesel oil consumption and national emission factors. Soil N_2O effluxes were assumed to be equal to the 1.25% of the N input in the soil. The results of the study indicated that the productivity was higher in SRF compared to PP (+36% and +56% compared to PPH and PPL, respectively), but the intensity of the cultural practices in SRC enhanced the overall GHG emissions by approximately +33% and +55% compared to PPH and PPL. The production and use of the fertilizers proved to be the major contributors to overall GHG emissions. PPL showed a 30% drop in GHG emissions compared to PPH. The most significant reductions proved to be those of pesticide use and energy consumption during irrigation. No remarkable difference in terms of efficiency in the energy use between the SRF and PP cultivation systems was found. However, in both cases a comparable amount of only 5%–7% of the energy produced from biomass was expended in the cultivation practices. For every ton of CO_2 equivalent expended during the cultivation, up to 16.8, 16.9 and 18.1 tons of CO_2 (in PPH, PPL and SRF, respectively) were avoided by substituting fossil fuel with biomass in energy production. To further reduce the GHG emission and production costs, irrigation and fertilization appear to be the fundamental factors [46–47].

The capacity to permanently accumulate C by SRF is related to the land use change (LUC) of the soil. LUC from forest to grassland and agricultural crops has induced a significant reduction of the soil organic carbon (SOC); about 30% of C contained in soil (until a depth of 100 cm) is lost during the first 30–50 years after the land conversion. However the C accumulation in poor soils can be achieved with suitable agricultural practices [48–49].

In SRF cultivations, initially a loss of SOC has been observed in several studies as a result of the field operations to prepare the land for the new crop; however this initial reduction is frequently followed by an increase after several years of cultivation [50–51].Interesting results, on SOC dynamics in SRF have been obtained in a long term study at the Research Centre for Industrial Crops (CRA-CIN) in Italy: two perennial herbaceous species miscanthus (*Miscanthus giganteus*) and giant reed (*Arundo donax* L.), were compared with SRF of poplar, black locust and willow; the herbaceous species were fertilized every year with 120 kg ha^{-1} of N and with 120 kg ha^{-1} of P_2O_5, while poplar and willow received an identical quantity of both fertilizers only once every two years after the biomass was harvested. The black locust, that has N-fixing *Rhizobium* on its roots, only received phosphorus fertilizer. Irrigation and soil cultivation were carried out only during the first vegetative season for the purpose of promoting the rapid establishment of the various species.

At the end of the study, soil samples collected at 0–20 cm and 20–40 cm reveal statistically significant increase in SOC. After seven years of cultivation, poplar increased SOC by 46% and willow by 53% (at 0–20 cm) compared to arable land crop, while no significant change were detected in the deeper layer. The trend to improve SOC was not correlated with biomass production. Despite its favorable effect on SOC, willows showed the lowest biomass productivity compared to others crops investigated. Yet, the higher biomass productivity of giant reed was not associated with significant SOC increments. However, the cultivation of perennial species allowed higher SOC storage compared to permanent grassland. The higher amount of C provided by roots and microclimatic conditions induced by the taller canopy (temperature and soil moisture) of the perennial crops might have played roles in SOC accumulation, as observed in other cases [52].

6.3   Biofuel Production and SRF

Liquid biofuel for the transports sector currently represents approximately only 1% of the sector's total fuel consumption [53]. At the present, the EU's policy is to support the production of 2nd generation biofuels. EU Directive 2009/28/EC sets mandatory targets for the use of energy from renewable sources: overall, renewable energy must comprise 20% of the EU's energy production by 2020, with a 10% share for renewable energy in the transport sector. At the same time, an amendment to Directive 98/70/EC ("the Fuel Quality Directive") [54] introduced a further mandatory target for 2020, requiring a 6% reduction in the GHG emission intensity from fuels used in road transport and non-road mobile machinery. Biofuels are expected to make a significant contribution to meeting these targets. On the other hand, increasing concern about the effects of indirect land use change on GHG emissions [55] has led the EU (COM 2012/0288 [56]) to limit the amount of biofuel that can be produced from food crops, with the use of food-based biofuels to meet the renewable energy target being limited to 5%. These biofuels are made from lignocellulosic material and thus provide high GHG savings with low risk of causing indirect land use change. They also do not compete directly for agricultural land with the food and animal feed markets. In this context, one might expect a considerable increase in the area of dedicated lignocellulosic bioenergy crops such as those in SRF. In addition to the support of the EU policy, some private companies have developed the technologies to produce 2nd generation biofuels on a large scale. Recently, in Italy, Chemtex (the Mossi and Ghisolfi Group) has developed a system for the production of approximately $40,000$ tons of 2nd generation bioethanol per year and it will serve as a basis for the future commercial scale-up of the process (single-line systems to produce $150,000$–$200,000$ tons of bioethanol per year) [57]. Currently, large amounts of residue from agricultural and forestry sources (22 and 2.14 Mt of dry matter, respectively) are available in Italy and in an initial phase, they could be utilized for the production of biofuels. However, to ensure a long-term and sustainable supply of biomass for renewable energy production, it is necessary to establish and grow new perennial, dedicated energy crops, particularly on marginal agricultural lands. Preliminary research has identified several perennial species that have the potential for energy production including a number of perennial grasses, poplars, and willows. These species have been studied as dedicated energy crops since the early 1980s in Italy. In the last ten years, SRF crops have been inserted in the cultural plans of several farms, particularly in northern Italy, that take advantage of their low input requirements and the added possibility of exploiting set-aside areas. The total surface area reached 6,700 ha in 2010 and poplar is the most commonly chosen species for SRF because it is already cultivated in the same area at a low planting density of 270–330 plants per hectare for the production of wood-based panels and paper. Recently, regional governments have supported this crop to induce farmers to grow SRF. Nonetheless, this type of support will not last forever; therefore, to increase the extent of SRF and to ensure a long-term supply of biomass, it is necessary to develop these crops on marginal lands in an economically sustainable way to make SRF sustainable without public grants. The profitability for farmers of SRF depends largely on public grants because the farmers must sell the biomass itself, not the bioenergy, to energy companies.

In Italy, further developments of these crops will be particularly connected with the implementation of SRF characterized by high density plantations, which were already used at the beginning of the 20th century for the production of wood for pulp and paper mills and then abandoned after the Second World War. This cultural model is now suitable for biofuel feedstock (pellets and 2nd-generation bioethanol), particularly if associated with phytoremediation. For this reason, it will be important to expand the number of clones and species that adapt to different site conditions. Another key factor that should be considered is the wood: the large variability in the specific gravity of the wood's poplar clones (from 0.50 g cm^{-3} to less than 0.30 g cm^{-3}) represents important criteria for the selection for biofuel production because, in recent studies on poplar genotypes, highly significant negative genetic correlations were observed between plant growth and lignin content, whereas there was a positive correlation with cellulose [58–59]. It will be interesting to see if the same result can be found with willow. Further research is needed to study the wood chemistry of new clones to verify the possibility of using low pre-treatment inputs to separate the wood components and for the conversion to biofuels. The increased biomass yields obtained by some clones of white poplar and willow could contribute to a long-term, gradual sustainable replacement of fossil fuels in Mediterranean regions. Multiple environmental benefits and socio-economic externalities also should be considered in addition to producing energy with no net addition of CO_2 to the atmosphere compared to other agricultural products and fossil fuel-based energy systems.

6.4   Conclusions

The environmental benefits provide by the SRF plantations represents a valuable contribution that can no longer be ignored. Besides the contribution that these crops can make in reducing the fossil fuel demand, control of soil erosion, or wastewater disposal and in light of the principle of sustainable development adopted by the EU, it appears to be important to consider the possibility of the economic compensation of the environmental benefits derived from SRF crops.

Although the cultivation of SRF could be a very promising renewable energy option for the future, its actual implementation in Italy and in most of the countries in Europe is still limited. Several authors have discussed the financial viability of SRF for bioenergy in a number of countries, but with varying conclusions. Mitchell et al. [60] argued that government incentives and a stable market for woody chips are indispensable for SRF to compete with conventional agricultural crops and to become profitable at a commercial scale in the UK. Ericsson et al. [61], on the other hand, found that willow is an economically feasible energy crop for relatively large farms in Poland as the production costs are significantly lower compared to western European countries because of the lower diesel, labor, and fertilizer costs. In a recent study, El Kasmioui and Ceulemans [62] analyzed the economic conditions of SRF in Belgium and highlighted a number of barriers for the widespread adoption of SRF by Belgian farmers. In order to convince farmers to establish SRF plantations, several conditions should be fulfilled: (i) SRF should be as profitable (with or without government incentives) as traditional agricultural crops, such as corn, wheat, sugar beets, etc.; (ii) a well-performing market should be present for the produced woody biomass chips; and (iii) the farmers should be confident that the special equipment to plant, to cultivate (e.g., specially designed line cultivators for energy crops), and to harvest the energy crops is available within a reasonable distance from the plantation site. Also, there is very little information about the current trading structures since, in most countries, there is very limited information on biomass from SRF.

References

[1] European Commission Communication from the European Commission on the Biomass Action Plan. 2005; COM 628.

[2] European Environment Agency. How much bioenergy can Europe produce without harming the environment? EEA Report No. 7. 2006; 67.

[3] Facciotto G, Bergante S, Lioia C, Mughini G, Nervo G, Giovanardi R. Short Rotation Forestry in Italy with poplar and willow. Proceedings of the 14th European Biomass Conference & Exhibition, Biomass for energy, industry and climate protection. Paris. France. 2005; 320–323.

[4] Lindroth A, Bath A. Assessment of regional willow coppice yield in Sweden on basis of water availability. Forest Ecol. Manag. 1999; 121: 57–65.

[5] Ceulemans R, Deraedt W. Production Physiology and growth potential of poplars under short rotation forestry culture. Forest Ecol. Manag. 1999; 121: 9–24.

[6] Di Muzio Pasta V, Negri M, Facciotto G, Bergante S. Maggiore TM. Growth dynamics and biomass production of 12 poplar and 2 willow clones in a short rotation coppice in Northern Italy.Proceedings of the 15th European Biomass Conference & Exhibition, Biomass for energy, industry and climate protection. Berlin. Germany. 2007; 749–754.

[7] Nervo G, Facciotto G. Activities on woody biomass for energy purposes in Italy. Proceedings of the 15th European Biomass Conference & Exhibition, Biomass for energy, industry and climate protection. Berlin. Germany. 2007; 691–694.

[8] Sixto H, Hernández MJ, Barrio M, Carrasco J, Cañellas I. Plantaciones del genero Populus para la producción de biomasa con fines energéticos: revisión, Investigación Agraria. Sistemas y Recursos Forestales. Spain 2007; 16: 277–294.

[9] Deckmyn G, Laureysens I, Garcia J, Muys B, Ceulemans R. Poplar growth and yield in short rotation coppice: model simulation using the process model SE-CRETS. Biomass & Bioenerg. 2004; 26: 221–227.

[10] Bergante S, Facciotto G, Minotta G. Identification of the main site factors and management intensity affecting the establishment of Short-Rotation-Coppices (SRC) in Northern Italy through Stepwise regression analysis. Central European Journal of Biology. 2010; 5: 522–530.

[11] Council Directive 91/676/EEC concerning the protection of waters against pollution caused by nitrates from agricultural sources.1991. (Accessed September, 2013 at http://ec.europa.eu/environment/water/water-nitrates/index_en.html).

[12] Directive 2000/60/EC of the European Parliament and of the Council establishing a framework for the Community action in the field of water policy. 2000. (Accessed September, 2013 at http://ec.europa.eu/environment/water/water-framework/index_en.html).

[13] Ongley DE. Control of water pollution from agriculture. Food and Agriculture Organization of the United Nations. Rome. 1996; ISBN 92-5-103875-9.

[14] Clean Coastal Waters: Understanding and Reducing the Effects of Nutrient Pollution. National Research Council Washington (USA) Press. 2000; ISBN 978-0-309-06948-9.

[15] EEA European Environment Agency. The European Environment State and outlook. Integrated Assessment Part A. EEa Report. 2005; 1.

[16] Scarascia Mugnozza G, Paris P. Nuovi impieghi ambientali per il pioppo. Il libro bianco della pioppicoltura. Italy. National poplar commission Press, 2007.

[17] Dimitriou I, Aronsson P. Willows for energy and phytoremediation in Sweden. Unasylva. 2005; 221: 47–50.

[18] Boz B, Gumiero B, Cornelio P. Rimozione dell'azoto diffuso: analisi dell'efficacia e dei processi di rimozione in un'area tampone boscata (sito sperimentale "Nicolas")—Proceedings of "Direttiva nitrati: dalla ricerca alla gestione del territorio inquinamento diffuso e aree tampone". Legnaro Italy. 2011.

[19] Gumiero, B, Boz B, Cornelio P. and Casella S. Shallow groundwater nitrogen and denitrification in a newly afforested, subirrigated riparian buffer. Journal of Applied Ecology. 2011; 48: 1135–1144.

[20] Gumiero B, Boz B, Cornelio P. Il sito sperimentale "Nicolas"—Efficacia delle fasce tampone nella riduzione dei carichi di azoto. Monitoraggio e sperimentazione presso l'azienda pilota e dimostrativa "Diana" di Veneto Agricoltura. 2008. (Accessed September, 2013 www.venetoagricoltura.org).

[21] Ceotto E, Di Candilo M. Sostenibilità ambientale delle colture da energia. Sherwood-Foreste ed Alberi oggi. 2012; 2: 94–97.

[22] Pellegrino E, Di Bene C, Tozzini C, Bonari E. Impact on soil quality of a 10-year-old short-rotation-coppice poplar stand compared with intensive agricultural and uncultivated systems in a Mediterranean area. Agriculture, Ecosystems & Environment. 2011; 140: 245–254.

[23] Haycock NE, Pinay G. Groundwater nitrate dynamics in grass and poplar vegetated riparian strips during the winter. Journal of Environmental Quality 1992; 2: 273–278.

[24] Volk TA, Verwijst T, Tharakan PJ, Abrahamson LP, White EH. Are short rotation woody crops sustainable? Proceedings of 2nd World Conference on Biomass for Energy, Industry and Climate Protection. Rome Italy. 2004; 34–39.

[25] Weih M. Intensive short rotation forestry in boreal climates: present and future perspectives. Canadian Journal of Forest Research. 2004; 34: 1369–1378.

[26] Sage R, Cunningham M, Boatman N. Birds in willow short-rotation coppice compared to other arable crops in central England and a review of bird census data from energy crops in the UK. Ibis. 2006; 148: 184–197.

[27] Berg Å. Breeding birds in short rotation coppice on farmland in central Sweden— the importance of *Salix* height and adjacent habitats. Agriculture, Ecosystems & Environment. 2002; 90: 265–276.

[28] Sage R, Tucker K. Invertebrates in the canopy of willow and poplar short rotation coppices. Aspects of Applied Biology. 1997; 49: 105–111.

[29] Dulla M, Vietto L, Chiarabaglio PM, Cristandi L. Conservazione di *Populus nigra* L. e *Populus alba* L. nell'ambito di attività di riqualificazione fluviale: il caso dell'isola Colonia di Palazzolo Vercellese. Proceedings of "Atti del Seminario Nazionale 'Il ruolo della vegetazione ripariale e la riqualificazione dei corsi d'acqua". Quaderni di tutela del territorio. Italy 2008; 3: 101–110.

[30] Labrecque M, Teodorescu TI. Influence of plantation site and wastewater sludge fertilisation on the performance and foliar nutrient status of two willow species grown under SRIC in southern Quebec (Canada). Forest ecology and management. 2001; 150: 223–239.

[31] Perttu KL, Kowalik PJ. *Salix* vegetation filters for purification of waters and soils. Biomass and Bioenergy. 1997; 12: 9–19.

[32] Ecosse A, Paris P, Mareschi L, Olimpieri G, Scarascia Mugnozza G. Smaltimento di reflui zootecnici in piantagioni da biomassa (SRFsrf): prime esperienze in un sito sperimentale di pioppo nel centro-Italia. Proceedings of "6° Congesso Nazionale SISEF" Arezzo Italy. 2007.

[33] Cavanagh A, Gasser MO, Labrecque M. Pig slurry as fertilizer on willow plantation. Biomass and Bioenergy. 2011; 35: 4165–4173.

[34] Guidi Nissim W, Labrecque M. Il salice nel controllo del degrado ambientale. Sherwood-Foreste e Alberi Oggi. 2012; 184: 40–45.

[35] Canesin C. Short Rotation Forestry in Nord-Italia. Limiti e potenzialità del suo impiego sotto differenti scenari: produzione di biomassa e prove di fertilizzazione. PhD thesis at University of Padova. Italy. 2010; 107.

[36] Santori F, Cicalini AR, Zingaretti A, Facciotto G, Bergante S. SRF of poplar fertirrigated with olive mill wastewater in demo plant. Proceedings of 15th European Biomass Conference & Exhibition. Berlin, Germany. 2007; 687–690.

[37] Doran WJ, and. Parking BT. Defining and Assessing Soil Quality. In: Doran, JW; DC Coleman; Bezdicek, DF; and Stewart. Defining Soil Quality for a Sustainable Environment. Madison, WI. Soil Sci. Soc. Am. 1994; 35: 3–21.

[38] Benedetti A, Mughini G, Alianiello F, Mascia MG. Uso di colture energetiche per il ripristino ambientale. Proceedings of Colture a scopo energetico e ambiente. Sostenibilità, diversità e conservazione del territorio. Rome Italy. 2006; 59–65.

[39] European Commission. Thematic Strategy for soil protection, 2006; COM 231: 2–12.

[40] Ranney JW, Mann LK. Environmental considerations in energy crop production. Biomass & Bioenergy. 1994; 6; 211–228.

[41] European Climate Change Programme (ECCP). Working Group Sinks Related to Agricultural Soils. Final Report. 2002.

[42] Liebig MA, Schmer MR, Vogel KP, Mitchell RB. Soil carbon storage by switchgrass grown for bioenergy. Bioenerg. Res. 2008; 1: 215–222.

[43] Crutzen PJ, Mosier AR, Smith KA, Winiwarter W. N_2O release from agro-biofuel production negates global warming reduction by replacing fossil fuels. Atmos. Chem. Phys. 2008; 8: 389–395.

[44] Ballarin Denti A, Giannella S, Lapi M. Progetto Kyoto. Lombardy foundation for the environment press. Milan Italy. 2008 ISBN 978-88-8134-067-5.

[45] Tedeschi V, Federici S, Zenone T, *et al.* Greenhouse gases balance of two poplar stands in Italy: a comparison of a short rotation coppice and a standard rotation plantation. Proceedings of 14th European Conference & Exhibition, Biomass for Energy, Industry and Climate Protection. Paris France. 2005; 2014–2016.

[46] Facciotto G, Bergante S, Ceotto E, Di Candilo M. Bilancio del Carbonio e dei gas serra. National research council for agriculture. Rome Italy. Pari L, Editor. 2012; 213–232.

[47] Gera M. Poplar culture for speedy carbon sequestration in India: a case study from Terai Region of Uttarakhand. Envis Forestry Bulletin. 2012; 12: 75–83.

[48] Lagomarsino A, De Angelis P, Moscatelli MC, Grego S, Scarascia Mugnozza G. Accumulo di C nel suolo di una piantagione di *Populus* spp. in condizioni di elevata CO_2 atmosferica e fertilizzazione azotata. Forest. 2009; 6: 229–239.

[49] Coleman M, Isebrand JG, Tolsted DN, Tolbert VR. Comparing Soil Carbon of Short Rotation Poplar Plantations with Agricultural Crops and Woodlots in North Central United States. Environmental Management. 2004; 33: 299–308.

[50] Grigal DF, Berguson WE. Soil carbon changes associated with short-rotation system. Biomass and Bioenergy. 1998; 14: 371–377.

[51] Don A, Osborne B, Carter MS, et al. Land-use change to bioenergy production in Europe: implications for the greenhouse gas balance and soil carbon. Global Change Biology Bioenergy. 2011; 4: 372–391.

[52] Ceotto E, Librenti I, Di Candilo M. Can bioenergy production and soil carbon storage be coupled? A case study on dedicated bioenergy crops in the low Po valley (Northern Italy). Proceedings of 18th European Biomass Conference & Exhibition. Lyon, France. 2010; 2261–2264.

[53] FAO. 2008. The State of Food and Agriculture 2008. Biofuels: Prospects, Risks and Opportunities. Rome, Italy.

[54] DIRECTIVE 98/70/EC of the European Parliament relating to the quality of petrol and diesel fuels and amending Council Directive 93/12/EEC L 350/58 Official Journal of the EC. 1998.

[55] Edwards R. Indirect Land Use Change from increased biofuels demand. European Commission Joint Research Centre Institute for Energy. 2010. ISBN 978-92-79-163913.

[56] Directive of the European Parliament amending Directive 98/70/EC relating to the quality of petrol and diesel fuels and amending Directive 2009/28/EC on the promotion of the use of energy from renewable sources. COM(2012) 595 final 2012/0288 (COD).

[57] Giordano D. Second generation bio-ethanol: M&G and its innovative technology. Special abstracts. J Biotechnol 2010; 150: 1–576.

[58] Novaes E, Kirst M, Chiang V, Winter-Sederoff H, Sederoff R. Lignin and biomass: a negative correlation for wood formation and lignin Content in trees. Plant Physiol. 2010; 154: 551–561.

[59] Sannigrahi P, Ragauskas AJ, Tuskan GA. Poplar as a feedstock for biofuels: a review of compositional characteristics. Biofuels, Bioprod Biorefin. 2010; 4: 209–226.

[60] Mitchell CP, Stevens EA. Watters M.P. Short-rotation forestry operations, productivity and costs based on experience gained in the UK. Forest Ecology & Management. 1999; 121: 123–136.

[61] Ericsson K, Rosenqvist H, Ganko E, Pisarek M, Nilsson L. An agro-economic analysis of willow cultivation in Poland. Biomass & Bioenergy. 2006; 30: 16–27.

[62] Kasmioui O, Ceulemans R. Financial analysis of the cultivation of short rotation woody crops for bioenergy in Belgium: barriers and opportunities. Bioenergy Resource. 2012; 6: 336–350.

According to Ericsson *et al.* [8], the calculated energy crop production costs were also found to be consistently lowest for SRC (4 to 5 € GJ^{-1}), followed by perennial grasses (6 to 7 € GJ^{-1}), and highest for annual straw crops (6 to 8 € GJ^{-1}). Moreover, the production costs of SRC and perennial grasses have the potential to be decreased to approximately 3 to 4 € GJ^{-1} and 5 to 6 € GJ^{-1}, respectively, as a consequence of economies of scale (an increase in the total cultivation area). In contrast, the production costs for annual straw crops have little potential for cost reductions in the future.

Finally, SRC have positive impacts on their surroundings from the socioeconomic prospects and environmental quality. Apart from energy independence and security, growing SRC enhance rural economies, biodiversity, site nutrient capture and retention, soil protection, water and air quality, and carbon sequestration [4, 15]. Wood chips from SRC show the best performance as biofuel of all of the raw materials tested, including winter rape, sugarcane, sorghum, soy, and oil palm, with respect to total environmental impact and greenhouse gas (GHG) emissions [16]. SRC are generally assigned as "environmentally friendly" crops since their management is usually less intensive than that needed for food crops [17]. This low input management is characterized by soil preparation and planting only during or prior to the year of establishment, mechanical weed control during the first one to three seasons (depending on the timing of canopy closure), a reduced number of harvests, and typically zero or low amounts of fertilizers and pesticides [18–19]. For example, low rates of nitrate leaching observed under SRC led to recommendations for their use as suitable crops for nitrate-sensitive areas or for groundwater protection zones around water supply boreholes [17]. In addition, since the fast growths of poplars and willows are linked with high water usage and high nutrient uptake, both enhanced by potentially deep root systems, SRC have been found to be the most suitable vegetation to grow in landfills and other waste disposal areas [20–21] or for the applications of wastewater, sewage sludge, or landfill leachate [22–23]. To conclude, the early successional genera of *Populus* and *Salix* are characterized by high productivity, vigorous juvenile growth, easy propagation, good coppice potential, and adaptation to a wide range of environmental conditions, making them ideal candidates for bioenergy production using SRC cultures.

7.2 Water Use of SRC

A recent process-based modeling analysis estimated future (2050) global bioenergy production potential comprising all biomass sources at 130 to 270 EJ yr^{-1}, equivalent to 15% to 25% of the World's future energy demand [24]. Dedicated

Chapter 7

Populus and *Salix* Grown in a Short-rotation Coppice for Bioenergy: Ecophysiology, Aboveground Productivity, and Stand-level Water Use Efficiency

Milan Fischer, Régis Fichot, Janine M. Albaugh, Reinhart Ceulemans, Jean Christophe Domec, Miroslav Trnka, and John S. King

7.1 Introduction

The term of short-rotation coppice (SRC) covers any high-yielding, fast-growing, mainly hardwood species managed in a coppice (cutting back to the ground level) system [1]. SRCs are grown commercially for heat and power generation [1–2] or wood products [3–5]. Depending on the primary purpose, these "crops" are harvested on 1-to 15-year rotation periods and remain viable for 15 to 30 years [4–7]. SRC plantations in temperate and subtropical latitudes have been mainly based on poplar (*Populus*) and willow (*Salix*) species grown in medium- to high-density cultures (c. 1,000 to 40,000 trees per ha) on abandoned, underused, or contaminated arable land. These poplar and willow SRC have been viewed as promising and relatively inexpensive sources of bioenergy over the last few decades [1, 7–10].

The replacement of fossil fuels by biomass in the generation of C-neutral energy and heat has recently become an important strategy promoted by the European Union (EU) to mitigate the effects of climate change [11–12]. Further, biomass and, in particular, energy crops, have attracted attention as promising renewable and local energy sources. Locally produced bioenergy could help to reduce the dependency on external energy sources as well as enhance the security of the supply, support the diversification of economic sectors, and avoid the depletion risk [9, 13–14].

[47] Gera M. Poplar culture for speedy carbon sequestration in India: a case study from Terai Region of Uttarakhand. Envis Forestry Bulletin. 2012; 12: 75–83.

[48] Lagomarsino A, De Angelis P, Moscatelli MC, Grego S, Scarascia Mugnozza G. Accumulo di C nel suolo di una piantagione di *Populus* spp. in condizioni di elevata CO_2 atmosferica e fertilizzazione azotata. Forest. 2009; 6: 229–239.

[49] Coleman M, Isebrand JG, Tolsted DN, Tolbert VR. Comparing Soil Carbon of Short Rotation Poplar Plantations with Agricultural Crops and Woodlots in North Central United States. Environmental Management. 2004; 33: 299–308.

[50] Grigal DF, Berguson WE. Soil carbon changes associated with short-rotation system. Biomass and Bioenergy. 1998; 14: 371–377.

[51] Don A, Osborne B, Carter MS, et al. Land-use change to bioenergy production in Europe: implications for the greenhouse gas balance and soil carbon. Global Change Biology Bioenergy. 2011; 4: 372–391.

[52] Ceotto E, Librenti I, Di Candilo M. Can bioenergy production and soil carbon storage be coupled? A case study on dedicated bioenergy crops in the low Po valley (Northern Italy). Proceedings of 18th European Biomass Conference & Exhibition. Lyon, France. 2010; 2261–2264.

[53] FAO. 2008. The State of Food and Agriculture 2008. Biofuels: Prospects, Risks and Opportunities. Rome, Italy.

[54] DIRECTIVE 98/70/EC of the European Parliament relating to the quality of petrol and diesel fuels and amending Council Directive 93/12/EEC L 350/58 Official Journal of the EC. 1998.

[55] Edwards R. Indirect Land Use Change from increased biofuels demand. European Commission Joint Research Centre Institute for Energy. 2010. ISBN 978-92-79-163913.

[56] Directive of the European Parliament amending Directive 98/70/EC relating to the quality of petrol and diesel fuels and amending Directive 2009/28/EC on the promotion of the use of energy from renewable sources. COM(2012) 595 final 2012/0288 (COD).

[57] Giordano D. Second generation bio-ethanol: M&G and its innovative technology. Special abstracts. J Biotechnol 2010; 150: 1–576.

[58] Novaes E, Kirst M, Chiang V, Winter-Sederoff H, Sederoff R. Lignin and biomass: a negative correlation for wood formation and lignin Content in trees. Plant Physiol. 2010; 154: 551–561.

[59] Sannigrahi P, Ragauskas AJ, Tuskan GA. Poplar as a feedstock for biofuels: a review of compositional characteristics. Biofuels, Bioprod Biorefin. 2010; 4: 209–226.

[60] Mitchell CP, Stevens EA. Watters M.P. Short-rotation forestry operations, productivity and costs based on experience gained in the UK. Forest Ecology & Management. 1999; 121: 123–136.

[61] Ericsson K, Rosenqvist H, Ganko E, Pisarek M, Nilsson L. An agro-economic analysis of willow cultivation in Poland. Biomass & Bioenergy. 2006; 30: 16–27.

[62] Kasmioui O, Ceulemans R. Financial analysis of the cultivation of short rotation woody crops for bioenergy in Belgium: barriers and opportunities. Bioenergy Resource. 2012; 6: 336–350.

bioenergy crops (non-food plant biomass or ligno-cellulosic biomass) could account for 20% to 60% of this, depending on scenarios of land availability and use of irrigation [24]. However, full exploitation of the dedicated bioenergy has become a source of controversy, since it is very water-intensive and may contribute to regional water shortages and soil salinization [24–26]. Likewise, in spite of the list of positive environmental and socioeconomic impacts assigned to SRC, the effects on the water cycle of their large-scale deployment remains a subject for discussion [4, 27–34]. A number of studies suggested that water availability constitutes one of the main constraints for the biomass yields and profitability of SRC grown on arable land [6, 28, 31, 35–36]. In addition, according to experimental and modelling studies, the water use of SRC is substantially higher than that of traditional agriculture crops and C3 or C4 grass species. Therefore, it has been hypothesized that the large-scale production of SRC will have detrimental impacts on the regional water budget linked with decreased recharges of aquifers and surface water [4, 17, 29–30, 37]. Other studies have provided conflicting results with comparable or lower water consumption of SRC to that of grasslands and reference crop evapotranspiration (ET) [8, 19, 32, 38–40].

These conflicting results highlight the need for more research into the water use of SRC. On the one hand, this diversity of results may be explained by the large physiological plasticity of each genus [32, 41]. On the other hand, stand dimension and structure plays an important role, where generally smaller and more heterogeneous stands are better ventilated and exposed to the possible advection of the sensible heat flux from the surrounding fields and, thus, lose more water in the process of ET [17, 19, 29, 42]. Knowledge of the species-specific ecophysiology and atmospheric coupling based on stand structure is necessary to decrease evaporative water losses of SRC bioenergy production. In order to mitigate the negative environmental impacts of large-scale dedicated bioenergy production, it is necessary to engineer and manage bioenergy cropping systems to be as water efficient as possible [34]. Further, maximizing bioenergy system water use efficiency will be important given the projected increases in the frequencies and intensities of drought events expected in the coming decades [34, 43].

7.3   Water Use Efficiency of SRC

A convenient way to assess how water use is coupled with biomass production is the concept of water use efficiency (WUE). The term WUE describes the rate of CO_2 uptake or plant dry matter production for a given rate of water loss. WUE is therefore a key trait in making the link between C and water cycles. Understanding plant ecophysiological controls over WUE is necessary for designing climatically appropriate bioenergy production systems [34]. Practically, there are many ways of describing WUE depending on scientific discipline and different measurement approaches. First, at the leaf-level, WUE can be defined as the ratio of CO_2 uptake and transpiration in the process of photosynthesis [44–47]. This is referred to as photosynthetic water use efficiency (WUE_{ph}) or instantaneous water use efficiency (WUE_{inst}) [45, 48–53]. Intrinsic WUE, *i.e.*, the ratio of the net CO_2 assimilation rate to stomatal conductance to water vapor, is also widely reported in the literature and can be easily calculated based on the gas exchange measurement [54–56]. For ecological, agricultural, and forestry purposes, however, the ratio of dry matter production to water consumption over longer periods and larger spatial scales is more informative than leaf-level gas exchange ratios. Hellriegel [57] and Maximow [58] are considered to be among the first who carried out the calculations on the relationship between the increase in the dry matter and water requirement. By dividing biomass productivity—expressed as organic dry matter—by water lost by transpiration or total ET, of productivity (WUE_P) or long-term water use efficiency (WUE_L) is obtained [32, 51–52, 59–60]. Sometimes, the reciprocal of the transpiration, ET, or assimilation ratio is used to describe the water use per unit of growth [61–62]. To estimate the WUE of whole ecosystems, geoscientists and ecologists commonly use the ratios of the main ecosystem fluxes such as gross primary production (GPP), gross ecosystem production (GEP), or net primary (ecosystem) production (NPP, NEP) to water loss by ET, and thus $WUE_{GPP(GEP,NPP,NEP)}$ is obtained [63–64]. Depending on the scale of analysis and integration, factors such as the diurnal variation in root and soil respiration, relative C allocation to roots and shoots, and turnover of fine roots and leaves will influence the resulting WUE [51, 62, 66].

Generally, the WUE_P of SRC lies within the higher range of the broadleaved species reaching up to 30–50 kg mm^{-1} (the unites refer to kg of dry matter biomass at 1 ha per 1 mm of ET throughout the whole chapter) [32, 51, 66–67]. However, as was recently demonstrated [34], there is still a very large gap in the ecophysiological database of the water use of SRC for bioenergy, despite the relatively large number of studies. We summarize, here, the current knowledge on poplar and willow SRC's water uses and WUEs. To this end, we first carefully review the available data in the peer-reviewed literature, focusing on field experiments, which are generally considered more representative compared to potometric or lysimetric studies. Then, we present a case study of our own *Populus* SRC experiment in the Bohemian-Moravian Highlands (Czech Republic) focusing on water use and biomass productivity from 2008 to the present.

7.4 WUE and Related Ecophysiological Variables Literature Surveys

Detailed knowledge of the ecophysiology and water relations of the major bioenergy crops under realistic field conditions is critically needed in order to match appropriate crops to prevailing agroclimatic conditions [43]. This will aid in the designs of bioenergy cropping systems with high WUE and improve model parameterizations aimed at predicting bioenergy SRC's responses to future climates. On a broad scale, the selection of energy crop species in conditions similar to those of their origins may be the first solution [68]. However, the degree of variation in water use traits among species and also among genotypes within a species is still poorly documented, masking opportunities to increase WUE and sustainability. Therefore, it is necessary to have insight into the relationships between water use and biomass productivity for particular poplars and willows species and their hybrids.

Recently, King *et al.* [34] summarized the literature on the representative species of the major groups (grasses, trees) of so-called "2nd generation" bioenergy crops—those based on cell-wall or "ligno-cellulosic" technologies. Overall, the best represented taxa were *Populus* (87 studies) and *Salix* (60 studies), while most others were poorly represented.

In the current chapter, we provide an in-depth analysis on the ecophysiological attributes controlling water use as related to biomass production, specifically for the genera of *Populus* and *Salix*. Further, since the genera of *Populus* and *Salix* are very broad and characterized by large and positive heterosis effects for growth (*i.e.*, inter-specific hybrids with superior vigor as compared to the parents) [56, 69], we analyzed the differences between pure species and their hybrids. We tried to normalize management regimes across studies by using data only from moderate to low-input systems but, occasionally, data from high-input systems were included.

Data related to leaf-level ecophysiology, stand-level water use, and biomass production in *Populus* and *Salix* pure species and hybrids were widely available in the literature when considered separately (Tabs. 7.1, 7.2 and 7.3). Nevertheless, only two groups of experiments reported simultaneous ecophysiological data and a complete water balance for a SRC culture system based on at least one year of measurements of the full hydrologic cycle. Early on, it was the very broad research from Sweden carried out mostly on *S. viminalis* including almost all of the available ecophysiological methods [27, 32, 35, 51, 70]. The second set of studies was from England, dedicated mainly to *P. trichocarpa* × *P. deltoides*, *P. deltoides* × *P. nigra*, and *S. burjatica* and primarily using sap flow, porometry, and modeling [17, 29–30, 71–73], which advanced our understanding of SRC productivity, water use, and site hydrology. Both sets of studies, in prin-

Tab. 7.1 Physiology, productivity, and water use of *Populus* (*P.*) and *Salix* (*S.*) pure species and their hybrids.

Taxa	A_{net} (μmol $m^{-2}s^{-1}$)	g_s (mmol $m^{-2}s^{-1}$)	WUE_i (μmol $mmol^{-1}$)	$\delta^{13}C$ (‰)	$ANPP_{wood}$ (Mg ha^{-1} yr^{-1})	ET_{us} (mm yr^{-1})
P. pure spp.	13.7	303	0.044	−28.0	9.9	117.0
P. hybrids	12.7	343	0.044	−28.1	8.1	142.5
S. pure spp.	13.7	316	0.045	−27.9	8.5	240.1
S. hybrids	20.6	378	0.040	−	8.5	119.4

Taxa	T (mm yr^{-1})	WUE_P (kg mm^{-1})	WUE_{bfg} (MJ m^{-3})	MAT (°C)	MAP (mm)	MAP/ET (mm)
P. pure spp.	419.2	23.7	39.6	11.4	607	1.1
P. hybrids	439.1	18.5	30.9	10.2	753	1.3
S. pure spp.	403.8	21.0	35.1	5.7	661	1.0
S. hybrids	324.9	26.1	43.7	6.6	703	1.6

Note: particular symbols and acronyms are explained in the list of abbreviations. Modified after King *et al.* [34].

ciple, agreed that SRC based on *Populus* or *Salix* are, despite the high WUE, great water consumers, and that water availability and water security will be the most critical issues related to the possible large-scale exploitation of SRC. However, these studies considered only a relatively narrow range of genotypes and environmental conditions specific for England and Sweden. It is therefore necessary to install representative field experiments that allow testing, improvement, and modeling of the productivity potential and water use of SRC across various genotypes and environmental conditions.

We provide the main trends of mean physiological rates, aboveground wood net primary production ($ANPP_{wood}$), water balance, and climatological variables, including all *Populus* and *Salix* pure species and hybrid data, separately for which data were available in the literature (Tab. 7.1). The highest net photosynthetic assimilation rates (A_{net}) were observed for *Salix* hybrids whereas the lowest were in *Populus* hybrids. Nevertheless, the very high value of A_{net} (20.6 μmol m^{-2} s^{-1}) by *Salix* hybrids comes from only one study on *S. schwerinii* × *S. viminalis* [74] making even such a broad comparison not representative. The A_{net} for pure species was practically identical, despite the general distribution of the *Salix* species in colder environmental conditions. Colder and similarly wetter conditions of *Salix* lead to lower evaporative demand (lower vapor pressure deficit) and lower risk of drought. This likely explains the slightly higher

Tab. 7.2 Averages of the main ecophysiological rates and hydrologic and climatic variables across the genus *Populus*.

Taxa	A_{net} (μmol m^{-2} s^{-1})	g_s (mmol m^{-2} s^{-1})	WUE_i (μmol mmol^{-1})	$\delta^{13}C$ (‰)	$ANPP_{wood}$ (Mg ha^{-1} yr^{-1})	ET_{us} (mm yr^{-1})	T (mm yr^{-1})	ET (mm yr^{-1})	WUE_P (kg mm^{-1})	WUE_{bfg} (MJ m^{-3})	MAT (°C)	MAP (mm)	MAP/ET (mm)
Pure species													
1. *P. alba*	14.1	276	0.049		15.81						14.0	714.3	
2. *P. angustifolia*	11.1	341	0.036	−28.83							5.7	425.0	
3. *P. balsamifera*					6.71								
4. *P. ciliata*				−28.90							9.3	929.0	
5. *P. deltoides*	16.2	422	0.043	−28.36	8.13		246.5		32.98	55.18	14.7	791.6	3.21
6. *P. euphratica*	11.9	225	0.057	−27.90				738.48			10.9	66.9	0.09
7. *P. fremontii*	13.2	294	0.038	−28.74			725.0	1345.0			16.7	239.0	0.33
8. *P. hopiensis*				−25.66									
9. *P. nigra*	15.0	391	0.049	−28.43	12.36			441.0			10.5	714.9	1.29
10. *P. tremula*		120			7.47						9.2	570.0	
11. *P. tremuloides*	10.6	323	0.038	−27.85	9.38	117.0	286.0	403.0	32.80	54.87	4.6	539.5	1.34
12. *P. trichocarpa*	14.4	306	0.049	−27.22	7.17						10.3	744.8	
Hybrids													
1. *P. deltoides* × *P. nigra*	13.1	360	0.044	−27.94	9.11			448.0			11.8	725.8	1.62
2. *P. deltoides* × *P. trichocarpa* (and vice versa)	10.9	289		−28.02	6.93	138	297.5	561.5	23.28	38.95	10.6	817.4	1.88
3. *P. trichocarpa* × *P. balsamifera* (and vice versa)	6.9	271			6.81						9.4	939.2	
4. *P. nigra* × *P. maximowizcii* (and vice versa)	9.5				12.40		679.75		18.24	30.52	9.2	806.4	1.19
5. *P. trichocarpa* × *P. nigra*	10.5				0.48						11.4	649.5	

162 — Milan Fischer et al.

(Continued)

Taxa	A_{net} (μmol m^{-2}s^{-1})	g_s (mmol m^{-2}s^{-1})	WUE_i (μmol mmol^{-1})	$\delta^{13}C$ (‰)	$ANPP_{wood}$ (Mg ha^{-1} yr^{-1})	ET_{us} (mm yr^{-1})	T (mm yr^{-1})	ET (mm yr^{-1})	WUE_P (kg mm^{-1})	WUE_{bfg} (MJ m^{-3})	MAT (°C)	MAP (mm)	MAP/ET (mm)
6. *P. balsamifera*×*P. simonii*	15.2	410	0.037	−28.77							3.3	380.0	
7. *P. deltoides*×*P. ×petrowskiana*	19.3	406	0.047	−28.2				298.0			3.3	364.5	1.22
8. *P. tremula*×*P. tremuloides*					4.55						8.0	702.0	
9. *P. balsamifera*×*P. tremula*					8.1						7.3	590.6	
10. *P. maximowizcii*×*P. balsamifera*					9.18							1050.0	
11. *P. maximowizcii*×*P. trichocarpa*					3.79	147.0	241.0	520.0	15.73	26.31	9.0	659.0	1.70
12. *P. trichocarpa*×*P. koreana*					9.7						7.3	590.6	
13. *(P. deltoides*×*P. nigra)*×*P. maxi-mowizcii*					7.55							1050.0	
14. *(P. deltoides*×*P. trichocarpa)* × *P. nigra*					8.66						13.3	620.4	
15. *(P. deltoides*×*P. trichocarpa)* × *P. trichocarpa*					12.55						13.1	698.8	
16. *P. nigra*× *(P. deltoides*×*P. nigra)*					14.87						11.2	644.0	
17. *P. maximowizcii*×*P. ×berolinensis*					10.2						7.3	590.6	
Unknown genetic background					9.78			552.1			11.7	649.9	1.18

Note: particular symbols and acronyms are explained in the list of abbreviations.

Tab. 7.3 Averages of the main ecophysiological rates and hydrologic and climatic variables across the genus *Salix*. Particular symbols and acronyms are explained in the list of abbreviations.

Taxa	A_{net} (μmol m^{-2} s^{-1})	g_s (mmol m^{-2} s^{-1})	WUE_i (μmol mmol^{-1})	δ^{13}C (‰)	$ANPP_{wood}$ (Mg ha^{-1} yr^{-1})	ET_{us} (mm yr^{-1})	T (mm yr^{-1})	ET (mm yr^{-1})	WUE_P (kg mm^{-1})	WUE_{bfg} (MJ m^{-3})	MAT (°C)	MAP (mm)	MAP/ET (mm)
Pure species													
1. *S. arctica*	19.1	328	0.058								−11.4	123	
2. *S. babylonica*						520.5	738.5	1300				404	0.31
3. *S. burjatica*		125											
4. *S. caprea*					4.48								
5. *S. cinerea*					3.83						6.4	777	
6. *S. dasyclados*	12.6	348.5	0.030		9.71						6.4	954	
7. *S. discolor*					9.42						6.4	954	
8. *S. eriocephala*					12.62							580	
9. *S. exigua*	20.6		0.049									900	
10. *S. glauca*	10.1	418	0.021								−3.0	900	
11. *S. gooddingii*	14.2	481	0.047					1764.0			20.3	208	0.12
12. *S. gordejevii*	14.5	238	0.049	−28.5							1.7	312.5	
13. *S. integra*	13.5	324											
14. *S. matsudana*	11.6	170	0.069								1.7	350	
15. *S. microstachya*	14.7	229	0.064								1.7	350	
16. *S. miyabeana*					13.17	505			26.08	43.63	7.7	986	
17. *S. monticola*	11.3										−3.0	900	
18. *S. planifolia*	12.2	542	0.023								9.0	1017	
19. *S. purpurea*						523							
20. *S. sachalinensis*					15.59	494			31.56	52.80	7.7	986	
21. *S. schwerinii*					2.99								
22. *S. sitchensis*					10.12								
23. *S. spaethii*					6.70								
24. *S. triandra*					7.67						7.7	581	
25. *S. viminalis*	13.2	360	0.043		8.78						7.2	706	1.59
						99.9	296.2	444.8	29.64	49.59			

164 — Milan Fischer et al.

(Continued)

Taxa	A_{net} (μmol m⁻² s⁻¹)	g_s (mmol m⁻² s⁻¹)	WUE_i (μmol mmol⁻¹)	$\delta^{13}C$ (‰)	$ANPP_{wood}$ (Mg ha⁻¹ yr⁻¹)	ET_{us} (mm yr⁻¹)	T (mm yr⁻¹)	ET (mm yr⁻¹)	WUE_P (kg mm⁻¹)	WUE_{bfg} (MJ m⁻³)	MAT (°C)	MAP (mm)	MAP/ET (mm)
Hybrids													
1. *S. burjatica*× *S. dasyclados*		306			10.59		275		38.49	64.39		600	
2. *S. burjatica*× *S. viminalis* (and vice versa)					9.55								
3. *S. dasyclados*× *S. aquatica*					2.72								
4. *S. dasyclados*× *S. caprea*					2.36								
5. *S. interior*× *S. eriocephala* (and vice versa)					7.76						6.4	954	
6. *S. matsudana*× *S. alba*					11.31							1000	
7. *S. sachalinensis*× *S. miyabeana*							533				9.0	1017	
8. *S. schwerinii*× *S. aquatica*	20.6	415	0.040		9.38								
9. *S. schwerinii*× *S. viminalis* (and vice versa)					10.49	148.0	266.0	480.0	39.45	66.00	6.6	645	1.34
10. *S. triandra*× *S. viminalis*					7.47	132.5	214.0	378.0	34.90	58.38		586	1.55
11. *S. viminalis*× *S. aquatica*					6.84								
12. *S. viminalis*× *S. candida*					8.14								

(Continued)

Taxa	A_{net} (μmol m^{-2}s^{-1})	g_s (mmol m^{-2}s^{-1})	WUE_i (μmol mmol^{-1})	$\delta^{13}C$ (‰)	ANPP$_{wood}$ (Mg ha^{-1} yr^{-1})	ET$_{us}$ (mm yr^{-1})	T (mm yr^{-1})	ET (mm yr^{-1})	WUE$_P$ (kg mm^{-1})	WUE$_{bfg}$ (MJ m^{-3})	MAT (°C)	MAP (mm)	MAP/ET (mm)
Hybrids													
13. *S. viminalis*× *S. caprea* (and vice versa)					7.14	92.5	402.5	555.0	17.74	29.69	5.8	554	1.00
14. *S. viminalis*× *S. dasyclados*					13.20	104.5	318.0	479.0	41.51	69.45	5.0	641	1.34
15. *S. viminalis*× *S. purpurea*					15.10						7.1	450	
16. *S. aurita*× *S. cinerea*×*S. viminalis*					6.60								
17. *S. caprea*× *S. cinerea*×*S. viminalis*					7.20								
18. *S. schwerinii*× *S. viminalis*× *S. dasyclados*					9.66								
Unknown genetic background					9.48			516.4				746	1.44

Note: particular symbols and acronyms are explained in the list of abbreviations.

stomatal conductance (g_s) of the *Salix* species and hybrids. Higher g_s together with comparable levels of A_{net} is, on the other hand, linked with the lowest intrinsic water use efficiency (WUE_i) at the leaf level, which is not consistent with the WUE at the stand level, mainly due to the lower stand transpiration rather than high $ANPP_{wood}$ reported in literature. This inconsistency may be explained by the fact that there are, practically speaking, two kinds of eco-physiological studies focused on different scales, *i.e.*, leaf-level and stand-level, which are unfortunately very rarely combined. This makes data comparison between scales (*e.g.*, leaf-level and stand-level) potentially fraught with error and uncertainty; although, some useful generalizations may emerge. This in-dicates the importance of combining basic measurements at the leaf and stand levels in future experiments in the context of site water balance under various environmental conditions to make the results more broadly applicable. Rates of $ANPP_{wood}$ were within the range of 8–10 Mg ha^{-1} yr^{-1}, with the highest found for pure *Populus* species (9.9 Mg ha^{-1} yr^{-1}) and the lowest for *Populus* hybrids (8.1 Mg ha^{-1} yr^{-1}) (Tab. 7.1). This contrasted with the general idea that hybrid cultivars are more productive than those of pure species. Conversely to $ANPP_{wood}$, the ranking in the stand transpiration (T) for the pure and hy-brid poplars was the opposite leading to the lowest WUE_P of 18.5 kg mm^{-1} for the latter. This suggests that *Populus* hybrids are more profligate with water, which may be explained by the different climatic conditions: hybrids across the studies were spread across more humid and colder conditions and thus periodic drought episodes may have been responsible for the lower stand T and higher WUE_P of the pure species.

Detailed data on ecophysiological and hydrological rates within the two gen-era and their hybrids have been compiled and provided as the averages for each particular species or hybrid (Tabs. 7.2 and 7.3). For both genera, traits related to physiological process rates (A_{net}, g_s, WUE_i) and $ANPP_{wood}$ were by far the most often reported. Leaf carbon isotopic composition ($\delta^{13}C$) was relatively well-represented in the genus *Populus*, but very rarely in *Salix*. Only a few studies reported hydrological parameters (sap flow-based estimates of T, micrometeoro-logically derived ET, and measured precipitation). Very little data was reported on understory evapotranspiration (ET_{us}) or stand-level water use efficiency of productivity (WUE_P). None of the studies provided so-called bioenergy WUE at the bioenergy cropping systems farm gate (WUE_{bfg}), as recently proposed by King *et al.* [34].

Although the genus *Salix* has a much broader ecological amplitude, *Populus* was generally better represented in the literature (87 studies). The most rep-resented genotypes were hybrids *P. deltoides* × *P. nigra* (29 studies) and *P. del-toides* × *P. trichocarpa* (26 studies). The rest of the *Populus* hybrids were repre-sented with less than ten studies, whereas eight were found for *P. trichocarpa* × *P. balsamifera* and eight for *P. nigra* × *maximowiczii*. The last mentioned hybrid, commonly known as clone NM5 or NM6 in the USA, has been recently relatively widely extended across Europe as clone J-105 or J-104. However, its physiology has been, so far, very rarely described, providing good potential for new and useful research. Most work on pure species has been dedicated to *P. trichocarpa* (15 studies), *P. deltoides* (13 studies), *P. nigra* (12 studies), and *P. alba* (7 studies). From these, four hybrids and three pure species were the most represented. The highest ANPP$_{wood}$ values were recorded for *P. alba* (15.81 Mg ha^{-1}), *P. nigra* × *maximowiczii* (12.40 Mg ha^{-1}), and *P. nigra* (12.36 Mg ha^{-1}) (Tab. 7.2). Stand and bioenergy WUE was highest for *P. deltoides* with respective values of 32.98 kg mm^{-1} and 55.18 MJ m^{-3} and lowest for *P. nigra*×*maximowiczii* (18.24 kg mm^{-1} and 30.52 MJ m^{-3}, respectively). The relatively low WUE of *P. nigra*×*maximowiczii* might be caused by lower temperature and higher precipitation compared to the rest of the studies examined and also by unrealistically high sap flow-based transpiration rates reported by Zalesny *et al.* [20]. If we exclude this study from the analysis, the WUE$_P$ for this hybrid is 26.14 kg mm^{-1} and WUE$_{bfg}$ is 43.73 MJ m^{-3}, which seems more realistic. The relatively low latitudes of Spain and Italy, represented by mean annual temperatures of 14°C, may also explain the highest ANPP$_{wood}$ by *P. alba*. Likewise, generally higher ANPP$_{wood}$ values reported for pure species may result from their geographic location in warmer climates.

In comparison to *Populus*, *Salix* was not as well-described (60 studies), with relatively good representation of leaf-level physiology for pure species but almost no data available for hybrids (Tab. 7.3). Conversely, *Salix* hybrids were better represented at the stand level, including ANPP$_{wood}$, ET$_{us}$, stand T, ET, and stand and bioenergy WUE. The most well-represented variable was ANPP$_{wood}$, with a mean of 8.50 Mg ha^{-1} y^{-1} for both pure species and hybrids. In contrast to *Populus*, the genus *Salix* was better represented by pure species. Overall, *S. viminalis* was described in 32 studies with gas exchange rates close to the averages for the other pure species, but with slightly higher g$_s$ and lower WUE$_i$. ANPP$_{wood}$ of *S. viminalis* was 8.78 Mg ha^{-1} and the stand and bioenergy WUE were 29.64 kg mm^{-1} and 45.59 MJ m^{-3}, respectively, ranking this species in the high range of WUE across all genera used for bioenergy [34]. It is noteworthy that *S. viminalis*, in particular, and the whole genus *Salix*, in general, are distributed in colder and more humid climates, although exceptions such as *S. gooddingii* have been reported [75]. The rest of the *Salix* pure species were poorly described, with eight studies for *S. dasyclados*, six studies for *S. burjatica*, and the remaining species described by less than five (typically one to three) studies. The *Salix* hybrids were the most represented by *S. schwerinii* × *S. viminalis* (12 studies), *S. triandra* × *S. viminalis* (6 studies), and *S. viminalis* × *S. caprea*. The rest of the hybrids were typically represented by one to three studies. *S. schwerinii* × *S. viminalis* had the highest A$_{net}$ of 20.6 µmol m^{-2} s^{-1}, which was associated with relatively high g$_s$ and low WUE$_i$. The ANPP$_{wood}$ of this hybrid was 10.49 Mg ha^{-1}, which was linked to high stand and bioenergy WUEs of 39.45 kg mm^{-1} and 66 MJ m^{-3}, respectively.

In order to provide a summary about the ecophysiological, hydrological, and climatic data and their relationships and suitability of such data backgrounds for general conclusions, we present simple correlation matrices relating all of the investigated variables. The correlation matrices (Tabs. 7.4 and 7.5) show generally similar trends for both *Populus* and *Salix*. Correlations were often not significant ($p > 0.05$), perhaps due to the limited data availability. However, we can still highlight several interesting points. A_{net} was positively correlated with g_s in *Populus* as expected from theory, but not in *Salix* due to high g_s (481 and 542 mmol m^{-2} s^{-1}) and moderate A_{net} (10.1 and 12.2 µmol m^{-2} s^{-1}) for *S. glauca* and *S. planifolia*, respectively, as reported by Reich *et al.* [76]. For both genera, g_s was negatively related with WUE$_i$ as expected, suggesting that better stomatal control (lower g_s) generally leads to higher WUE$_i$. By omitting the low g_s of 125 mmol m^{-2} s^{-1} for *P. fremontii* [77] from the analysis, the average g_s increases to 378 mmol m^{-2} s^{-1} and the relationship becomes statistically significant for *Populus*. Surprisingly, WUE$_i$ positively correlated with δ^{13}C for both genera, and moreover, in the case of *Populus* this relationship was even significant ($p < 0.05$), which is contradictory to the theory about leaf C isotope composition and WUE$_i$ [78]. The poor correlation we observed probably lies in the fact that gas exchange rates and δ^{13}C, on which correlations are based, do not necessarily originate from the same studies. However, the reliability of δ^{13}C as a proxy for WUE$_i$ has already been widely demonstrated in tree species including poplar [54–56]. For this reason, δ^{13}C has often been used as a cheap alternative for identifying species with favorable WUE to decrease the water use of bioenergy systems [79]. Nevertheless, δ^{13}C was at least negatively related to WUE at the stand level, though the data is very limited. Apart from the limited database and the nature of its averaging, this indicates inconsistency between WUE$_i$ and WUE$_P$ or WUE$_{bfg}$ at the leaf and stand levels which may be explained as an effect of the low canopy coupling to the atmosphere [80–81]. Similar relationships with stronger data backgrounds including various bioenergy tree species were also described [34], suggesting that atmospheric decoupling may play an important role.

Interestingly, ANPP$_{wood}$ for both genera were not significantly correlated with any other variables, though marginally correlated with stand T ($p = 0.055$ and 0.054 for *Populus* and *Salix*, respectively), as expected according to the basic concept of WUE describing that water loss is proportional to biomass productivity [53]. Although the relationship between ANPP$_{wood}$ and WUE$_P$ or WUE$_{bfg}$ was poor, it was positive for both genera. By omitting *P. nigra* $\times$ *P. maximowiczii* with high ANPP$_{wood}$ and low WUE, the relationship became significant. Obviously, data availability did not allow correlation of WUE$_i$ with stand and

Tab. 7.4 The correlation matrix for the *Populus* genus comprising the Pearson correlation coefficient and particular *p*-values (1-tailed test) for water- and biomass-related ecophysiological properties.

	A_{net}	g_s	WUE_i	$\delta^{13}C$	$ANPP_{wood}$	T	ET	WUE_P	WUE_{bfg}	MAT	MAP	MAP/ET
A_{net}	1.000	**0.663 / 0.007**	0.206 / 0.272	–	0.314 / 0.188	0.324 / 0.250	0.327 / 0.368	0.648 / 0.176	0.648 / 0.176	0.387 / 0.252	0.142 / 0.087	0.268 / 0.261
g_s		1.000	0.108	0.390 / 0.117	0.182 / 0.319	0.282 / 0.101	0.359 / **0.034**	0.709 / 0.249	0.709 / 0.249	0.227 / 0.055	0.383 / 0.485	0.503 / 0.102
WUE_i			1.000	**0.589 / 0.036**	0.381 / 0.228	0.792 / 0.055	0.267	-0.209 / 0.433	-0.209 / 0.433	0.433 / 0.221	0.100 / 0.379	-0.309 / 0.276
$\delta^{13}C$				1.000	-0.565 / 0.121	–	–	0.228 / 0.356	0.228 / 0.356	0.175 / 0.387	0.420 / 0.179	0.016 / 0.487
$ANPP_{wood}$					1.000	–	–	-0.410 / 0.246	-0.410 / 0.246	0.224	0.089 / 0.335	-0.296 / 0.238
T						1.000	**0.981 / 0.010**	-0.757 / 0.227	-0.757 / 0.227	**0.811 / 0.004**	0.335 / 0.186	**-0.735 / 0.048**
ET							1.000	–	–	0.070 / 0.455	0.181	**-0.660** / 0.494
WUE_P								1.000	**1.000 / 0.000**	0.070 / 0.455	–	0.199 / 0.494
WUE_{bfg}									1.000	–	–	0.199 / 0.064
MAT										1.000	–	0.426
MAP											1.000	**0.776 / 0.002**
MAP/ET												1.000

The data shown in bold character are statistically significant relations ($p < 0.05$). Particular symbols and acronyms are explained in the list of abbreviations.

Tab. 7.5 The correlation matrix for the *Salix* genus comprising the Pearson correlation coefficient and particular *p*-values (1-tailed test) for water- and biomass-related ecophysiological properties.

	Anet	gs	WUEi	δ¹³C	ANPPwood	T	ET	WUEp	WUEbfg	MAT	MAP	MAP/ET
Anet	1.000	0.052	0.308	–	0.804	–	–	–	–	0.418	0.149	0.238
gs	0.440	1.000	0.178	0.482	0.202	–	0.381	0.097	0.097	–	**0.753**	0.424
WUEi	–	–	1.000	0.747	–0.021	0.404	0.106	0.469	0.469	0.485	**0.002**	**0.894**
δ¹³C	–	0.482	**0.001**	1.000	–0.247	–	0.917	–	–	–	**0.001**	0.148
ANPPwood	0.804	0.202	–0.021	0.489	1.000	0.610	0.445	0.302	0.234	0.190	0.222	**–0.856**
T	0.610	0.054	0.357	–	0.073	1.000	**0.982**	–0.565	–0.565	**0.740**	0.218	**–0.968**
ET	0.073	0.445	–	0.131	–	**0.000**	1.000	–0.513	–0.513	**0.985**	0.259	**–0.959**
WUEp	–	0.097	0.469	–	0.302	0.072	0.189	1.000	**1.000**	**0.018**	0.161	0.089
WUEbfg	–	0.097	0.469	–	0.234	0.072	0.189	**0.000**	1.000	0.985	0.222	0.433
MAT	0.312	0.278	0.485	–	0.190	**0.001**	**0.018**	–	0.985	1.000	0.184	**–0.897**
MAP	0.334	0.205	**0.001**	0.874	0.205	0.218	0.259	0.161	0.222	0.212	1.000	**0.932**
MAP/ET	0.542	0.172	0.172	–	**0.000**	0.172	–	0.542	0.542	**0.019**	**0.000**	1.000

The data shown in bold character are statistically significant relations ($p < 0.05$). Particular symbols and acronyms are explained in the list of abbreviations.

bioenergy WUE, which confirms the already mentioned inconsistency of leaf- and stand-level experiments. This was well-demonstrated by the lack of correlation or even inverse relationships between leaf physiological rates and stand T within the *Populus* genus and there was almost no data for *Salix*. Similarly, $\delta^{13}C$ was not available for any *Salix* hybrids for which more stand-level experiments were typical and thus no correlation was possible. Due to the lack of ET_{us} data, we completely excluded it from this analysis. Stand T was positively related to mean annual temperature (MAT) in the cases of both genera. However, in the case of *Populus*, the correlation was not possible due to only one study on *P. deltoides* in India, which was not climatologically representative (characterized by a MAT of 25.3°C and mean annual precipitation (MAP) of 970 mm [82]). The relationship with MAT was much stronger for ET, which is generally beyond plant control. For both genera, ET was negatively related to MAP, whereas for stand T such a relationship was evident only for *Populus*. This was, however, due to the atypical results for *P. euphratica* [83–84], *P. fremontii*, and *S. gooddingii* [75], located in almost rain-free conditions with access to underground water or *S. babylonica* growing next to the river [85]. After exclusion of these studies, the relationship between T or ET and MAP was strong and positive. So, together with MAP/ET ratios relatively close to unity, it is suggested that *Populus* and *Salix* consume most of the available water and, as a consequence, they constitute one of the most important factors for their natural distribution. Though the MAP/ET ratios for both *Populus* and *Salix* hybrids are usually higher than unity, water availability is undoubtedly a crucial factor for the economic viability and ecological sustainability of SRC-based bioenergy systems.

7.5 Case Study: *Populus* in the Bohemian-Moravian Highlands

7.5.1 Introduction

Bioenergy cropping systems in central Europe and mainly in the Czech Republic are relatively new and undeveloped. SRC for direct combustion are seen as a prospective way to partly substitute fossil fuels and increase the energy diversity and independence of the Czech Republic. Nevertheless, the areas planted in bioenergy SRC are still very small, amounting to about 1,300 to 1,500 ha. This area of bioenergy production is very low compared to many other EU countries, but there is large potential for expansion in the near future. To avoid food versus fuel issues, potential areas for SRC are intended to be established in so-called marginal agricultural areas where the climate or soil conditions are not optimal and farming is neither effective nor sustainable. Since marginal areas are typically rain-fed, the question arises as to whether SRC bioenergy production would be economically profitable and ecologically sustainable. This is also relevant to SRC tolerance to future climatic conditions, characterized by a drier climate with lower precipitation to the ET ratio [43]. Here, we present a case study of water use, WUE, and productivity of a *Populus* SRC grown in a marginal agricultural area of the Bohemian-Moravian Highlands.

7.5.2 Site and Stand Description

The study was conducted in a typical rain-fed area of the Bohemian-Moravian Highlands at the research locality Domanínek (Czech Republic, 49°31′N, 16°14′E and altitude 530 m a.s.l.) in the west part of the town of Bystřice nad Pernštejnem. The MAT at this site over the period of 1981–2010 was 7.2°C, the MAP was 609.3 mm and the mean annual reference ET (Allen *et al.* 1998) was 650 mm. The length of the growing season (daily mean air temperature above 5°C) is 217 days, beginning at the end of March and lasting until the beginning of November. The site is highly suitable for planting *Populus* SRC due to the deep soil profile, but of marginal value for agriculture [18]. The site is characterized by slightly undulating topography with mild slopes of 3–5° with an eastern aspect and is subject to a cool and relatively wet temperate climate typical for this part of central Europe, with both continental and maritime influences. Soil conditions at the location are representative of the wider region, with deep luvic Cambisol soils influenced by gleyic processes and with a limited amount of stones in the profile.

In April, 2002, the first of the investigated operational high-density mono-clonal plantation (hereafter, SRC 1) was established for verification of the performance of hybrid poplar clone J-105 (*Populus nigra* × *P. maximowiczii*) at the total area of 2.85 ha (49°31′25″N, 16°14′31″E and altitude 540 m a.s.l.) The plantation was planted on agricultural land previously cropped predominantly for cereals and potatoes. Following conventional tillage operations, the hardwood cuttings were planted in a double-row design with between-row and within-row spacing of 2.5 m and 0.7 m, respectively, accommodating a theoretical density of 9, 216 trees per ha. To mimic the most probable future SRC based low-input systems, no irrigation, fertilization, or herbicide treatments (except the local application of glyphosate on the most vigorous and tenacious weeds) were applied during the experiment. Subsequent mechanical weeding was carried out two times per growing season until canopy closure in 2005. As a consequence of the intense weed competition, the rotation period was lengthened to eight years and the first harvest was carried out at the end of winter 2009/2010. In April 2001, the second of the investigated coppice culture (hereafter SRC 2) was established less than 300 m south-westward up the slope from SRC 1 (49°31'15"N, 16°14'10"E and altitude 575 m) at place with similar soil conditions, similar slope angle (3.5°), and same land-use history. Note that the numbering of the SRC (1, 2) does not reflect their age but expresses the chronology of the investigations. The total area of SRC 2 consisted of a 1.2 ha monoclonal block of hybrid poplar clone J-105 and an additional 1.6 ha block with a mixture of four *Populus* species and three *Salix* species, described by Trnka *et al.* [18]. In the first rotation cycle, SRC 2 received the same management as SRC 1. However, after first coppicing during the winter of 2008/2009, a fertilization experiment was arranged in a randomized plot design within the block of J-105, which included four treatments (nitrogen, phosphorus, potassium inorganic fertilizer; a mixture of sewage sludge and ash; lime; and a control) replicated three times. Since the preliminary results did not reveal any significant effects of fertilization on biomass productivity of SRC 2, this treatment was dropped from the analysis. The main purpose for both plantations was to serve as field models, both in terms of size and composition, to eventual future plantations in the region.

7.5.3   Methods

The intensive measurement campaign started in June, 2008, when a 14 m-high aluminum mast was erected above SRC 1. The mast was equipped with a Bowen ratio energy balance (BREB) system (EMS Brno, Czech Republic) for estimating actual ET. Two years later, at the end of 2010, a similar BREB system was installed above the nearby canopy of SRC 2. The BREB method is based on the measurements of air temperature and humidity gradients between at least two vertical levels as well as above canopy net radiation and ground heat flux measurements. For our purposes, we used a similar BREB design, which was previously analyzed and validated [86]. The system is based on a fixed-positioned combination of thin-film polymer capacitive relative humidity and the adjacent resistance temperature sensor instruments EMS 33 (EMS Brno, Czech Republic). We used three vertical heights above SRC 1 (2+1 m distance) and two heights (2 m distance) above SRC 2. In both cases, we kept the lowest sensor just above the canopy top in order to minimize fetch/footprint issues [87], being aware of the good reliability of BREB employed in the roughness sub-layer if the canopy is dense and uniform [88–89]. We measured net radiation and soil heat flux using a net radiometer (Schenk 8110, Philipp Schenk GmbH Wien, Austria) and a plate sensor (HFP01, Hukseflux, The Netherlands), respectively. These core sensors of each system were supplemented with other instrumentation such as a diode pyranometer, rain gauge, wind speed and direction sensor, soil moisture and soil temperature probes, and soil water potential sensors—all connected to a solar energy supply and dataloggers with measuring frequencies of 1 min averaged every 10 min [19].

Apart from stand ET measured by BREB systems, we used the sap flow technique for deriving transpiration rates at the tree level during the season of 2009. Since we used the tissue heat balance method [90], which is limited to trees with diameters of at least 0.1 m, we were able to investigate only the tree cohort of the most dominant trees (Fig. 7.1). Therefore, we did not attempt scaling from tree to canopy and limited our results to tree-level water use and, in particular, tree-level WUE of the dominant diameter class.

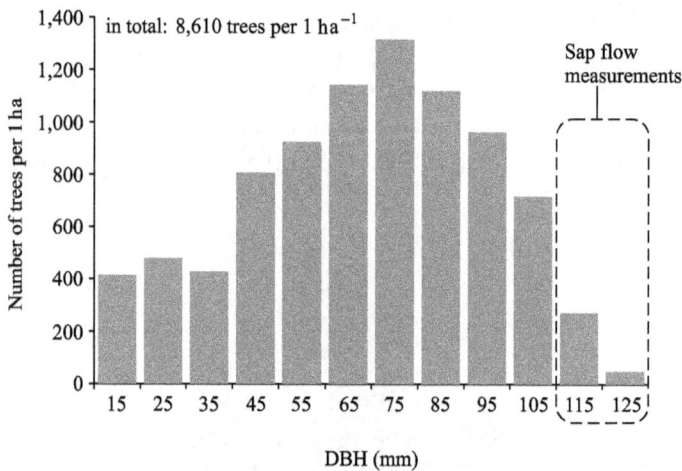

Fig. 7.1 The frequency distribution of breast height diameters of trees in the SRC 1 stand at the end of the first rotation cycle in 2009 expressed in absolute number per ha. The diameter range of the trees measured by the tissue heat balance method is marked in the dashed quadrangle.

For estimating biomass increments, an array of 15 mechanical (DB 20) and three automatic dendrometers (DRL 26, EMS Brno, Czech Republic) were installed in SRC 1 at the beginning of 2008. These measurements were updated by adding 15 DB 20 and 1 DRL 26 dendrometers at the start of the growing season in 2009. The DBs and DRLs were fixed to the trunks at breast height and were manually read weekly while the DRL 26 dataloggers were adjusted to an hourly measuring step. Incremental values obtained by these dendrometers are very useful because they can be converted to biomass increments through allometric equations [7, 52, 91]. Further, destructive measurements of 40 randomly chosen trees where the dendrometers were placed and another randomly selected 80 trees were carried out during the harvesting at the beginning of 2010—the end of first eight-year-long rotation period. This procedure followed the same methodology as described by Fajman *et al.* [92]. The allometric relationship between the diameter at breast height (DBH) and the dry matter content (DMC) of aboveground woody biomass (AB_{wood}) was described by a biometric power function in the form of $AB_{wood} = a * DBH^b$, where a and b are empirically obtained parameters. Furthermore, the DBHs of 100 randomly chosen trees from all diameter and height classes were measured with calliper at the start of June, 2009, and then again at the end of July, 2009. This two-month increment during the peak of the growing season was taken as representative for all DBH classes and served as a so-called reference increment. By relating the reference DBH increment to the DBH, a scaling power function in the form of $\Delta DBH_{ref} = a * DBH^b$ was determined. This relation followed the natural competition law, which favors the dominant trees at the expense of the suppressed ones, similarly to the process of self-thinning [93]. This parameterized function was applied to the DBHs of all 702 trees measured during regular stem inventory (Fig. 7.1) and thus their reference DBH increments were estimated. At the same time, the function was applied on the sampled trees where the DBH increments were measured regularly with dendrometers. By dividing the actual by the calculated (reference) increments, tree-specific scaling coefficients were obtained. This means that for each record from one dendrometer, there is one scaling coefficient enabling conversion of the calculated reference increment from all of the trees to the real DBH increment. Although there were 40 dendrometers in total, there were only 15 with uninterrupted data series for both consecutive years. Due to the homogeneity and comparability of both years, only these 15 measured trees were used for scaling purposes. For each of these sampled trees, one scaling coefficient was obtained, resulting in 15 slightly different increments for all 702 trees. This may be due to the r^2 value of 0.83. The mean of 15 variants of the calculated increments of 702 trees within the defined area were finally scaled up to the whole stand, assuming that the 702 trees were sufficient to provide a representative diameter distribution of the whole stand. The total stem increments were subsequently converted using the allometric equation for AB_{wood} to the biomass increment (DMC) per area (1 ha).

Further, the $ANPP_{wood}$ increment per 1 ha was divided by the amount of ET (mm) integrated to the periods corresponding to dendrometer readings and thus the long-term WUE_P was obtained. In this work, WUE is defined only from a part of the aboveground woody biomass (stems and branches – the growth of leaves and roots is not considered) divided by ET (not only transpiration). The term gross WUE_P is then used in order to emphasize the difference from typically calculated stand WUE. Similarly, the $ANPP_{wood}$ increment at the four sampled trees with sap flow instrumentation was related to the transpiration and thus the WUE_P of the particular trees was derived.

7.5.4 Results and Discussion

The BREB, like any other micrometeorological method, provides whole ET comprising overstory transpiration, understory ET (including soil evaporation), and evaporation of canopy interception. Using the BREB method only, there is no direct way to discriminate tree transpiration, which can subsequently be related to $ANPP_{wood}$ in order to obtain WUE_P. For this reason, we used notation of gross WUE_P to indicate combined overstory and understory ET. By relating $ANPP_{wood}$ to annual ET, we found that the WUE_P of SRC 1 and SRC 2 were quite conservative and very high when the canopy was completely closed, which is typically three years after planting or coppicing at our site (Tab. 7.6). During the early stages of canopy development (*i.e.*, during the first and second year after coppicing as indicated in Tab. 7.6 for SRC 1), understory ET played an important role and decreased gross WUE_P. In our previous study [19], we found that during the eighth year of the first rotation, transpiration comprised 70% of whole ET over the growing season, in good agreement with other studies conducted on SRC [17, 70]. During open canopy periods (*i.e.*, 2010 and beginning of 2011 for SRC 1), this portion of ET had to be much lower and soil evaporation played a more important role. Considering this, it is obvious that gross WUE_P (Tab. 7.6) was significantly underestimated compared to the transpiration-based WUE_P and thus the WUE_P presented in this case study lies at the high end of all the pure species and hybrids reported in the literature (Tabs. 7.2 and 7.3).

Tab. 7.6 Gross WUE_P based on $ANPP_{wood}$ to the total ecosystem ET ratio as a result from experimental locality Domanínek (Czech Republic). Particular symbols and acronyms are explained in the list of abbreviations.

Stand	Year	Stand/shoot age (yr)	$ANPP_{wood}$ (t ha^{-1}yr^{-1})	ET (mm yr^{-1})	Gross WUE_P (kg mm^{-1})	Gross WUE_{bfg} (MJ m^{-3})
SRC 1	2008	7/7	13.4	516.0	25.97	43.45
SRC 1	2009	8/8	16.5	549.2	30.04	50.26
SRC 1	2010	9/1	3.9	344.3	11.33	18.95
SRC 1	2011	10/2	6.9	519.2	13.23	22.14
SRC 2	2011	11/3	15.5	529.3	29.27	48.96

Gross WUE_P based on synchronous measurements of AB_{wood} increments (allometrically derived from stem circumference increments) and canopy ET (by the BREB method) during 2008 and 2009 at SRC 1 showed high seasonal variability in both years (Fig. 7.2). The seasonal patterns were typical with the highest rates of gross WUE_P at the beginning of the season reaching up to 67.53 kg mm^{-1}, decreasing with the progression of the growing season. Similar seasonal behavior was observed in tree WUE_P of the dominant DBH class based

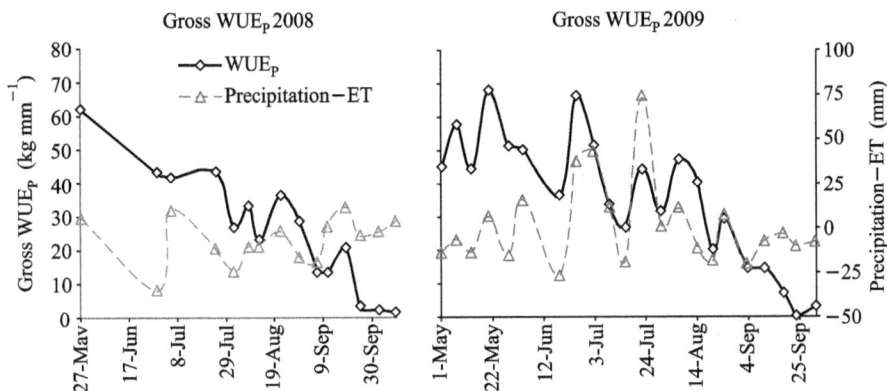

Fig. 7.2 Seasonal patterns of gross water use efficiency of productivity (WUE$_P$) of SRC 1 as the ratio of ET (mm) and allometrically derived aboveground woody biomass increments (kg of DMC per 1 ha) compared with the differences in total precipitation and actual ET (mm). All values were integrated, usually in weekly time steps, except the beginning of 2008, and curves were smoothed by a cubic spline fit.

on sap flow and stem increment measurements (Fig. 7.3). Lindroth *et al.* [51] described similar seasonal trends of WUE$_P$ in intensively managed (irrigated and fertilized) willow SRC (*Salix viminalis*) in Sweden, where the most marked fall of WUE$_P$ was linked to a reduction in leaf area index (LAI) at the end of the summer. Our results confirmed this link to LAI dynamics since the pronounced decline of gross WUE$_P$ was observed around mid-August, when LAI culminated at about 7 [19]. Further, Lindroth *et al.* [51] explained the maximum peak in WUE$_P$, defined as the ratio of measured ANPP increment (allometrically derived from shoot diameter) and modeled transpiration (physically based model KAUSHA tuned against BREB measurements), as an effect of rainy weather and thus with considerable amounts of evaporation of intercepted water and thus lower transpiration rates. However, in our research, which took into account the gross WUE$_P$ based on total ET and ANPP$_{wood}$, the maxima were also found during rain events. This is also in contrast with the results of Grelle *et al.* [94], who concluded that during precipitation and immediately after, lower WUE$_{GPP}$ is due to enhanced surface evaporation of water that was never part of plant metabolism.

The higher gross WUE$_P$ as a consequence of precipitation could be explained by a few reasons. Firstly, by using the long-term WUE based on the ANPP$_{wood}$ increment and not on CO_2 uptake or transpiration ratio, the relative C allocation between roots and aboveground biomass can play an important role [51, 65, 95–96]. During the period with reduced soil water availability, assimilated carbohydrates are directed away from shoots and toward root growth. After alleviation of such conditions by the replenishment of soil water status by rain

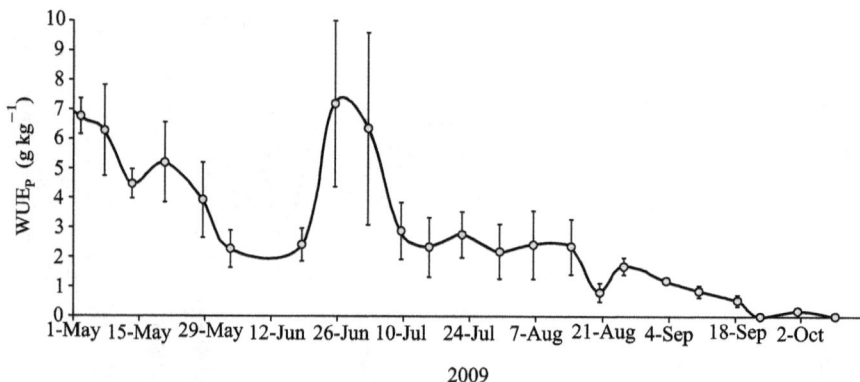

Fig. 7.3 Long-term water use efficiency of productivity (WUE$_P$) of three randomly selected trees from the dominant DBH class of SRC 1 (see Fig. 7.1). This tree WUE$_P$ is based on sap flow measurements (in kg of water) and simultaneous measurements of the DBH increment allometrically converted into aboveground woody biomass (in g of DMC). The sums of sap flow and biomass increments were usually integrated in a weekly time step (see the particular points) and the curve was smoothed by a cubic spline fit. The error bars indicate the standard deviation.

or irrigation, temporary C allocation to roots is compensated by a later increase in shoot growth [51, 97–99]. A similar effect caused by mobile C pools also influences the seasonal variation with high WUE$_P$ in the spring, characterized by the so-called "spring flush" (strong upward translocation of non-structural carbohydrates produced in assimilation during the end of the previous season) and, conversely, with the drop to zero at the end of the season linked with downward accumulation [65, 99–102]. The high accumulation of root reserves is very relevant for the management of SRC bioenergy systems. Secondly, growth is the biological phenomenon of an increase in size over time. Growth involves the formation, differentiation, and expansion of new cells, tissues, or organs. The sudden increase in tree diameter often observed after rain is not necessary due to growth but reflects the hydration of shrunk xylem and bark tissues after an extended dry period [103–104]. Finally, during dry periods when stem water deficits occur, new cells, which were recently created, do not immediately expand, but a release of low pressure conditions in the cambium suddenly enlarges the already existing cells to their mature size [98, 105–108]. This means that within a certain period of time and for a range of soil water deficits, growth is not inhibited but just delayed [108]. Within this context, the term WUE is disputable and using gross WUE$_P$ (either based on total ET or just on pure transpiration) seems to be a more reasonable metric of biomass production efficiency.

Further, the results of WUE$_{GPP}$ based on directly measured fluxes of CO_2 and ET by the eddy-covariance method across European forest ecosystems showed that WUE$_{GPP}$ increased with rising monthly precipitation and rising average monthly temperatures [109]. This might well be explained by a low vapor pressure deficit under warm but humid conditions leading to a higher CO_2 uptake to water loss ratio [109]. Finally, the higher productivity during these periods can result from an increase in diffuse radiation that might stimulate assimilation [110–111]. Analogous explanations can be forwarded for differences in sap flow (Fig. 7.3) versus BREB-based WUE (Fig. 7.2) if we consider that the dominant trees are more exposed to direct solar radiation and a higher vapor pressure deficit. This suggests that suppressed and co-dominant trees may have higher WUE$_P$ compared to dominant tree classes.

The gross WUE$_P$ was notably higher during 2009 as compared to 2008 (Tab. 7.6). A similar situation was also described by Lindroth *et al.* [51] where the authors attributed the contrast in the consecutive seasons to different-aged SRC and consequent changes in the root-to-shoot ratio. The decreasing root-to-shoot ratio with ontogenic aging is a well-known phenomenon in SRC as well as in plantations with other tree species [112–115] and could provide an interpretation of higher ANPP$_{wood}$ increment during 2009. However, the year 2009 was also abnormally wet (the annual precipitation sum was 778 mm compared to 520 mm in 2008) and the growing season in 2009 started approximately two weeks earlier due to the abnormally warm and dry April. The heterogeneity in C allocation during particular ontogenetic phases and also during particular parts of the season may cause difficulties in predicting yields with a simplified method based on ET and biomass relationships, but such information can still provide some general and gross estimation of SRC production.

Generally, the average WUE$_P$ of most coniferous and broadleaved trees of temperate zones ranges from 30 to 50 kg mm^{-1} of transpired water [53]. Research on WUE$_P$ of three different poplar clones (Beaupré, Trichobel, and Ghoy) growing in weighing lysimeters placed in a greenhouse indicated relatively constant values varying from 35 to 44 kg mm^{-1}, despite strongly fluctuating soil moisture during the season, leading to marked variation in root-to-shoot ratios [67]. However, Guidi *et al.* [22] demonstrated a strong influence of fertilization on WUE$_P$, where the effect of fertilizers increased WUE$_P$ from 4.3 to 21.4 kg mm^{-1}and from 6.8 to 24 kg mm^{-1} of evapotranspired water for willows and poplars, respectively, used as a vegetation filter. Relatively high values of mean seasonal WUE$_P$ of up to 41 and 55 kg mm^{-1} of transpired water for two consecutive years by fertilized and irrigated high-density willow stands, respectively, was reported by Lindroth *et al.* [51]. Even higher mean seasonal values of WUE$_P$ associated with high foliar N concentrations reaching up to 63 kg mm^{-1} of transpired water were described by Lindroth and Cienciala [35]. In our case, gross WUE$_P$ values ranged from 25.97 to 30.04 kg mm^{-1} under closed canopy conditions and from 11.33 to 13.23 kg mm^{-1} (open canopy) (Tab. 7.6). As total ET in mature closed canopy SRC is typically 30% higher than pure transpiration, the WUE of poplars is not only comparable but slightly higher than other broadleaved tree species of temperate climate zones [53].

An annual stemwood productivity of 10 to 12 Mg ha^{-1} is generally considered as the economic threshold [35]. According to the results presented in this study, such a yield will consume more than 450–500 mm per growing season. Therefore, it is assumed that a location with higher precipitation distributed across the growing season is a prerequisite for correct SRC site selection, especially in rain-fed areas such as the Bohemian-Moravian Highlands.

As demonstrated by Fischer [19], the ET of SRC at the experimental locality of Domanínek did not exceed the ET of the neighboring grasslands (unfertilized with one to two cuts per year) at annual or monthly levels during four years of investigations. At the same time, it was estimated that the aboveground biomass productivity of these grasslands was 3 to 4 t ha^{-1} yr^{-1} (dry mass). This suggests that the bioenergy systems based on *Populus* or *Salix* do not likely provide more competition for water than unmanaged native vegetation or other low-input systems in marginal agricultural areas and can provide a useful commodity such as bioenergy in areas that might not be suitable for agriculture.

7.6 Conclusions

We surveyed the literature on ANPP$_{wood}$, water use, and WUE and related physiological rates of several pure species and hybrids of the genera *Populus* and *Salix* as the most likely candidate tree species for bioenergy SRC cropping systems. Our analysis was restricted to operational field experiments. Although almost 150 studies were reviewed, we found several limitations of the resulting database. For one, there is a serious lack of experiments combing the inves-tigations at the leaf and stand levels. Since processes at these two scales can become decoupled or confounded, our ability to relate one level to the other remains limited. Another weakness of the database was the low coverage of var-ious environmental conditions together with a relatively narrow range of species investigated, despite the large genera of *Populus* and *Salix*. This limits our ability to take advantage of genetic variations in physiological traits that might be used to increase bioenergy WUE through tree improvement programs (*e.g.*, breeding). Therefore, future research involving SRC for bioenergy should test numerous genotypes, both pure species and their hybrids, across a wide range of soils and environmental conditions using an ecophysiological approach to quan-tify productivity and bioenergy WUEs. Such an approach will also allow for the development of process-based models that can be used to scale results to the ecosystem level and test hypotheses related to how the bioenergy productivity potential (and water use) will be affected by a changing climate. Finally, we presented a case study from the Bohemian-Moravian Highlands (Czech Republic), which provides an example of the kind of integrated research needed on water use and WUE from operational SRC based on *P. nigra* × *P. maximowiczii* grown under the rain-fed conditions of central Europe. This work has demonstrated that the water use of the *Populus* SRC in such conditions are not more water-demanding than reference grasslands while, at the same time, they maintain relatively high gross WUE, in the range of 11 to 30 kg per mm of water used by ET, which significantly increases with age and canopy closure.

References

[1] Aylott MJ, Casella E, Tubby I, Street NR, Smith P, Taylor G. Yield and spatial supply of bioenergy poplar and willow short-rotation coppice in the UK. New Phytologist 2008; 178: 358–370.

[2] Laureysens I, Bogaert J, Blust R, Ceulemans R. Biomass production of 17 poplar clones in a short-rotation coppice culture on a waste disposal site and its relation to soil characteristics. Forest Ecology and Management 2004; 187: 295–309.

[3] Hansen EA. Poplar woody biomass yields: a look to the future. Biomass and Bioenergy 191; 1: 1–7.

[4] Perry CH, Miller RC, Brooks KN. Impacts of short-rotation hybrid poplar plantations on regional water yield. Forest Ecology and Management 2001; 143: 143–151.

[5] Migliavacca M, Meroni M, Manca G, *et al*. Seasonal and interannual patterns of carbon and water fluxes of a poplar plantation under peculiar eco-climatic conditions. Agricultural and Forest Meteorology 2009; 149: 1460–1476.

[6] Deckmyn G, Laureysens I, Garcia J, Muys B, Ceulemans R. Poplar growth and yield in short rotation coppice: model simulations using the process model SECRETS. Biomass and Bionergy 2004; 26: 221–227.

[7] Al Afas N, Marron N, Van Dongen S, Laureysens I, Ceulemans R. Dynamics of biomass production in a poplar coppice culture over three rotations (11 years). Forest Ecology and Management 2008; 255: 1883–1891.

[8] Ericsson K, Roseqvist H, Nilsson, LJ. Energy crop production costs in the EU. Biomass and Bioenergy 2009; 33: 1577–1586.

[9] Njakou Djomo S, El Kasmioui U, Ceulemans R. Energy and greenhouse gas balance of bioenergy production from poplar and willow: a review. Global Change Biology Bioenergy 2011; 3: 181–197.

[10] Broeckx LS, Verlinden MS, Ceulemans R. Establishment and two-year growth of a bio-energy plantation with fast-growing *Populus* trees in Flanders (Belgium): effects of genotype and former land use. Biomass and Bioenergy 2012; 42: 151–163.

[11] International Energy Agency. IEA statistics-renewable information, IEA/OECD, Paris 2003.
[12] IPCC 2007. Climate change mitigation. In: Metz B, Davidson OR, Bosch PR, Dave R, Meyer LA. (Eds.) Contribution of Working Group 3 to the Fourth Assessment Report of the Intergovernmental Panel on Climate Change, Cambridge University Press, Cambridge, UK.
[13] De Vries BJM, Vvan Vuuren DP, Hoogwijk MM. Renewable energy sources: their global potential for the first-half of the 21st century at global level: an integrated approach. Energy Policy 2006; 35: 2590–2610.
[14] Gasol CM, Gabarrell X, Anton A, Rigola M, Carrasco J, Ciria P. Life Cycle, assessment of a *Brassica carinata* bioenergy system in southern Europe. Biomass and Bioenergy 2007; 31: 543–555.
[15] Isebrands JG, Karnonsky DF. Environmental benefits of poplar culture. In: Dickmann DI, Isebrands JG, Eckenwalder JE, Richardson J (Eds.). Poplar Culture in North America. Ottawa: National Research Council of Canada, 2001, 207–218.
[16] Scharlemann JPW, Laurance WF. How green are biofuels? Science 2008; 4: 43–44.
[17] Hall RL, Allen SJ, Rosier PTW, *et al.* Hydrological effects of short rotation energy coppice. Annual report 1993–1994. Wallingford. Institute of Hydrology and British Geological Survey. B/W5/00275/00/00, 1996; 89 pp.
[18] Trnka M, Trnka M, Fialová J, *et al.* Biomass production and survival rates of selected poplar clones grown under a short-rotation system on arable land. Plant Soil Environment 2008; 54: 78–88.
[19] Fischer M. Water balance of short rotation coppice. Mendel University in BrnoBrno, Ph.D thesis, 2012.
[20] Zalesny Jr RS, Wiese AH, Bauer EO, Riemenschneider DE. Sapflow of hybrid poplar (*Populus nigra* L. × *P. maximowiczii* A. Henry 'NM6') during phytoremediation of landfill leachate. Biomass and Bioenergy 2006; 30: 784–793.
[21] Mirk J, Volk TA. Seasonal sap flow of four *Salix* varieties growing on the Solvay wastebeds in Syracuse, NY, USA. Intrenational Journal of Phytoremediation 2010; 12: 1–23.
[22] Guidi W, Piccioni E, Bonari E. Evapotranspiration and crop coefficient of poplar and willow short-rotation coppice used as vegetation filter. Bioresource Technology 2008; 99: 4832–4840.
[23] Dimitriou I, Aronsson PA. Wastewater and sewage sludge application to willows and poplars grown in lysimeters—Plant response and treatment efficiency. Biomass and Bioenergy 2011; 35: 161–170.
[24] Beringer T, Lucht W, Schapoff S. Bioenergy production potential of global biomass plantations under environmental and agricultural constraints. GCB Bioenergy 2011; 3: 299–312.
[25] Tilman D, Fargione J, Wolff B, *et al.* Forecasting agriculturally driven global environmental change. Science 2001; 292: 281–284.
[26] Gerbens-Leenes W, Hoekstra AY, Van der Meer TH. The water footprint of bioenergy. Proceedings of the National Academy of Science 2009; 106: 10219–10223.

[27] Grip H, Halldin S, Lindroth A.Water use by intensively cultivated willow using estimated stomatal parameter values. Hydrological Processes 1989; 3: 51–63.

[28] Persson G.Comparison of simulated water balance for willow, spruce, grass ley and barley. Nordic Hydrology 1997; 28: 85–98.

[29] Hall RL, Allen SJ, Rosier PTW, Hopkins R. Transpiration from coppiced poplar and willow measured using sap-flow methods. Agricultural and Forest Meteorology 1998; 90: 275–290.

[30] Allen SJ, Hall RL, Rosier PTW. Transpiration by two poplar varieties grown as coppice for biomass production. Tree Physiology 1999; 9: 493–501.

[31] Lindroth A, Båth A. Assessment of regional willow coppice yield in Sweden on basis of water availability. Forest Ecology and Management 1999; 121: 57–65.

[32] Linderson ML, Iritz Z, Lindroth A. The effect of water availability on stand-level productivity, transpiration, water use efficiency and radiation use efficiency of field-grown willow clones. Biomass and Bioenergy 2007; 31: 460–468.

[33] Sevigne E, Gasola CM, Brunc F, *et al.* Water and energy consumption of *Populus* spp. bioenergy systems: a case study in Southern Europe. Renewable and Sustainable Energy Reviews 2011; 15: 1133–1140.

[34] King JS, Ceulemans R, Albaugh JM, *et al.* The challenge of ligno-cellulosic bioenergy in a water-limited world. BioScience 2013; 63: 102–117.

[35] Lindroth A, Cienciala E. Water use efficiency of short-rotation *Salix viminalis* at leaf, tree and stand scales. Tree Physiology 1996; 16: 257–262.

[36] Lasch P, Kollas C, Rock J, Suckow F. Potentials and impacts of short-rotation coppice plantation with aspen in Eastern Germany under conditions of climate change. Regional Environmental Change 2010; 10: 83–94.

[37] Petzold R, Schwärzel K, Feger KH.Transpiration of a hybrid poplar plantation in Saxony (Germany) in response to climate and soil conditions. European Journal of Forest Research 2010; 130: 695–706.

[38] Meiresonne L, Nadezhdina N, Cermak J, Van Slycken J, Ceulemans R. Measured sap flow and simulated transpiration from a poplar stand in Flanders (Belgium). Agricultural and Forest Meteorology 1999; 96: 165–179.

[39] Bungart R, Hütll RF. Growth dynamics and biomass accumulation of 8-year-old hybrid poplar clones in a short-rotation plantation on a clayey-sandy mining substrate with respect to plant nutrition and water budget. European Journal of Forest Research 2004; 23: 105–115.

[40] Tricker PJ, Trewin H, Kull O, Clarkson GJJ, Eensalu E, Tallis MJ, Colella A, Doncaster CP, Sabatti M, Taylor G. Stomatal conductance and not stomatal density determines the long-term reduction in leaf transpiration of poplar in elevated CO_2. Oecologia 2005; 143: 652–660.

[41] Ceulemans R, Impens I, Imler I. Stomatal conductance and stomatal behavior in *Populus* clones and hybrids. Canadian Journal of Botany 1988; 66: 1404–1414.

[42] Lindroth A, Iritz Z.Surface energy budget dynamics of short-rotation willow forest. Theoretical and Applied Climatology 1993; 47: 175–185.

[43] Trnka M, Olesen JE, Kersebaum KC, *et al.* Agroclimatic conditions in Europe under climate change. Global Change Biology 2011; 17: 2298–2318.

[44] Polster H. Die Physiologischen Grundlagen der Stofferzeugung im Walde. Bayericher Landwirtchaftsverlag Gmbh, Munchen, 1950; 96 pp.

[45] Bierhuizen JF, Slatyer RO. Effect of atmospheric concentration of water vapour and CO_2 in determining transpiration-photosynthesis relationships of cotton leaves. Agricultural Meteorology 1965; 2: 259–270.

[46] Holmgren P, Jarvis PG, Jarvis MS. Resistance to carbon dioxide and water vapour transfer in leaves of different plant species. Physiologia Plantarum 1965; 18: 557–573.

[47] Tanner CB, Sinclair TR. Efficient water use in crop production: Research or re-search? In: Taylor HM, Jordan WR, Sinclair TR. (Eds.). Limitations of Efficient Water Use in Crop Production. American Society of Agronomy, Madison, WI, 1983; 1–27.

[48] Zur B, Jones JW.Diurnal changes in the instantaneous water-use efficiency of soybean crop. Agriculture and Forest Meteorology 1984; 33: 41–51.

[49] Baldocchi DD, Verma SB, Anderson DE. Canopy photosynthesis and water use efficiency in a deciduous forest. Journal of Applied Ecology 1987; 24: 251–260.

[50] Denmead OT, Dunin FX, Wong SC, Greenwood EAN. Measuring water use efficiency of eucalypt trees with chambers and micrometeorological techniques. Journal of Hydrology 1993; 150: 649–664.

[51] Lindroth A, Verwijsta T, Halldin S. Water-use efficiency of willow: Variation with season, humidity and biomass allocation. Journal of hydrology 1994; 156: 1–19.

[52] Cienciala E, Černý M, Apltaure J, Exnerová Z. Biomass functions applicable to European beech. Journal of Forest Science 2005; 51: 147–154.

[53] Larcher W. Physiological Plant Ecology, 4th edition. Springer, Berlin, 2003.

[54] Ripullone F, Lauteri M, Grassi G, Amato M, Borghetti M. Variations in nitrogen supply changes wateruse efficiency of *Pseudotsuga menziesii* and *Populus × euroamericana*; a comparison of three approaches to determine water-use efficiency. Tree Physiology 2004; 24: 671–679.

[55] Monclus R, Dreyer E, Delmotte FM, *et al.* Impact of drought on productivity and water use efficiency in 29 genotypes *Populus deltoides × Populus nigra.* New Phytologist 2006; 169: 765–777.

[56] Fichot R, Chamaillard S, Depardieu C, Le Thiec D, Cochard H, Barigah TS, Brignolas F. Hydraulic efficiency and coordination with xylem resistance to cavitation, leaf function, and growth performance among eight unrelated *Populus deltoides × Populus nigra* hybrids. Journal of Experimental Botany 2011; 62: 2093–2106.

[57] Hellriegel H. Beiträge zu den naturwissenschaftlichen Grundlagen des Ackerbaus mit besonderer Berücksichtigung der agrikultur-chemischen Methode der Sandkultur. Friedrich Vieweg und Sohn, Braunschweig, 1883; 796 pp.

[58] Maximov NA. Physiologisch-ökologische Untersuchungen über die Dürreresistenz der Xerophyten. Jahrbücher für wissenschaftliche Botanik 1923; 62: 128–144.

[59] De Wit, CT.Transpiration and crop yields. Institute of Biological and Chemical Research on Field Crops and Herbage, No. 64.6, Wageningen, The Netherlands, 1958.

[60] Forrester DI, Theiveyanathan S, Collopy JJ, Marcar NE. Enhanced water use efficiency in a mixed *Eucalyptus globulus* and *Acacia mearnsii* plantation. Forest Ecology and Management 2010; 259: 1761–1770.

[61] Jones RJ, Mansfield TA. Effects of abscisic acid and its esters on stomatal aperture and the transpiration ratio. Physiologia Plantarum 1972; 26: 321–327.

[62] Masle J, Farquhar GD, Wogn SC. Transpiration ratio and plant mineral content are related among genotypes of a range of species. Australian Journal of Plant Physiology 1992; 19: 709–721.

[63] Law BE, Falge E, Gu L, *et al.* Environmental controls over carbon dioxide and water vapor exchange of terrestrial vegetation. Agricultural and Forest Meteorology 2002; 113: 97–120.

[64] Reichstein M, Tenhunen JD, Ourcival JM, *et al.* Severe drought effects on ecosystem CO_2 and H_2O fluxes at three Mediterranean evergreen sites: revision of current hypothesis? Global Change Biology 2002; 8: 999–1017.

[65] Dickmann DI, Isebrands JG, Terence JB, Kosola K, Kort J. Physiological ecology of poplars. In: Dickmann DI, Isebrands JG, Eckenwalder JE, Richardson J (Eds.). Poplar Culture in North America, Ottawa: National Research Council of Canada, 2001; 77–118.

[66] Fischer M, Trnka M, Kučera J, Fajman M, Zalud Z. Biomass productivity and water use relation in short rotation poplar coppice (*Populus nigra* × *P. Maximowiczii*) in the conditions of Czech Moravian Highlands. Acta Universities Agriculturae et Silivculturae Mendelianae Brunensis 2011; 59: 141–151.

[67] Souch CA, Stephens W. Growth, productivity and water use in three hybrid poplar clones. Tree Physiology 1998; 18: 829–835.

[68] Lewandowski I, Scurlock JMO, Lindvall E, Christou M. The development and current status of perennial rhizomatous grasses as energy crops in the US and Europe. Biomass and Bioenergy 2003; 25: 335–361.

[69] Larssson S. Genetic improvement of willow for short-rotation coppice. Biomass and Bioenergy 1998; 15: 23–26.

[70] Persson G, Lindroth A. Simulating evaporation from short-rotation forest: variations within and between seasons. Journal of Hydrology 1994; 156: 21–45.

[71] Hall RL, Allen SJ. Water use of poplar clones grown as short-rotation coppice at two sites in the UK. Annals of Applied Biology 1997; 49: 163–172.

[72] Hall JP. Sustainable production of forest biomass for energy. The Forestry Chronicle, May/June 2002, 78 (3): 1–6.

[73] Finch JW, Hall RL, Rosier PTW, *et al.* The hydrological impacts of energy crop production in the UK. Final report. London, UK, Department of Trade and Industry, 2004; 151pp. (CEH Project Number: C01937).

[74] Robinson KM, Karp A, Taylor G. Defining leaf traits linked to yield in short-rotation coppice *Salix*. Biomass and Bioenergy 2004; 26: 417–431.

[75] Hartwell S, Morino K, Nagler PL, Glenn EP. On the irrigation requirements of cottonwood (*Populus fremontii* and *Populus deltoides* var. *wislizenii*) and willow (*Salix gooddingii*) grown in a desert environment. Journal of Arid Environments 2010; 74: 667–674.

[76] Reich PB, Ellsworth DS, Walters MB, *et al.* Generality of leaf trait relationships: a test across six biomes. Ecology 1999; 80: 1955–1969.

[77] Busch DE, Smith SD. Mechanisms associated with decline of woody species in riparian ecosystems of the southwestern U.S. Ecological Monographs 1995; 65: 347–370.

[78] Farquhar GD, Ehleringer JR, Hubick KT.Carbon isotope discrimination and photosynthesis. Annual Review of Plant Physiology and Plant Molecular Biology 1989; 40: 503–537.

[79] Dillen SY, Monclus R, Barbaroux C, *et al.* Is the ranking of poplar genotypes for leaf carbon isotope discrimination stable across sites and years in two different full-sib families? Annals of Forest Science 2011; 68: 1265–1275.

[80] Jarvis PG, McNaughton KG. Stomatal control of transpiration: scaling up from leaf to region. Advances in Ecological Research 1986; 15: 1–49.

[81] Lindroth A. Aerodynamic and canopy resistance of short-rotation forest in relation to leaf area index and climate. Boundary-Layer Meteorology 1993; 66: 265–279.

[82] Singh B. Biomass production and nutrient dynamics in three clones of *Populus deltoides* planted on Indogangetic plains. Plant and Soil 1998; 203: 15–26.

[83] Khamzina A, Sommer R, Lamers JPA, Vlek PLG. Transpiration and early growth of tree plantations established on degraded cropland over shallow saline groundwater table in northwest Uzbekistan. Agricultural and Forest Meteorology 2009; 149: 1865–1874.

[84] Hou LG, Xiao HL, Si JH, Xiao SC, Zhou MX, Yang YG. Evapotranspiration and crop coefficient of *Populus euphratica* Oliv forest during the growing season in the extreme arid region northwest China. Agricultural Water Management 2010; 97: 351–356.

[85] Doody T, Benyon R. Quantifying water savings from willow removal in Australian streams. Journal of Environmental Management 2011; 92: 926–935.

[86] Savage MJ. Field evaluation of polymer capacitive humidity sensors for Bowen ratio energy balance flux measurements. Sensors 2010; 10: 7748–7771.

[87] Stannard DI. A theoretically based determination of Bowen-ratio fetches requirements. Boundary-Layer Meteorology 1997; 83: 375–406.

[88] Cellier P, Brunet Y. Flux-gradient relationships above tall plant canopies. Agricultural and Forest Meteorology 1992; 58: 93–117.

[89] Iritz Z, Lindroth A. Energy partitioning in relation to leaf area development of short-rotation willow coppice. Agricultural and Forest Meteorology 1996; 81: 119–130.

[90] Čermák J, Kučera J, Nadezhdina N. Sap flow measurements with some thermodynamic methods, flow integration within trees and scaling up from sample trees to entire forest stands. Trees 2004; 18: 529–546.

[91] King JS, Albaugh TJ, Allen HL, Kress LW. Stand-level allometry in Pinus taeda as affected by irrigation and fertilization. Tree Physiology 1999; 19: 769–778.

[92] Fajman M, Palát M, Sedlák P. Estimation of the yield of poplars in plantations of fast-growing species within current results. Acta Universitatis Agriculture et Silviculturae Mendelianae Brunensis 2009; 2: 25–36.

[93] Osawa A, Kurachi N. Spatial leaf distribution and self-thinning exponent of Pinus banksiana and Populus tremuloides. Trees 2004; 18: 327–338.

[94] Grelle A, Lundberg A, Lindroth A, Moren AS, Cienciala E. Evaporation components of a boreal forest: variations during the growing season. Journal of Hydrology 1997; 197: 70–87.

[95] Höll W. Seasonal fluctuation of reserve materials in the trunkwood of spruce *Picea abies* (L.) Karst. Journal of Plant Physiology 1985; 117: 355–362.

[96] Hoch G, Richter A, Korner C. Non-structural carbon compounds in temperate forest trees. Plant, Cell and Environment 2003; 26: 1067–1081.

[97] Hsiao TC, Acevedo E. Plant responses to water deficits, water-use efficiency, and drought resistance. Agricultural Meteorology 1974; 14: 59–84.

[98] Kramer PJ. Water relations of plants. Academic Press, Orlando, Florida, USA 1983.

[99] Barbaroux C, Breda N, Dufrene E. Distribution of aboveground and below-ground carbohydrate reserves in adult trees of two contrasting broad-leaved species (*Quercus petraea* and *Fagus sylvatica*). New Phytologist 2003; 157: 605–615.

[100] Teskey RO, Hinckley TM. Influence of temperature and water potential on root growth of white oak. Physiologia Plantarum 1981; 52: 363–369.

[101] Deans JD, Ford ED. Seasonal patterns of radial root growth and starch dynamics in plantation-grown *Sitka spruce* trees of different ages. Tree Physiology 1986; 1: 241–251.

[102] Lacointe A, Deleens E, Ameglio T, *et al.* Testing the branch autonomy theory: a ^{13}C/^{14}C double labelling experiment on differentially shaded branches. Plant, Cell and Environment 2004; 27: 1159–1168.

[103] Herzog KM, Häsler R, Thum R. Diurnal changes in the radius of a subalpine Norway spruce stem: their relation to the sap flow and their use to estimate transpiration. Trees 1995; 10: 94–101.

[104] Offenthaler I, Hietz P, Richter H. Wood diameter indicates diurnal and long-term patterns of xylem water potential in Norway spruce. Trees 2001; 15: 215–221.

[105] Hinckley TM, Lassoie JP. Radial growth in conifers and deciduous trees: a comparison. Mitteilungen der forstlichen Bundesversuchsanstalt Wien 1981; 142: 17–56.

[106] Barbaroux C, Breda N. Contrasting distribution and seasonal dynamics of carbo-hydrate reserves in stem wood of adult ring-porous sessile oak and diffuse-porous beech trees. Tree Physiology 2002; 22: 1201–1210.

[107] Steppe K, De Pauw DJW, Lemeur R, Vanrolleghem PA. A mathematical model linking tree sap flow dynamics to daily stem diameter fluctuations and radial stem growth. Tree Physiology 2006; 26: 257–273.

[108] Zweifel R, Zimmermann L, Zeugin F, Newbery DM. Intra-annual radial growth and water relations of trees—implications towards a growth mechanism. Journal of Experimental Botany 2006; 57: 1445–1459.

[109] Kuglitsch FG, Reichstein M, Beer C, *et al.* Characterisation of ecosystem water-use efficiency of European forests from eddy covariance measurements. Biogeo-sciences Discussion 2008; 5: 4481–4519.

[110] Alton PB, North PR, Los SO. The impact of diffuse sunlight on canopy light-use efficiency, gross photosynthetic product and net ecosystem exchange in three forest biomes. Global Change Biology 2007; 13: 776–787.

[111] Knohl A, Baldocchi DD. Effects of diffuse radiation on canopy gas exchange processes in a forest ecosystem. Journal of Geophysical Research 2008; 113, G0 2023, 17 pp.

[112] Ovington JD. Dry-matter production by *Pinus sylvestris* L. Annals of Botany 1957; 21: 287–314.

[113] Reynolds HL, D'Aantonio C. The ecological significance of plasticity in root weight ratio in response to nitrogen: opinion. Plant and Soil 1996; 185: 75–97.

[114] Coleman MD, Friend AL, Kern CC. Carbon allocation and nitrogen acquisition in a developing Populus deltoids plantation. Tree Physiology 2004; 24: 1347–1357.

[115] King JS, Giardina CP, Pregizer KS, Friend AL. Biomass partitioning in red pine (*Pinus resinosa*) along a chronosequence in the Upper Peninsula of Michigan. Canadian Journal of Forest Research 2007; 37: 93–102.

[116] Aasamaa K, Heinsoo K, Holm B. Biomass production, water use and photosynthesis of *Salix clones* grown in a wastewater purification system. Biomass and Bioenergy 2010; 34: 897–905.

[117] Aasamaa K, Sober A, Hartung W, Niinemets U. Rate of stomatal opening, shoot hydraulic conductance and photosynthetic characteristics in relation to leaf abscisic acid concentration in six temperate deciduous trees. Tree Physiology 2002; 22: 267–276.

[118] Aasamaa K, Sober A. Hydraulic conductance and stomatal sensitivity to changes of leaf water status in six deciduous tree species. Biologia Plantarum 2001; 44: 65–73.

[119] Alstad KP, Welker JM, Williams SA, Trlica MJ. Carbon and water relations of *Salix monticola* in response to winter browsing and changes in surface water hydrology: an isotopic study using $\delta^{13}C$ and $\delta^{18}O$. Oecologia 1999; 120: 375–385.

[120] Armstrong A, Johns C, Tubby I. Effects of spacing and cutting cycle on the yield of poplar grown as an energy crop. Biomass and Bioenergy 1999; 17: 305–314.

[121] Aronsson P, Dahlin T, Dimitriou I. Treatment of landfill leachate by irrigation of willow coppice—Plant response and treatment efficiency. Environmental Pollution 2010; 158: 795–804.

[122] Begley D, McCracken AR, Dawson WM, Watson S. Interaction in short rotation coppice willow, *Salix viminalis* genotype mixtures. Biomass and Bioenergy 2009; 33: 163–173.

[123] Bergante S, Facciotto G, Minotta G. Identification of the main site factors and management intensity affecting the establishment of short-rotation-coppices (SRC) in northern Italy through stepwise regression analysis. Central European Journal of Biology 2010; 5: 522–530.

[124] Bernacchi CJ, Calfapietra C, Davey PA, *et al*. Photosynthesis and stomatal conductance responses of poplars to free-air CO_2 enrichment (PopFACE) during the first growth cycle and immediately following coppice. New Phytologist 2003; 159: 609–621.

[125] Black TA, Den Hartog G, Neumann HH, *et al*. Annual cycles of water vapour and carbon dioxide fluxes in and above a boreal aspen forest. Global Change Biology1996; 2: 219–229.

[126] Boehmel C, Lewandowski I, Claupein W. Comparing annual and perennial energy cropping systems with different management intensities. Agricultural Systems 2008; 96: 224–236.

[127] Bonhomme L, Barbaroux C, Monclus R, *et al.* Genetic variation in productivity, leaf traits and carbon isotope discrimination in hybrid poplars cultivated on contrasting sites. Annals of Forest Science 2008; 65: 503.

[128] Bowman WD, Conant RT. Shoot growth dynamics and photosynthetic response to increased nitrogen availability in the alpine willow *Salix glauca.* Oecologia 1994; 97: 93–99.

[129] Brooks JR, Flanagan LB, Buchmann N, Ehleringer JR. Carbon isotope composition of boreal plants: functional grouping of life forms. Oecologia 1997; 110: 301–311.

[130] Bullard MJ, Mustill SJ, McMillan SD, Nixon PMI, Carver P, Britt CP. Yield improvements through modification of planting density and harvest frequency in short rotation coppice *Salix* spp. — 1. Yield response in two morphologically diverse varieties. Biomass and Bioenergy 2002; 22: 15–25.

[131] Cai T, Price DT, Orchansky AL, Thomas BR. Carbon, water, and energy exchanges of a hybrid poplar plantation during the first five years following planting. Ecosystems 2011; 14: 658–671.

[132] Calfapietra C, Gielen B, Galema ANJ, *et al.* Free-air CO_2 enrichment (FACE) enhances biomass production in a short-rotation poplar plantation. Tree Physiology 132; 23: 805–814.

[133] Calfapietra C, Tulva I, Eensalu E, *et al.* Canopy profiles of photosynthetic parameters under elevated CO_2 and N fertilization in a poplar plantation. Environmental Pollution 2005; 137: 525–535.

[134] Cao SK, Feng Q, Si JH, Su YH, Chang ZQ, Xi HY. Relationships between foliar carbon isotope discrimination with potassium concentration and ash content of the riparian plants in the extreme arid region of China. Photosynthetica 2009; 47: 499–509.

[135] Chen Y, Chen Y, Xu C, Li W. Photosynthesis and water use efficiency of *Populus euphratica* in response to changing groundwater depth and CO_2 concentration. Environmental Earth Sciences 2011; 62: 119–125.

[136] Cienciala E, Lindroth A. Gas-exchange and sap-flow measurements of *Salix viminalis* trees in short-rotation forest. II: Diurnal and seasonal variations of stomatal response and water use efficiency. Trees 1995; 9: 295–301.

[137] Cochard H, Casella E, Mencuccini M. Xylem vulnerability to cavitation varies among poplar and willow clones and correlates with yield. Tree Physiology 2007; 27: 1761–1767.

[138] De Lillis M, Matteucci G, Valentini R. Carbon assimilation, nitrogen, and photochemical efficiency of different Hymalayan tree species along an altitudinal gradient. Photosynthetica 2004; 42: 597–605.

[139] DeBell DS, Clendenen GW, Harrington CA, Zasada JC. Tree growth and stand development in short-rotation *Populus* plantings: 7-year results for two clones at three spacings. Biomass and Bioenergy 1996; 11: 253–269.

[140] Deckmyn G, Laureysens I, Garcia J, Muys B, Ceulemans R. Poplar growth and yield in short rotation coppice: model simulations using the process model SECRETS. Biomass and Bioenergy 2004; 26: 221–227.

[141] DesRochers A, Vvan Dden Driessche R, Thomas BR. NPK fertilization at planting of three hybrid poplar clones in the boreal region of Alberta. Forest Ecology and Management 2006; 232: 216–225.

[142] Dillen SY, Marron N, Koch B, Ceulemans R. Genetic variation of stomatal traits and carbon isotope discrimination in two hybrid poplar families (*Populus deltoides* 'S9-2' × *P. nigra* 'Ghoy' and *P. deltoides* 'S9-2' × *P. trichocarpa* 'V24'). Annals of Botany 2008; 102: 399–407.

[143] Donovan LA, Ehleringer JR. Ecophysiological differences among juvenile and reproductive plants of several woody species. Oecologia 1991; 86: 594–597.

[144] Fang S, Xu X, Lu S, Tang L. Growth dynamics and biomass production in short-rotation poplar plantations: 6-year results for three clones at four spacings. Biomass and Bioenergy 1999; 17: 415–425.

[145] Fichot R, Barigah ST, Chamaillard S, Le Thiec D, Laurans F, Cochard H, Brignolas F. Common trade-offs between xylem resistance to cavitation and other physiological traits do not hold among unrelated *Populus deltoides* × *Populus nigra* hybrids. Plant, Cell and Environment 2010; 33: 1553–1568.

[146] Fortier J, Gagnon D, Truax B, Lambert F. Biomass and volume yield after 6 years in multiclonal hybrid poplar riparian buffer strips. Biomass and Bioenergy 2010; 34: 1028–1040.

[147] French C, Dickinson NM, Putwain PD. Woody biomass phytoremediation of contaminated brownfield land. Environmental Pollution 2006; 141: 387–395.

[148] Funk JL, Jones CG, Lerdau MT. Leaf- and shoot-level plasticity in response to different nutrient and water availabilities. Tree Physiology 2007; 27: 1731–1739.

[149] Gazal RM, Scott RL, Goodrich DC, Williams DG. Controls on transpiration in a semiarid riparian cottonwood forest. Agricultural and Forest Meteorology 2006; 137: 56–67.

[150] Giampietro M, Ulgiati S, Pimentel D. Feasibility of large-scale biofuel production. BioScience 197; 47: 587–600.

[151] Gong JR, Zhang XS, Huang YM. Comparison of the performance of several hybrid poplar clones and their potential suitability for use in Northern China. Biomass and Bioenergy 2011; 35: 2755–2764.

[152] Gornall JL, Guy RD. Geographic variation in ecophysiological traits of black cottonwood (*Populus trichocarpa*). Canadian Journal of Botany 2007; 85: 1202–1213.

[153] Gruenewald H, Brandt BKV, Schneider BU, Bens O, Kendzia G, Hüttl RF. Agroforestry systems for the production of woody biomass for energy transformation purposes. Ecological Engineering 2007; 29: 319–328.

[154] Guo XY, Zhang XS. Performance of 14 hybrid poplar clones grown in Beijing, China. Biomass and Bioenergy 2010; 34: 906–911.

[155] Hofmann-Schielle C, Jug A, Makeschin F, Rehfuess KE. Short-rotation plantations of balsam poplars, aspen and willows on former arable land in the Federal Republic of Germany. I. Site-growth relationships. Forest Ecology and Management 1999; 121: 41–55.

[156] Horton JL, Kolb TE, Hart SC. Leaf gas exchange characteristics differ among Sonoran Desert riparian tree species. Tree Physiology 2001; 21: 233–241.

[157] Iritz Z, Tourula T, Lindroth A, Heikinheimo M. Simulation of willow short-rotation forest evaporation using a modified Shuttleworth-Wallace approach. Hydrological Processes 2001; 15: 97–113.

[158] Isebrands JG, Eckenwalder JE, Richardson J. Poplar Culture in North America. Ottawa: National Research Council of Canada Press, 2001; 207–218.

[159] Johansson T, Karacic A. Increment and biomass in hybrid poplar and some practical implications. Biomass and Bioenergy 2011; 35: 1925–1934.

[160] Kopp RF, Abrahamson LP, White EH, Burns KF, Nowak CA. Cutting cycle and spacing effects on biomass production by a willow clone in New York. Biomass and Bioenergy 1997; 12: 313–319.

[161] Labrecque M, Teodorescu TI. Influence of plantation site and wastewater sludge fertilization on the performance and foliar nutrient status of two willow species grown under SRIC in southern Quebec (Canada). Forest Ecology and Management 2001; 150: 223–239.

[162] Labrecque M, Teodorescu TI. High biomass yield achieved by *Salix* clones in SRIC following two 3-year coppice rotations on abandoned farmland in southern Quebec, Canada. Biomass and Bioenergy 2003; 25: 135–146.

[163] Labrecque M, Teodorescu TI. Field performance and biomass production of 12 willow and poplar clones in short-rotation coppice in southern Quebec (Canada). Biomass and Bioenergy 2005; 29: 1–9.

[164] Laureysens I, Pellis A, Willems J, Ceulemans R. Growth and production of a short rotation coppice culture of poplar. III. Second rotation results. Biomass and Bioenergy 2005; 29: 10–21.

[165] Leffler AJ, Evans AS. Physiological variation among Populus fremontii populations: short- and long-term relationships between $\delta^{13}C$ and water availability. Tree Physiology 2001; 21: 1149–1155.

[166] Letts MG, Phelan CA, Johnson DRE, Rood SB. Seasonal photosynthetic gas exchange and leaf reflectance characteristics of male and female cottonwoods in a riparian woodland. Tree Physiology 2008; 28: 1037–1048.

[167] Liberloo M, Calfapietra C, Lukac M, et al. Woody biomass production during the second rotation of a bio-energy Populus plantation increases in a future high CO_2 world. Global Change Biology 2006; 12: 1094–1106.

[168] Liberloo M, Dillen SY, Calfapietra C, et al. Elevated CO_2 concentration, fertilization and their interaction: growth stimulation in a short-rotation poplar coppice (EUROFACE). Tree Physiology 2005; 25: 179–189.

[169] Liberloo M, Gielen B, Calfapietra C, et al. Growth of a poplar short rotation coppice under elevated atmospheric CO_2 concentrations (EUROFACE) depends on fertilization and species. Annals of Forest Science 2004; 61: 299–307.

[170] Liberloo M, Tulva I, Raïm O, Kull O, Ceulemans R. Photosynthetic stimulation under long-term CO_2 enrichment and fertilization is sustained across a closed Populus canopy profile (EUROFACE). New Phytologist 2007; 173: 537–549.

[171] Lindegaard KN, Parfitt RI, Donaldson G, Hunter T. Comparative trials of elite Swedish and UK biomass willow varieties. Aspects of Applied Biology 2001; 65: 183–192.

[172] Liu MZ, Jiang GM, Li YG, et al. Gas exchange, photochemical efficiency, and leaf water potential in three *Salix* species. Photosynthetica 2003; 41: 393–398.

[173] Lowthe-Thomas SC, Slater FM, Randerson PF. Reducing the establishment costs of short rotation willow coppice (SRC)—A trial of a novel layflat planting system at an upland site in mid-Wales. Biomass and Bioenergy 2010; 34: 677–686.

[174] Manzanera JA, Martinez-Chacon MF. Ecophysiological competence of Populus alba L., Fraxinus angustifolia Vahl., and Crataegus monogyna Jacq. used in plantations for the recovery of riparian vegetation. Environmental Management 2007; 40: 902–912.

[175] McCracken AR, Dawson WM, Bowden G. Yield responses of willow (*Salix*) grown in mixtures in short rotation coppice (SRC). Biomass and Bioenergy 2001; 21: 311–319.

[176] McCracken AR, Walsh L, Moore PJ, *et al.* Yield of willow (*Salix* spp.) grown in short rotation coppice mixtures in a long-term trial. Annals of Applied Biology 2011; 159: 229–243.

[177] Meinzer FC, Hinckley TM, Ceulemans R. Apparent responses of stomata to transpiration and humidity in a hybrid poplar canopy. Plant, Cell and Environment 1997; 20: 1301–1308.

[178] Merilo E, Heinsoo K, Kull O, Söderbergh I, Lundmark T, Koppel A. Leaf photosynthetic properties in a willow (*Salix viminalis* and *Salix dasyclados*) plantation in response to fertilization. European Journal of Forest Research 2006; 125: 93–100.

[179] Mirck J, Volk T. Seasonal sap flow of four *Salix* varieties growing on the solvay wastebeds in Syracuse, NY, USA. International Journal of Phytoremediation 2010; 12:1–23.

[180] Monclus R, Villar M, Barbaroux C, *et al.* Fichot R, Delmotte FM, Delay D, Petit JM, Bréchet C, Dreyer E, Brignolas F. Productivity, water-use efficiency and tolerance to moderate water deficit correlate in 33 poplar genotypes from a Populus deltoides × Populus trichocarpa F1 progeny. Tree Physiology 2009; 29: 1329–1339.

[181] Nagler P, Jetton A, Fleming J, *et al.* Evapotranspiration in a cottonwood (Populus fremontii) restoration plantation estimated by sap flow and remote sensing methods. Agricultural and Forest Meteorology 2007; 144: 95–110.

[182] Nassi O Di Nasso N, Guidi W, Ragaglini G, Tozzini C, Bonari E. Biomass production and energy balance of a 12-year-old short-rotation coppice poplar stand under different cutting cycles. Global Change Biology Bioenergy 2010; 2: 89–97.

[183] Niu S, Jiang G, Wan S, Li Y, Gao L, Liu M. A sand-fixing pioneer C3 species in sandland displays characteristics of C4 metabolism. Environmental and Experimental Botany 2006; 57: 123–130.

[184] Niu SL, Jiang GM, Li YG, *et al.* Comparison of photosynthetic traits between two typical shrubs: legume and non-legume in Hunshandak Sandland. Photosynthetica 2003; 41: 111–116.

[185] Paris P, Mareschi L, Sabatti M, *et al.* Comparing hybrid Populus clones for SRF across northern Italy after two biennial rotations: survival, growth and yield. Biomass and Bioenergy 2011; 35: 1524–1532.

[186] Persson G, Jansson PE. Simulated water balance of a willow stand on clay soil. In: Perttu KL., Kowalik PJ (Eds.): Energy Forestry Modelling—Growth, Water Relations and Economics. Simulation Monographs. 30. Pudoc, Wageningen, 1989; 147–162.

[187] Persson G. Willow stand evapotranspiration simulated for Swedish soils. Agricultural water management 1995; 28: 271–293.

[188] Proe MF, Craig J, Griffiths J, Wilson A, Reid E. Comparison of biomass production in coppice and single stem woodland management systems on an imperfectly drained gley soil in central Scotland. Biomass and Bioenergy 1999; 17: 141–151.

[189] Proe MF, Griffiths JH, Craig J. Effects of spacing, species and coppicing on leaf area, light interception and photosynthesis in short rotation forestry. Biomass and Bioenergy 2002; 23: 315–326.

[190] Quaye AK, Volk TA, Hafner S, Leopold DJ, Schirmer C. Impacts of paper sludge and manure on soil and biomass production of willow. Biomass and Bioenergy 2011; 35: 2796–2806.

[191] Rae AM, Robinson KM, Street NR, Taylor G. Morphological and physiological traits influencing biomass productivity in short-rotation coppice poplar. Canadian Journal of Forest Research 2004; 34: 1488–1498.

[192] Rowland DL, Beals L, Chaudhry AA, Evans AS, Grodeska LS. Physiological, morphological, and environmental variation among geographically isolated cottonwood (Populus deltoides) populations in New Mexico. Western North American Naturalist 2001; 61: 452–462.

[193] Schreiber SG, Hacke UG, Hamann A, Thomas BR. Genetic variation of hydraulic and wood anatomical traits in hybrid poplar and trembling aspen. New Phytologist 2011;190: 150–160.

[194] Sims REH, Maiava TG, Bullock BT. Short rotation coppice tree species selection for woody biomass production in New Zealand. Biomass and Bioenergy 2001; 20: 329–335.

[195] Sixto H, Salvia J, Barrio M, Ciria MP, Cañellas I. Genetic variation and genotype-environment interactions in short rotation Populus plantations in southern Europe. New Forests 2011; 42: 163–177.

[196] Souch CA, Martin PJ, Stephens W, Spoor G. Effects of soil compaction and mechanical damage at harvest on growth and biomass production of short rotation coppice willow. Plant and Soil 2004; 263: 173–182.

[197] Sparks JP, Ehleringer JR. Leaf carbon isotope discrimination and nitrogen content for riparian trees along elevational transects. Oecologia 1997; 109: 362–367.

[198] Stephens W, Hess T, Knox J. Review of the effects of energy crops on hydrology. Institute of Water and Environment, Cranfield University, Bedford, 2001; 59 pp.

[199] Stolarski M, Szczukowski S, Tworkowski J, Klasa A. Productivity of seven clones of willow coppice in annual and quadrennial cutting cycles. Biomass and Bioenergy 2008; 32: 1227–1234.

[200] Su H, Li Y, Lan Z, et al. Leaf-level plasticity of *Salix gordejevii* in fixed dunes compared with lowlands in Hunshandake Sandland, North China. Journal of Plant Research 2009; 122: 611–622.

[201] Sullivan PF, Welker JM. Variation in leaf physiology of *Salix arctica* within and across ecosystems in the High Arctic: test of a dual isotope (δ^{13}C and δ^{18}O) conceptual model. Oecologia 2007; 151: 372–386.

[202] Szczukowski S, Stolarski M, Tworkowski J, Przyborowski, Klasa A. Productivity of willow coppice plants grown in short rotations. Plant Soil and Environment 2005; 51: 423–430.

[203] Tanaka-Oda A, Kenzo T, Koretsune S, Sasaki H, Fukuda K. Ontogenic changes in water-use efficiency (δ^{13}C) and leaf traits differ among tree species growing in a semiarid region of the Loess Plateau, China. Forest Ecology and Management 2010; 259: 953–957.

[204] Telenius BF. Stand growth of deciduous pioneer tree species on fertile agricultural land in southern Sweden. Biomass and Bioenergy 1999; 16: 13–23.

[205] Thomas BR, Macdonald SE, Dancik BP. Variance components, heritabilities and gain estimates for growth chamber and field performance of Populus tremuloides: gas exchange parameters. Silvae Genetica 1997; 46: 309–317.

[206] Tozawa M, Ueno N, Seiwa K.Compensatory mechanisms for reproductive costs in the dioecious tree *Salix integra*. Botany 2009; 87: 315–323.

[207] Vande Walle I, Van Camp N, Van de Casteele L, Verheyen K, Lemeur R. Short-rotation forestry of birch, maple, poplar and willow in Flanders (Belgium). I — Biomass production after 4 years of tree growth. Biomass and Bioenergy 2007; 31: 267–275.

[208] Vose JM, Wayne L, Swank T, Harvey GJ, Clinton BD, Sobek C. Leaf water relations and sapflow in eastern cottonwood (*Populus deltoides* Bartr.) trees planted for phytoremediation of a groundwater pollutant. International Journal of Phytoremediation 2000; 2: 53–73.

[209] Wang HL, Yang SD, Zhang CL. The photosynthetic characteristics of differently shaped leaves in Populus euphratica Olivier. Photosynthetica 1997; 34: 545–553.

[210] Wilkinson JM, Evans EJ, Bilsborrow PE, Wright C, Hewison WO, Pilbeam DJ. Yield of willow cultivars at different planting densities in a commercial short rotation coppice in the north of England. Biomass and Bioenergy 2007; 31: 469–474.

[211] Wilske B, Lu N, Wei L, *et al.* Poplar plantation has the potential to alter the water balance in semiarid Inner Mongolia. Journal of Environmental Management 2009; 90: 2762–2770.

[212] Zabek LM, Prescott CE. Biomass equations and carbon content of aboveground leafless biomass of hybrid poplar in Coastal British Columbia. Forest Ecology and Management 2006; 223: 291–302.

[213] Zhang H, Morison JIL, Simmonds LP. Transpiration and water relations of poplar trees growing close to the water table. Tree Physiology 1999; 19: 563–573.

Part IV

Biofuels and Natural Resource Management

Chapter 8

Afforestation of Salt-affected Marginal Lands with Indigenous Tree Species for Sustainable Biomass and Bioenergy Production

Yashpal Singh, Gurbachan Singh, and Dinesh K. Sharma

8.1 Introduction

The unrelenting shortage of fuel wood and petroleum reserves in different parts of the world and the impact of environmental pollution necessitated the afforestation on marginally productive salt-affected soils. About 831×10^6 ha of land, which is about 20% of the world's irrigated land, is salt-affected in the world. Salinization of arable land will result in 30% to 50% land loss in the next 25 years until the year 2050 if remedial measures are not taken. Increasing population pressure on land and a consequent reduction in the land/man ratio poses a serious threat. The current gap between demand and the supply of food, fuel, fodder, and timber is likely to worsen in the near future as a consequence of the continuing degradation of lands and reduced per capita land availability. Scarcities of fuel wood in third world countries have initiated several programs for developing renewable energy sources. In India, availability of fuel wood is nearly four times less than its demand in household sector of rural areas and small-scale industries in urban areas (brick kiln, bakery, pottery, soap industries, etc.). Until the recent past, there was no exclusive program to raise fuel wood output, which was derived as an incidental or a byproduct from timber or industrial wood harvesting. Since the 1980s, social forestry projects were initiated to take care of the fuel and forage needs of society; yet, the allocation of degraded wastelands for this purpose could not generate a sufficient quantity to accomplish the emerging needs of the ever-growing population. Continuous illicit extraction of fuel wood from the natural forests could not be restricted on account of such a high population pressure. As a consequence, only 19.3% of the land surface has been left under forest cover leading to decreased and erratic rainfall patterns throughout the country. Most of the natural disaster and environmental catastrophes have been developed in this way. Recurrent floods and droughts, global warming and greenhouse effects, desertification, soil erosion, and river silting are some of the major consequences of the large-scale deforestation of tropical forests [1]. It is, therefore, imperative to create new forest resources on salt-affected wastelands.

Indo-Gangetic plains lying between 21°55' to 32°39' N and 73°45' to 88°25' E comprise the states of Punjab, Haryana, Uttar Pradesh, and part of Bihar (north), West Bengal (south), and Rajasthan (north), having 2.7 m ha salt-affected soils constitute the country's old barren sodic soils without any land use system [2]. It has an ample scope for afforestation to generate fuel wood and also to have an ideal quality of environment in this biographic region of the country. These soils have been regarded as unfit for agriculture on account of a high concentration of soluble salts capable of producing alkaline hydrolysis products such as Na_2CO_3, $NaHCO_3$, and sufficient exchangeable sodium to impart poor soil physical conditions to the soil. The presence of $CaCO_3$ concretions at various depths (caliche bed) causes physical impedance for root proliferation, therefore making it difficult for tree establishment. Dedicated efforts were made to rehabilitate these inhospitable soils under tree cover over the past four decades [3–4]. Some indigenous species established on these soils suffer with stunted growth and poor yield [5–6]. With the scarcity of fuel wood in many developing countries, various programs of short rotation forestry were launched in the past two decades to meet this basic need of rural communities [7–9]. In India, sodic lands were generally allocated to the poor and landless peoples under a poverty alleviation program for rehabilitation and simultaneous improvement of the area [5–6, 10]. Despite their slow growth and low productivity, afforestation on salt-affected soils produces a good amount of biomass and has reclaimed the soils significantly [11–12]. The salt-affected soils are poor in fertility and it is not ascertained whether nutrients removed from the soil during fuel wood production would be replenished naturally or would require fertilization. Very little information on biomass as well as bioenergy production and nutrient concentration of tree species growing in an environment of soil sodicity are available. An attempt is made, here, to collate the available information on the afforestation of salt-affected soils for biomass and bioenergy production.

8.2 Origin and Distribution of Salt-affected Soils in India

Salt-affected soils are commonly found in the Indo-Gangetic plains of Uttar Pradesh, Punjab, Haryana, Rajasthan, Bihar, and West Bengal. There are various regions associated with the formation of salt-affected soils that are both natural and anthropogenic. The geological deposition of clay minerals comprises quartz, feldspars (orthoclase and plagioclase), muscovite, biotite, chloritised biotite, tourmaline, zircon, and hornblende in their sand fractions [13–14]. Quartz and feldspars occur distinctly in the salt fraction. However, illite, mixed layer minerals, vermiculite, and chloride are common to both the silt and clay fractions. The mixed-layer minerals of vermiculites and smectite in these soils originate from biotite mica. Different workers have reported variable estimates of salt-affected soils in India. According to the latest estimation in India, salt-affected soils occupy about 6.73 million ha of land, which is 2.1% of the geographical area of the country [2]. Out of 584 districts in the country, 194 have salt-affected soils. Out of the total 6.73 million ha of land, 2.96 million ha are saline and the remaining 3.77 million ha are sodic. Out of the total 2.35 million ha salt-affected soils in the Indo-Gangetic plains, 0.56 million ha are saline and 1.79 million ha are sodic (Tab. 8.1).

Tab. 8.1 State-wise extent of salt-affected soils in India (million ha).

State	Saline	Sodic	Total
Andhra Pradesh	0.78	1.97	2.75
Andaman & Nicobar	0.08	0.00	0.08
Bihar	0.47	1.06	1.53
Gujarat	1.68	0.54	2.22
Haryana	0.49	1.83	2.32
Karnataka	0.02	1.48	1.50
Kerala	0.20	0.00	0.20
Madhya Pradesh	0.00	1.40	1.40
Maharashtra	1.84	4.23	6.07
Orissa	1.47	0.00	1.47
Punjab	0.00	0.15	0.15
Rajasthan	0.20	0.18	0.38
Tamil Nadu	0.01	0.35	0.36
Uttar Pradesh	0.22	1.35	1.57
West Bengal	0.44	0.00	0.44
Total	2.96	3.77	6.73

Source: [2].

There are three distinct categories of salt-affected soils from a soil characterization point of view. However, from the point of view of clay mineralogy, micaceous/illite and smectite are the two classes into which the salt-affected soils can be broadly classified. From the point of view of soil characterization, the salt-affected soils are categorized in three distinct categories, *i.e.*, alkali or sodic, saline, and acid sulphate saline. The distribution of salt-affected soils in different geo-climatic regions of India is given in Table 8.2.

Tab. 8.2 Distribution of salt-affected soils in different geo-climatic regions of India.

S. No.	Main characteristics	Rainfall (mm yr^{-1})	Distribution
1.	Alkali soils of Indo-Gangetic alluvial plain, developed on less calcareous alluvium		
	High pH, EC, ESP and preponderance of sodium bicarbonate and carbonates	600–1,000	Parts of Punjab, Haryana, UP, South Bihar, Palwama and Badgam districts of Kashmir, Jammu region and Rajasthan
2.	Alkali soil of Indo-Gangetic alluvial plain developed on fine, highly calcareous alluvium		
	High pH, EC, ESP and preponderance of sodium bicarbonate and carbonates	1,000–1,400	North Bihar and parts of Western UP
3.	Inland saline soils of arid and semi-arid regions		
	Neutral to alkaline pH, high EC and preponderance of chlorides and sulphates	< 500	Part of Punjab, Haryana, UP, Rajasthan, Gujarat, Leh district of Jammu and Kashmir
4.	Inland saline soils of sub-humid regions		
	Neutral to alkaline pH, high EC and preponderance of chlorides and sulphates	1,000–1,400	North Bihar
5.	Inland salt-affected deep black soils (vertisols)		
	Neutral to highly alkaline pH, variable EC and preponderance of chlorides and sulphate with miner amounts of sodium carbonate and smectitic mineralogy	700–1,000	Parts of Madhya Pradesh, Maharashtra, Rajasthan, Andhra Pradesh, Gujarat, Karnataka and Tamil Nadu

(Continued)

S. No.	Main characteristics	Rainfall (mm yr^{-1})	Distribution
6. Medium to deep black soils of deltaic and costal semi-arid regions			
	Neutral to highly alkaline pH, high EC, preponderance of chlorides and sulphates with or without sodium bicarbonate smectitic mineralogy	700–900	Saurashtra region of Gujarat, and deltas of Godavari, Krishna and Cauveri river in Andhra Pradesh and Tamil Nadu
7. Saline micaceous, deltaic alluvium of humid regions			
	Neutral to slightly acid pH, high pH and preponderance of chlorides	1,400–1,600	Sunderban delta in West Bengal and parts of Mahanadi delta in Orissa
8. Saline humic and acid sulphate soils of humid tropical region			
	Acid pH, high EC, presence of humus (Organic) horizon and preponderance of chlorides and sulphates, sulpher and pyritic material	200–3,000	Malabar coast of Kerala and parts of Sunderban delta in West Bengal
9. Saline marsh of the Rann of Kachh			
	Neutral to slightly alkaline pH, high EC and preponderance of chlorides and sulphates	< 300	Rann of Kachh of Gujarat

Source: [15].

8.3 Properties of Salt-affected Soils

Salt-affected soils differ from arable soils with respect to two important properties, i.e., the soluble salt and the soil reaction. A buildup of soluble salts in the soil may influence its behavior for crop production through changes in the proportions of exchangeable cations, soil reaction, physical properties, and the effects of osmotic and specific ion toxicity. Salt-related properties of soils are subject to rapid change. Salt-affected soils in India are broadly placed into two broad groups: (i) sodic (alkali) soils and (ii) saline soils. There are certain specific situations where saline-sodic soils also exist. Since the management of saline-sodic soils will be more similar to that of the sodic soils, they are generally grouped into the sodic soil category. The salt-affected soils are broadly grouped as saline and alkaline/sodic soils [16–17].

Saline soils with white salt encrustation on the surface have predominantly chlorides and sulphates of Na, Ca, and Mg. The soils with neutral soluble salts have a saturation paste pH <8.2. The electrical conductivity of the saturation extract (ECe) of saline soils is generally more than 4 dS m^{-1} at 25°C. Such soils invariably have a sodium adsorption ration (SAR) of the soil solution >15. In saline soils, the excess of neutral salts restricts normal plant growth. The soil properties of a typical saline soil profile are given in Table 8.3.

Tab. 8.3 Soil characteristics of a typical saline soil of the Indo-Gangetic plain region.

Soil depth (cm)	Silt (%)	Clay (%)	Bulk density (g cm^{-3})	pH	ECe (dS m^{-1})	Cation content (mmol)			Anion content		SAR
						Na$^+$	Ca $^{2+}$	Mg $^{2+}$	Cl$^-$	HCO$_3^-$	
0–30	15.5	11.2	1.69	7.2	36.4	85.5	141.0	66.5	495	1.15	6.0
30–60	13.3	17.3	1.55	7.2	23.5	64.0	84.5	38.5	300	1.30	5.8
60–120	12.8	19.6	1.51	7.1	21.4	72.0	56.5	33.5	233	1.15	8.0

Source: [18].

Sodic soils suffer from varying levels of degradation in structural, chemical, nutritional, hydrological, and biological properties. These soils are compact and heavy with a high bulk density and silty clay loam texture (Typic Natrustalf). They also have a higher proportion of sodium in relation to other cations in soil solution and on the exchange complex. The sodic soils of the Indo-Gangetic plain are generally gypsum-free (CaSO$_4$, 2H$_2$O) but are calcareous, with CaCO$_3$ increasing with depth, which is present in an amorphous form, in a concretionary form, or even as an indurate bed at about 1 m of depth. A high pH (>10) and high exchangeable sodium percent (ESP) of more than 60 imbalances the ionic equilibrium of the soil solution, which leads to abnormal nutrient physiology. The high ratio of Na:Ca and low ratio of C:N cannot sustain the vegetation. The growth of most crops on sodic soils is adversely affected because of the impairment of physical conditions, disorder in nutrient availability, and suppression of biological activity due to high pH. Deficiency of some micronutrients (Zn, Fe, Cu, Mn) and toxicity of other elements (Na, B, Mo) further aggravate the situation for stressed growth of whatever plants exist on such land. Poor water permeability (hydraulic conductivity and infiltration rate) due to interlocked pore space as well as compactness impedes the root development of plants. In water-logging conditions, root respiration is inhibited under oxygen stress. A wide range of microbial populations and diversity do not exist in sodic soils due to hostile conditions, which retards the rate of litter decomposition and nutrient mineralization leading to poor nutrient availability in the growing plants. The soil properties of a typical sodic soil profile are given in Table 8.4.

Tab. 8.4 Physico-chemical properties of a typical sodic soil profile of the Indo-Gangetic plain region.

Soil parameter	Soil depth (cm)				
	0–11	11–29	29–88	88–118	118–148
pH	10.70	10.80	9.33	9.75	9.85
ECe (dS m^{-1})	10.67	10.25	1.20	1.17	2.56
Ca+Mg (meq L^{-1})	1.60	2.10	2.10	1.60	1.60
Na (meq L^{-1})	141.00	138.00	9.40	9.60	24.50
K (meq L^{-1})	0.11	0.10	0.06	0.04	0.03
CO$_3$ (meq L^{-1})	118.00	110.00	4.00	4.00	6.00
HCO$_3$ (meq L^{-1})	24.00	21.00	6.00	6.50	15.00
Cl (meq L^{-1})	7.00	7.00	2.50	2.00	4.00
SO$_4$ (meq L^{-1})	0.00	2.00	0.00	0.00	0.00
CaCO$_3$ <2.0 mm	0.40	0.50	0.70	3.30	10.80
CaCO$_3$ >2.0 mm	0.00	0.00	0.00	60.0	30.00
Sand (%)	62.80	53.00	45.80	46.50	53.70
Silt (%)	19.50	25.50	26.00	20.00	19.90
Clay (%)	17.70	21.50	28.20	32.70	26.40
Textural class	1	sil	sicl	cl	cl
CEC (cmol kg^{-1})	9.00	10.80	12.80	14.00	12.50
O.M (%)	0.17	0.15	0.12	0.12	0.12
ESP	73.30	86.10	54.60	64.20	69.60
Ca (me 100g^{-1})	0.80	0.60	3.60	2.20	2.00
Mg (me 100g^{-1})	0.60	0.30	2.00	1.60	1.20
Na (me 100g^{-1})	6.60	9.30	7.00	9.00	8.70
K (me 100g^{-1})	0.50	0.20	0.90	0.80	0.50

Source: [19].

8.4   Natural Vegetation on Salt-affected Soils

On the basis of the fidelity class, certain selective indicator species of trees such as *Prosopis juliflora*, *Acacia nilotica*, *Clerodandrum phlomidis*, *Prosopis cineraria*, and *Mimosa hamata*, climbers such as *Asparagus racemosus*, *Coculus pendulus*, *Momordica dioica*, *Mukia maderaspatana*, and *Cyranthes aspera*, and grasses such as *Sporobolus maderaspatanus*, *Sporobolus marginatus*, *Desmostachya bip-*

*innata, Sacharum spontanium, Aristida abbscendens, Calotropis procera, Puli-
caria crispa, Eragrostis tennella,* and *Fimbristylis dchotroma* are exclusively
found in barren sodic soils and do not occur in semi-reclaimed nor non-sodic
soil sites. Some other species like *Cyperus triceps, Sporobolus diander, Dacty-
loctenium aegypticum,* etc. are found selectively in sodic lands but occasionally
in semi-reclaimed soils. *Paspalun vaginatum* and few species of *Cyperus* and
Panicum are also found in such soils, although they are restricted to the rainy
season only.

8.5 Management Practices for Afforestation on Salt-affected Soils

Management of both saline and sodic soils for afforestation differs because of
different physical and chemical characteristics of the both soils. Management of
salt-affected soil requires effective management practices, developing appropriate
techniques for planting, and the selection of the most suitable plant species for
the particular environment. To create a favorable environment for proper tree
establishment in a salt-affected environment, diagnosis and knowledge of the
magnitude of salt-affected soils is a precondition. Management practices for
creating a favorable root environment and suitability of tree species varied for
both saline and sodic soils.

8.5.1 Selection of Tree Species

The initial establishment including germination and initial growth of tree seedlings
in saline and sodic environments is a difficult task for researchers. The selection
of suitable tree species for high biomass and bioenergy production in salt-affected
soils depends upon the tolerance of the species to salinity and sodicity, suitability to local agro-climate, and purpose of plantation. Several studies have been
conducted to evaluate the performances of a large number of tree species in
saline and sodic conditions in India. Initial trials on the afforestation of sodic
soils, as early as 1931, were conducted by the Uttar Pradesh Forest Department,
but no significant results were obtained until the 1960s, with the exception of
some success on the establishment of *Prosopis juliflora* in such soils. Yadav [4]
suggested several afforestation techniques and stressed that species like *Prosopis
juliflora, Eucalyptus tereticornis,* and *Acacia nilotica* can grow better on sodic
soils. Some preliminary studies have been done to select salt tolerant species

through the pot culture experiments; in which six tree species, *i.e.*, *Casurina equisetifolia*, *Eucalyptus tereticornis*, *Acacia nilotica*, *Dalbergia sissoo*, *Pongamia pinnata*, and *Araucaria cunninghammii* were evaluated [20]. All of these species failed to grow above 61.4 ESP. However, the successful growth was observed at 30.6 ESP for *Acacia nilotica*, *Eucalyptus tereticornis*, and *Casurina equisetifolia* and at 15.2 ESP for *Dalbergia sissoo*, *Pongamia pinnata*, and *Araucaria cunninghammii*. Seedlings of several tree species tested in pot culture by maintaining different salinity levels revealed that *Acacia nilotica* and *Eucalyptus camaldulensis* could grow with 50% growth reduction on 5 dS m^{-1} salinity and *Acacia nilotica* at a relatively high salinity level. *Casurina equisetifolia* showed a moderate salt tolerance. Singh and Yadav [21] and Yadav and Singh [22] reported a 50% reduction in the growth of *Acacia nilotica* and *Eucalyptus camaldulensis* at 5.0 dS m^{-1} salinity in clay soil, but they grew satisfactorily at ECe 10.0 dS m^{-1} in sandy soil. However, *Acacia auriculiformis* could not survive beyond ECe 2.5 dS m^{-1}. Similarly, Gupta *et al.* [23] observed a significant reduction in dry plant weight at ECe 2.5 dS m^{-1} in *Leucaena leaucocephala* and *Pallophorma pterocarpum*, at ECe 5.0 dS m^{-1} in *Eucalyptus tereticornis* and *Albizzia lebbek*, and at ECe 7.0 dS m^{-1} in *Acacia indica*. Bandhopadhyay *et al.* [24] reported that *Casurina equisetifolia* did not germinate and showed a reduction in the growth of seedlings at ECe 8.0 dS m^{-1}. *Dalbergia sissoo*, with its threshold of ECe of 2.2 dS m^{-1} and slope of 8.9 per unit increase in ECe above 2.2 dS m^{-1}, was moderately sensitive during the establishment stage [25].

In a field trial of the Biomass Research Centre at Banthra, Lucknow, *Prosopis juliflora*, *Acacia nilotica*, and *Terminalia arjuna* were found to have promising growth on sodic soils [5]. *Casurina equisetifolia* and *Acacia nilotica* could grow on sodic soils with ESP 30.6, whereas *Pongamia pinnata* and *Delbergia sissoo* survived only up to ESP 15.2 [21–22]. Based on the performance of tree saplings planted in soils of different pH (7–12), the relative tolerance followed the order: *Prosopis juliflora*> *Acacia nilotica*> *Haplophragma adenophyllum*> *Albizzia lebbek* > *Syzygium cumini* [26]. 30 forest tree species were evaluated at high sodicity (pH>10.0). After seven years of planting, only 13 out of 30 species survived. Out of these 13 surviving species, only *Prosopis juliflora*, *Tamarix articulate*, and *Acacia nilotica* were found to be suitable for such soils. *Eucalyptus tereticornis* showed good survival and height but no meaningful biomass was observed. However, *Dalbergia sissoo*, *Pithecellobium dulce*, *Terminalia arjuna*, *Kigali pinnata*, *Parkinson aculeate*, and *Cordial Rothay* showed more than 70% survival but could not attain economically suitable biomass [27]. Singh *et al.* [28] evaluated the performance of ten tree species in sodic soils having ESP 89. After ten years of field studies, only three species, *Prosopis juliflora*, *Acacia nilotica*, and *Casuarina equisetifolia* recorded survival rates of >90% and attained economical biomass. *Eucalyptus tereticornis* showed good performance during the initial four years, but its growth rate declined thereafter. *Azadirachta indica*,

Melia azadirach, and *Dalbergia sissoo* were poor performers. On the basis of the available information, a short list of consistently better-performing species that could be recommended for saline and alkali soils of Indo-Gangetic plains are given in Table 8.5.

Tab. 8.5 Recommended tree species for the afforestation on salt-affected soils.

Soil parameter	Firewood/timber/fruit species (common name)
Alkali soils (pH$_2$ to 1.2 m)	
> 10.0	*Acacia nilotica* (Kikar), *Butea Monosperma* (dhak), *Casuarina equisetifolia* (Casurina, saru), *Prosopis juliflora* (mesquite, pahari kikar), *Prosopis cinerraria* (khejri, jand)
9.0–10.0	*Albizzia lebbeck* (siris), *Cassia siamea* (cassia), *Eucalyptus teriticornis* (mysore gum, safeda), *Tamarix articulata* (faransh), *Terminalia arjuna* (arjun)
8.6–9.0	*Azardirachta indica (neem)*, *Dalbergia sissoo* (shisham, tahli), *Grevillia robusta (silver oak)*, *Hardwickea binnata (anjan)*, *Kajellea pinnata* (balam khira), *Morus alba* (mulberry, shehtoot), *Moringa olifera* (sonjna), *Mangifera indica* (mango), *Pyris communis* (pear, nashpati), *Populus delteoides (poplar)*, *Tectona grandis* (teak, saguan), *Syzium cumuni* (jamun)
Saline and waterlogged soils (ECe (dS m^{-1}) below 0.3 m)	
20–30	*Acacia farnesiana* (pissi babul), *Prosopis juliflora* (mesquite, pahari kikar), *Parkinsonia aculeate* (Jerusalem thorn, parkinsonia), *Tamarix aphylla* (faransh)
14–20	*Acacia nilotica* (desi kikar), *A. pennatula* (kikar), *A. tortilis* (Israeli Kikar), *Callistemon lanceolatus* (bottle brush), *Casuarina glauca* (casuarinas, saru), *C. obese*, *C. equisetifolia*, *Eucalyptus camaldulensis* (river-red gum, safeda), *Ferronia limonia* (kainth, kabit), *Leucaena leucophala* (subabul), *Ziziphus jujube*(ber)
10–14	*Casuarina canninghamiana* (casuarinas, saru), *Eucalyptus teriticornis* (mysore gum, safeda), *Terminalia arjuna* (arjun)
5–10	*Albizia caribaea* (albizia, tantacayo), *Dalbergia sissoo* (shisham), *Gauzuma ulmifolia* (guacima), *Pongamia Pinnata* (papri), *Samanea saman* (rain tree)
< 5	*Acacia auriculiformis* (Australian kikar, akash mono), *A. deamii* (zarza), *A. catechu* (khair), *Syzygium cumini* (jamun), *Salix* spp. *(willow, salix)*, *Tamarindus indica* (imli)

Source: [29–30].

8.5.2   Pre-planting Management Strategies

Afforestation on salt-affected soils is a challenging task because salt-affected soils are poor in organic matter and fertility and contain the presence of a calcium carbonate layer within one meter depth of the soil profile. Land preparation for plantation in sodic soils is slightly different from that in saline soils. In addition to organic matter, sodic soils are deficient in nitrogen (N), available zinc, and calcium. Therefore, application of N fertilizers and zinc sulphates is considered essential at the time of transplanting for better establishment of the saplings. However, the application of N fertilizers may be avoided for N-fixing trees. Application of N at 100 g urea per plant in split doses at the onset of the monsoon season aids better plant growth [31].

The root systems of plants take two to six weeks to establish on planted sites, depending on the soil conditions. The monsoon season is generally considered the best time for all types of plantations. Suitable drainage must be maintained if a plantation is out of the rainy season. Six- to twelve-month-old saplings grown in normal soils are suitable for plantation on salt-affected soils. Most of the earlier attempts failed due to the lack of suitable technology, proper interest and care, and scarcity of funds [3–4]. To provide a more congenial environment around sapling roots and faster establishment and growth of saplings, adequate nutrition and a proper soil-filling mixture are required. A proper filling mixture is also necessary to minimize the effect of high pH and ESP. Gypsum ($CaSO_4$, $2H_2O$) is mostly used as a chemical amendment for mixing with the excavated salt-affected soils. The quantity of gypsum required will depend on the amount of soil to be amended and the pH of the excavated soil. An eight-year study conducted to evaluate the effects of different filling mixtures along with site preparation methods showed that mixing of 3 to 6 kg of gypsum with the original soils significantly improved the survival of eucalyptus and acacia trees [32]. Singh et al. [33] also reported that 3 kg gypsum + 8 kg FYM (farmyard manure) per auger hole of 45 cm diameter at the surface, 20 cm diameter at the bottom, and 120 cm deep are optimal for the survival and growth of Prosopis juliflora (Tab. 8.6).

Plant-to-plant and row-to-row spacing are governed by the growth habits of the planted tree species and the purpose for which trees are being grown. In salty soils, it is desirable to plant with relatively closer spacing, e.g., 2 m × 2 m, 2 m × 3 m, or 3 m × 3 m, to ensure good tree stand against mortality due to salinity or sodicity stresses. Excess canopies can be pruned and the pruned biomass will add an additional fuel biomass and opportunities to retain vigorous plants. Singh et al. [33–34] did not observe any effect from spacing (2, 3, and 4 m) on the tree height of Prosopis juliflora. However, the pruned material was increased under closer spacing. Pruning is done to produce well-shaped and

Tab. 8.6 Effects of different filling mixtures on the performances of eucalyptus and acacia.

Filling mixture (kg)		Augerhole/ pit Size (cm × cm)	Eucalyptus			Acacia		
Gypsum	FYM		SP (%)	Height (cm)	DSH (cm)	SP (%)	Height (cm)	DSH (cm)
Experiment I								
Nil	Nil(OS)	15 × 120	0			38	74	
3	–	15 × 120	50	142	6.8	94	227	6.8
6	–	15 × 120	69	199	8	100	247	8.2
3	8	15 × 120	100	292	11.1	100	257	10.1
3	8 (sand)	15 × 120	100	341	13.4	100	295	11.3
Experiment II								
4	3	10 × 120	100	35	12.9	100	309	12.2
6	3	10 × 180	100	424	15.8	100	318	14.6
8	3	15 × 120	100	349	13.2	100	307	14.3
12	3	15 × 180	100	466	17.9	100	334	15.6
24	12	90 × 90	100	385	15.4	100	348	16.6

Source: [32].

clear bolls and, in the thorny tree species, to provide access to plantations for growing crops under such trees. The pruned branches also provide for firewood, and sometimes foliage can be used for fodder. Singh *et al.* [35] reported that the mean plant height of *Prosopis juliflora* was significantly increased in the pruned treatments grown in high-alkali soils.

8.5.3 Planting Techniques

Tree growth is adversely affected in saline soils due to the reduced water availability caused by the excess of salts along with periodic water-logging and poor aeration, especially during the monsoon season. For the successful establishment of tree species in high-saline soils, appropriate techniques are needed to improve soil conditions. To provide better aeration and avoid excessive salinization, planting on high ridges or mounds was often considered beneficial for establishing tree plantations [36]. This method was compared with the subsurface planting method. Substantially higher salts accumulated in the ridges that resulted in poor survival and sapling growth (Tab. 8.7). Difficulties in conserving rainwater on the ridge top and sides were other disadvantages observed

Tab. 8.7 Effects of planting methods on tree growth in saline soils (ECe 35–45 dS m^{-1}).

Tree species	Sub-surface			Surface		
	Height (m)	DSH (cm)	PS (%)	Height (cm)	DSH (cm)	PS (%)
After 9 years of planting						
Acacia nilotica	6.41	44.6	50			0
Acacia tortilis	5.31	34.3	56	3.11	10.8	25
Leucaena leucocephala	6.91	36.7	50			0
Prosopis juliflora	8.06	55.9	100	6.40	42.5	100
	Sub surface			SPFIM*		
After 27 months of planting						
Acacia auriculiformis	1.43	13	2.42			65
Acacia nilotica	3.21	69	2.89			95
Casuarina equisetifolia	2.13	46	3.0			95
Eucalyptus camaldulensis	2.24	50	3.78			95
Terminalia arjuna	1.83	81	2.00			90

*Subsurface planting and furrow irrigation method; Source: [37].

with ridge planting. The performance of trees was better when planted with the subsurface method but the additional need for spot irrigation was the main problem. The method was then improved by planting in the sole of furrow (60 cm wide and 20 cm deep), which was subsequently used for the irrigation of the tree saplings. Besides the uniform application of irrigation water and the reduction in application costs, the subsurface planting and the furrow irrigation method helped to create a low-salinity zone below the sills of the furrows. Creation of such niches favored the establishment of young seedlings from the trees [37].

The tree growth in sodic soils is constrained due to the inability of their roots to proliferate through the hard kankar (calcite) pan existing usually at depths below 50–75 cm from the surface. Therefore, even the earlier afforestation attempts resorted to the replacement of excavated sodic soils (50 cm deep pits) with normal soils [38] to improve upon their drainage by digging holes (90–150 cm deep) and refilling the holes with a filling mixture of good soils, farmyard manure, and gypsum before planting tree saplings [39]. This technique was introduced in 1895 and named as the "deep thala system" or "panchali system" of plantation. Keeping in view the expenditures involved in the replacement of the soil, it was suggested in the early 1960's that the soil conditions must be ameliorated *in situ* by the appropriate amendments [40], but Pande [41] again suggested the plantation of trees in pits (90 cm depth and 90 cm diameter) by replacing the original soils with normal soils from elsewhere. Later, it was concluded that the addition of gypsum (50%GR (gypsum requirement)) and FYM at 25 kg per pit (90 cm × 90 cm) was comparable to the replacement of the original sodic soil (pH 10.0) with normal soil with the growth of sapling and their survival [3, 42–43]. In the experiments conducted in Cherat, Aligarh, and Kusheri near Unnao in Uttar Pradesh, Ghosh [44] observed that sapling survival and their growth were better when planted in pits (120 cm × 120 cm), where the lower half of the pit soil was replaced by normal soil and the upper half was amended with vermiculite, gypsum, and FYM. The pit planting technique suffers from the disadvantages of high requirements of amendments, laborious pit-digging operation involving more earth work, and impedance to roots through the calcic horizon (hard pan). Keeping these limitations in view, the planting technique has been improved with the 'auger hole technique' at the Central Soil Salinity Research Institute in Karnal, India [34, 45–48]. Here, the auger holes with 100–140 cm deep and 20–25 cm diameter were dug with a tractor-operated auger and sapling were planted after suitably amending the dug-out soils. The performance of trees planted with this method, as opposed to the routine pit or trench methods, has been quite satisfactory in field trials (Tab. 8.8). This method has succeeded very well to the piercing of hard kankar layer. The advantages of this technique include the encouraging and training of deeper rooting. Thus, the trees are able to probe deeper soil layers for water and nutrients to sustain their growth.

Tab. 8.8 Comparative performance of trees planted with pit, trench and auger whole techniques in highly alkali soils.

Dimension		Eucalyptus tereticornis (8 years)		Acacia nilotica (8 years)		Prosopis juliflora (6 years)	
Depth (cm)	Width (cm)	Height (m)	DSH* (cm)	Height (m)	DSH (cm)	Height (m)	DSH (cm)
Pit							
90	90	9.2	17.3	7.5	19.6		
30	30					7.0	8.4
Trench							
30	30					6.9	7.9
Augerhole							
120	10	8.2	14.2	6.7	14.6		
120	10	8.9	16.2	7.7	16.2		
90	15					7.7	9.6
180	15	8.4	14.8	6.8	16.6		
180	15	9.1	16.6	7.4	17.6		

*DSH: Diameter at stump height; Source: [32, 34].

8.5.4   Post-planting Management Strategies

Salt-affected soils exist mostly in arid, semi-arid, and hot sub-humid regions of India where 70%–80% of the total rainfall is received during July to September. Generally, transplanting of trees is completed during the monsoon season, but these saplings may suffer for want of water during the post-monsoon periods, especially the summers. The Central Soil Salinity Research Institute of Karnal, Haryana, India has been conducting several long-term experiments on irrigation management for the proper establishment of saplings in salt-affected soils. For successful plantation in saline soils, utilization of rainwater to the maximum possible extent and keeping the salt concentration in the active root zone at minimal level are important for minimizing the adverse effects of the high salinity of the soil. Similarly, in sodic soils, provisions of supplemental irrigation during the early establishment period are very essential. A marked response to irrigation in terms of survival and biomass yield of *Prosopis juliflora* was recorded in sodic soils [49]. In low rainfall areas (30–35 cm per annum), survival and biomass yields were significantly higher and plantations continued to respond to irrigation up to four to five years of planting. In experiments conducted in India from 1985 to 1988, Singh *et al.* [33] found that during the first two years, the growth of *Prosopis* was far better when irrigated compared to the plantation, which depended on rainfall alone. Within two years of planting, 36% of rain-fed *Prosopis* died, while the mortality rate was only 9% among the irrigated plants. The water use efficiency (WUE) was also higher under irrigated conditions. Irrigation brought very little change in the chemical compositions of different plant parts but significantly decreased root zone soil sodicity. The application of irrigation during first two years of planting is absolutely necessary. However, after two years, irrigation may be withdrawn as plant roots can meet their requirements from groundwater.

Sometimes, there is saline water in the underlying soil in salt-affected areas. Afforestation under such conditions can only succeed if poor-quality water is utilized suitably for irrigating tree saplings. Research efforts in this direction have to be made to develop the surface planting and furrow irrigation method (SPFIM) system, which, in addition to optimizing the water regimes of the rooting zone, also helps to better control salinity. This system not only saves irrigation time and labor but also leads to the addition of less salt to the soil profile [50]. Weeding and hoeing around the planted saplings remain quite useful for a successful attempt, which may be carried out at least three to four times in a year. Excess water should be removed out of the field as soon as possible.

8.6 Biomass Production

8.6.1 Saline Soils

The ultimate aim of any afforestation program is to get maximum biomass per unit of time. Like suitable planting techniques, saline soils require proper selection of trees for high biomass production. More than 40 native and exotic tree species of arid and semi-arid areas were evaluated at Research Farm Sampla by Tomar *et al.* [18]. Based upon periodical observations for the survival, height, and girth of experimental plants, woody species like *Acacia farnesiana, Parkinsonia aculeate, Prosopis juliflora,* and *Tamarix articulate* have been rated as the most tolerant to salinity and could be grown satisfactorily on soils with salinity levels up to 50 dS m^{-1} in their root transmission zones. Tree species like *Acacia nilotica, Acacia tortilis, Casuarina gluca, Casuarina abesa,* and *Casuarina equisetifolia* could be grown on sites with ECe varying from 10–25 dS m^{-1}. The performance of some important tree species after nine years of growth has been compared when these were grown with different methods of plantation. The data on biomass indicated that *Prosopis juliflora* and *Casuarina gluca* cv. *13987* were the highest (98 and 89 Mg ha^{-1}) followed by *Acacia nilotica* (52–67 Mg ha^{-1}) and *Acacia tortilis* (41 Mg ha^{-1}) when planted with the subsurface or furrow techniques (Tab. 8.9).

Tab. 8.9 Biomass estimation of trees harvested from the experimental site.

Species	Method of planting	Range of soil salinity at 0–120 cm depth EC (dS m^{-1})	Estimated biomass (Mg ha^{-1})
Acacia nilotica	Subsurface Furrow	10.6–25.3	52
		11.1–21.0	67
Acacia tortilis	Subsurface Ridge	6.8–28.1	41
		19.7–29.1	6
Eucalyptus camaldulensis	Furrow	10.0–17.9	28
Prosopis juliflora	Subsurface Ridge	10.3–24.0	98
		23.5–57.5	65
Casuarina equisetifolia	Furrow	5.6–20.7	28
Casuarina gluaca 13987	Furrow	6.5–33.9	96
Casuarina obesa 27	Furrow	9.0–19.5	38
Leucaena leucocephala	Subsurface	6.9–23.9	30
Tamarix sp.	Furrow	8.2–21.3	12

Source: [18].

8.6.2   Sodic Soils

Estimates on biomass production for plantations of 8–10 years in sodic soils followed the order: *Prosopis juliflora*>*Acacia nilotica*>*Eucalyptus tereticornis* [51]. If the plantations on sodic lands are carefully attended, 60%–80% survival of different species has been noticed [4]. The growth and productivity was highest in *Prosopis juliflora* followed by *Acacia nilotica* and *Terminalia arjuna*. The growth and productivity of *Eucalyptus tereticornis* has also been found to be excellent at Banthra, Lucknow, India [52]. In high-density energy plantations, *Prosopis juliflora* produced about 69 Mg ha^{-1} of biomass in eight years [5]. Under the same conditions, *Populus deltoids* could produce 49 Mg ha^{-1} total biomass [53]. *Terminalia arjuna* in two sodic soil sites produced about 114 and 280 Mg ha^{-1} total biomass at 12 and 15 years, respectively, whereas a mixed forest of several species consisted of 342 Mg ha^{-1} of total biomass (Tab. 8.10).

Tab. 8.10   The biomass productivity of some tree species on sodic soil in Banthra, Lucknow, India.

Species	Age (yr)	Population density (number ha^{-1})	Mean girth* (cm)	Basal area (m^2 ha^{-1})	Total biomass (Mg ha^{-1})
Leucaena leucocephala	5	3,990	29.2	29.5	195
Terminalia arjuna	12	2,078	29	15.5	114
Terminalia arjuna	35	138	134.2	21.8	280
Populus deltoids	10	791	40.5	11.8	48
Acacia nilotica	15	619	60.5	22.8	161
Acacia nilotica	30	167	93.5	11.7	202
Acacia auriculiformis	15	2,593	28.8	22.2	130
Pithecellobium dulce	15	2,410	22.1	12.9	66
Casuarina glauca	8	2,227	25.3	13.0	51
Prosopis juliflora	10	3,013	28.9	25.6	157
Eucalyptus camaldulensis	35	570	82.5	38.4	405
Mixed forest	35	554	63.9	29.2	342

*At 1.3 m height; Source: [54].

To find high biomass-producing tree species for sodic soils, long-term experiments were conducted on highly sodic soils (pH >10.0) in the Saraswati Range forest site in Haryana, India.   30 forest tree species were planted with deep augers piercing the concretion layer and shallow augers not piercing the concretion layer.   After seven years of planting, Dagar *et al.* [54] reported that among the 30 species planted with two methods *Prosopis juliflora, Acacia nilot-*

ica, Tamarix articulate, and *Eucalyptus tereticornis* were found to be the most suitable and highest biomass-producing tree species for sodic soils (Tab. 8.11).

Tab. 8.11 Average air-dried biomass of some trees species after seven years of growth.
(Unit: Mg ha^{-1})

Species	Deep auger hole	Shallow auger hole
Tamarix articulate	97.33	37.71
Acacia nilotica	69.78	39.09
Prosopis juliflora	51.27	22.06
Eucalyptus tereticornis	14.38	5.20
Pithecellobium dulce	3.96	2.14
Terminalia arjuna	2.68	1.76
Dalbergia sissoo	1.75	1.18
Cordia rothii	1.48	0.62
Kigelia pinnata	1.17	0.49
Parkinsonia aculeate	1.15	0.90

LSD ($P \leqslant 0.05$): Between species=5.94; Between auger depths=1.17; Interactions (auger $\times$ species) =3.70

Source: [54].

In another experiment conducted at the Shivri Research Farm of the Central Soil Salinity Research Institute, *Prosopis juliflora* gave the maximum dry biomass (56.50 Mg ha^{-1}) with about 96% biomass allocated to stem and branch woods followed by *Acacia nilotica* (50.75 Mg ha^{-1}) with 95% biomass in the wood components [28] (Tab. 8.12). This is because of their fast growth and

Tab. 8.12 Dry biomass production of different tree species in sodic soils.

Species	Tree biomass (Mg ha^{-1})			
	Stem	Branch	Leaf	Total
Terminalia arjuna	23.78	10.70	7.13	41.62
Azadirachta indica	11.17	6.21	1.84	19.22
Prosopis juliflora	27.73	26.60	2.17	56.50
Pongamia pinnata	9.05	14.45	3.10	26.60
Casuarina equisetifolia	28.60	9.15	4.35	42.10
Prosopis alba	14.70	11.10	1.95	27.75
Acacia nilotica	22.15	26.14	2.46	50.75
Eucalyptus tereticornis	24.40	5.27	2.10	31.77
Pithecellobium dulce	23.50	6.81	1.94	32.25
Cassia siamea	14.30	5.65	1.70	21.65
LSD ($P \leqslant 0.05$)	2.43	4.63	1.21	5.42

Source: [28].

higher yields in sodic soils [9, 34]. The highest portions of dry biomass (76.8% and 72.9%) in the stem were recorded with *Eucalyptus tereticornis* and *Pithecellobium dulce*, respectively, because of the smaller number of branches, whereas the share of dry biomass through the branches was higher (54.3%) in *Pongamia pinnata* and *Terminalia arjuna* showed relatively high proportions of foliar biomass (7.13 Mg ha^{-1}) because of broad laminar morphology.

The highest annual increment in biomass yield (13.78 Mg ha^{-1}) was observed in *Prosopis juliflora* between stand age of two and four years, followed by *Acacia nilotica* (9.44 Mg ha^{-1}), *Terminalia arjuna* (9.40 Mg ha^{-1}), and *Casuarina equisetifolia* (8.98 Mg ha^{-1}), whereas it increased linearly in *Pithecellobium dulce* up to the age of eight years. The annual increments of other species like *Azadirachta indica*, *Pongamia pinnata*, *Pithecellobium dulce*, and *Cassia siamea* suffered heavily during the early growth periods due to high sodicity and low soil fertility, which, in general, led to stressed growth (Tab. 8.13).

Tab. 8.13 Annual biomass increments of tree species.

Species	Annual increment (Mg ha^{-1})					r^2
	0–2 years	2–4 years	4–6 years	6–8 years	8–10 years	
Terminalia arjuna	4.99	9.40	5.18	1.93	0.63	0.996
Azadirachta indica	2.33	4.3	3.02	1.46	0.61	0.991
Prosopis juliflora	9.23	13.78	7.46	2.89	1.00	0.996
Pongamia pinnata	2.4	5.18	4.84	3.08	1.65	0.997
Casuarina equisetifolia	3.97	8.98	7.69	4.43	2.14	0.999
Prosopis alba	3.54	6.12	4.07	1.89	0.77	0.996
Acacia nilotica	6.66	9.44	7.28	4.25	2.18	0.996
Eucalyptus tereticornis	5.09	7.43	4.61	2.08	0.83	0.996
Pithecellobium dulce	0.44	3.01	6.58	7.69	6.39	0.999
Cassia siamea	2.61	5.96	3.65	1.46	0.50	0.996

Source: [28].

8.7   Bioenergy Production

Afforestation on salt-affected soils, apart from meeting several of the needs of mankind, conserve the environment to a great extent. A study carried out in a 40-year-old man-made forest established on sodic soils at Banthra, Lucknow, India indicated 8.94 TJ ha^{-1} (Tera jule) of energy conservation and 168 Mg ha^{-1} C sequestration in the vegetation (Tab. 8.14). It was observed that a

Tab. 8.14 Role of a 40-year-old rehabilitation forest on sodic wastelands for bioenergy production.

Vegetation strata	Biomass (mg ha^{-1})	Energy production (*GJ ha^{-1})	C sequestration (Mg ha^{-1})	N standing state (kg ha^{-1})
Overstory	343±27	8836±778	166±10	3444±200
Understory	3.8±0.4	99±11	2.3±0.3	69±21
Ground layer	0.04±0.02	0.85±0.5	0.02±0.01	0.8±0.5
Total	347±27	8936±790	168±10	3514±221

$*$GJ=Giga Joule.

Tab. 8.15 Aboveground biomass (t ha^{-1}) and energy content (in parenthesis as GJ ha^{-1}) of a 3.5-year-old species grown on an alkaline wasteland under different treatments.

Species	Treatments			Means for species LSD ($P \leqslant 0.05) = 1.207$
	Gypsum	Pyrite	Control	
Prosopis juliflora	12.05	3.38	1.06	5.49
	(242.11)	(67.91)	(21.30)	(110.44)
Acacia nilotica	8.89	2.01	0.69	3.86
	(181.41)	(41.02)	(14.08)	(78.84)
Terminalia arjuna	3.69	1.39	0.17	1.75
	(77.95)	(29.36)	(3.59)	(36.96)
Pongamia pinnata	0.43	0.27	0.10	0.26
	(8.71)	(5.47)	(2.02)	(5.40)
Means for treatments	6.26	1.76	0.54	
LSD ($P \leqslant 0.05) = 1.045$	(127.54)	(35.94)	(10.25)	
LSD ($P \leqslant 0.05) = 2.09$ to compare partial means				

Source: [6].

forest having 300–400 Mg ha^{-1} of biomass may convert 2%–3% of usable solar radiation in a year with 10–15 t ha^{-1} yr^{-1} of carbon fixations in the biomass.

The aboveground biomass production and energy contents were estimated after 3.5 years of field establishment. The maximum biomass over the period and the energy content were recorded for *Prosopis juliflora* treated with gypsum. *Pongamia pinnata* produced the lowest dry weight and energy content when no soil treatment was applied (Tab. 8.15).

Through a long-term study conducted at the Shivri, Lucknow Research Farm, Singh *et al.* [28] reported that the leaves had slightly higher heats of combustion

(21.40–23.71 MJ kg^{-1}), whereas it was lowest in the stems (20.45–23.23 MJ kg^{-1}) (Tab. 8.16). The calorific values of the stems and branches exhibited fewer variations, with *Acacia nilotica* having the highest heat combustion in both stems and branches (23.23 and 24.24 MJ kg^{-1}), respectively. The differences in total energy production and its allocation to different plant parts, led to variation between biomass yield and its allocation to stem, branch and leaves per hectare. *Prosopis juliflora* gave the highest energy harvest of 1,267.75 GJ ha^{-1}, followed by *Acacia nilotica* with 1,206.32 GJ ha^{-1}, and the lowest of *Azadirachta indica* (520.66 GJ ha^{-1}).

Tab. 8.16 Energy values of different plant components in ten tree species.

Species	Calorific values (MJ kg^{-1})			Total energy (GJ ha^{-1})
	Stem	Branch	Leaf	
Terminalia arjuna	22.57	21.60	23.24	933.53
Azadirachta indica	20.60	20.54	21.42	520.66
Prosopis juliflora	22.53	23.20	23.71	1267.75
Pongamia pinnata	21.60	21.60	22.34	576.85
Casuarina equisetifolia	22.20	22.14	22.21	934.11
Prosopis alba	21.46	22.20	23.21	607.13
Acacia nilotica	23.23	24.24	23.64	1206.32
Eucalyptus tereticornis	20.45	22.43	21.40	662.12
Pithecellobium dulce	21.50	21.60	22.64	696.26
Cassia siamea	21.40	21.68	22.58	466.89

Source: [28].

8.8   Soil Amelioration

Various studies have been conducted to monitor the soil dynamics due to the afforestation of salt-affected soils. As the tree grows, a large amount of litter is shed on the ground, which, during decomposition, releases several weak acids (humic and fumic) to lower the soil pH and EC. Singh *et al.* [28] observed that the litter production after ten years of tree growth by *Prosopis juliflora*, *Casuarina equisetifolia*, *Acacia nilotica*, *Terminalia arjuna*, and *Pongamia pinnata* was 6.1 Mg ha^{-1}, 5.7 Mg ha^{-1}, 5.4 Mg ha^{-1}, 5.1 Mg ha^{-1}, 5.0 Mg ha^{-1}, respectively. The winter months accounted for 40%–55% of the total litter fall that was composed of about 75.80% foliage (Tab. 8.17).

Tab. 8.17 Comparative litter fall yield and its composition.

Tree species	Litter fall yield (Mg ha^{-1})	Nutrient content (%)				
		N	P	K	Ca	Mg
Terminalia arjuna	5.1	0.84	0.13	0.15	0.67	0.52
Azadirachta indica	2.8	0.84	0.14	0.32	0.54	0.28
Prosopis juliflora	6.1	1.70	0.10	0.86	0.62	0.36
Pongamia pinnata	5.0	1.55	0.15	0.63	0.46	0.32
Casuarina equisetifolia	5.7	0.85	0.16	0.42	0.51	0.30
Prosopis alba	2.0	1.10	0.12	0.59	0.53	0.32
Acacia nilotica	5.4	1.14	0.10	0.28	0.43	0.26
Eucalyptus tereticornis	1.3	0.88	0.14	0.16	0.73	0.36
Pithecellobium dulce	2.4	0.86	0.16	0.43	0.52	0.36
Cassia siamea	1.3	0.78	0.14	0.40	0.86	0.84

Source: [55].

Garg and Jain [11] measured about 5–6 Mg ha^{-1}yr^{-1} litter fall in a young plantation (eight years) and 10–12 Mg ha^{-1}yr^{-1} in an old forest (35–40 years). Singh et al. [55] reported the highest litter fall yield at the ten-year tree growth stage under Prosopis juliflora (6.1 Mg ha^{-1}) followed by Casuarina equisetifolia (5.7 Mg ha^{-1}), Acacia nilotica (5.4 Mg ha^{-1}), Terminalia arjuna (5.1 Mg ha^{-1}), and Pongamia. pinnata (5.0 Mg ha^{-1}) on high-alkali soils. In addition to the recycling of nutrients, the decomposition of litter leads to the evolution of CO_2, which helps mobilize the inherent calcium (Ca). The released Ca can hasten the reclamation by replacing the exchangeable sodium (Na) from the soil, thus reducing the soils sodicity and pH levels. Singh et al. [28] reported that after ten years of plantation, the highest improvement in terms of soil pH, electrical conductivity, and exchangeable sodium percentage in 0–15 cm of soil depth was recorded under Prosopis juliflora, followed by Acacia nilotica, and Pongamia pinnata (Fig. 8.1). Singh and Gill [56], Singh and Gill [10], and Mishra et al. [10] have also reported higher soil amelioration in terms of decreased soil pH with Prosopis juliflora. The increase in the organic C content of the surface soil (0–15 cm) in a span of ten years was about four-fold under Prosopis juliflora and Pongamia pinnata and about three-fold in other species. Tripathi and Singh [12] also reported an overall higher improvement in the soil organic matter under Prosopis juliflora. Fisher [58] advocated five hypotheses for the mechanisms that regulate soil amelioration by trees. These mechanisms involve (i) an increase in the soil organic matter content as a result of C fixation in photosynthesis and its transfer via leaf fall and root turn to the soil [59–60], (ii) leguminous trees fixing atmospheric N and resulting in an increase in soil N under the tree canopy [61], (iii) rhizosphere effects of trees on soils resulting in enhanced N mineralization and increased microbial biomass [62–63], (iv) microclimate modification by tree

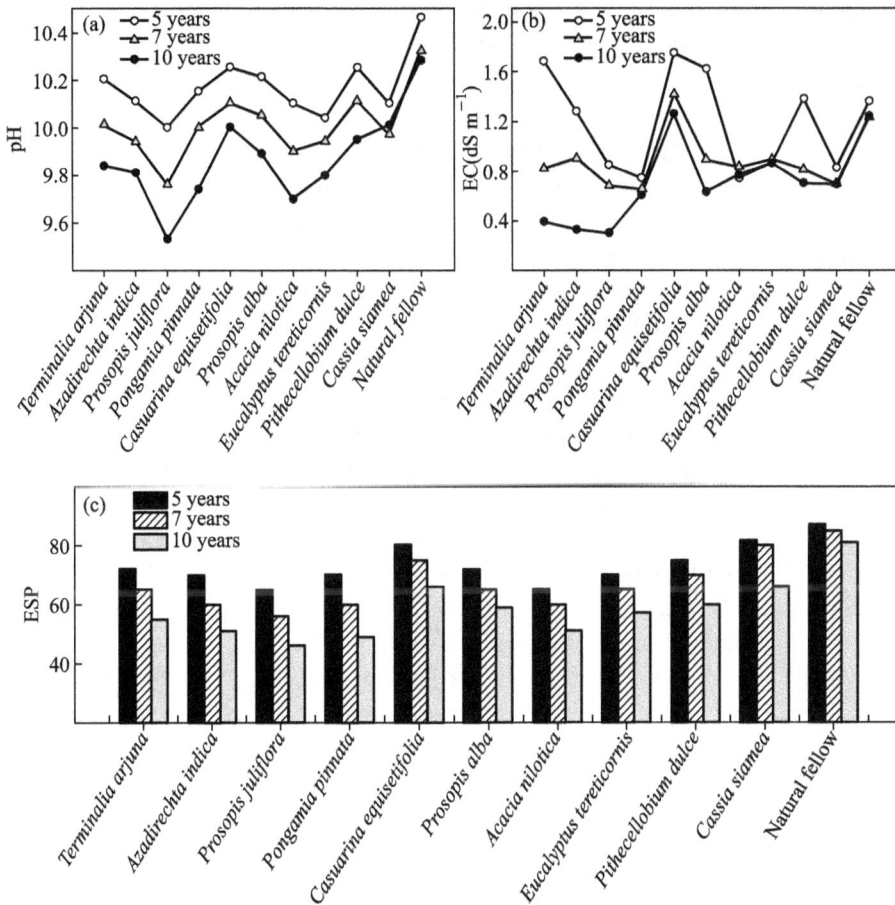

Fig. 8.1 Changes in (a) soil pH, (b) electrical conductivity, and (c) ESP of a sodic soil after 5, 7, and 10 years of plantation at 0–15 cm depth (a significant difference in soil pH at 5, 7, and 10 years of growth stages at a 5% level of significance is 0.16, 0.18, and 0.21, respectively. A significant difference in soil EC at 5, 7, and 10 years of growth stages at a 5% level of significance is 0.06, 0.04, and 0.04, respectively. A significant difference in soil ESP at 5, 7, and 10 years of growth stages at a 5% level of significance is 4.63, 3.68, and 7.32, respectively.

canopies that moderate soil and air temperatures and soil moisture regimes [64], and (v) nutrient pumping reflecting the uptake of nutrients from greater depths by tree roots and accumulation in a smaller volume of surface soil as a result of litter fall [64].

After ten years of plantation, a significant improvement in the physical properties of the sodic soil was recorded in an experiment conducted at the Shivri Research Farm in Lucknow, India. Bulk density in the 0–75 mm soil layer decreased significantly over the control, whereas the porosity and infiltration rates increased. A maximum reduction in bulk density was recorded under *Casuarina equisetifolia* (1.21 Mg m^{-3}) followed by *Pithecellobium dulce* (1.25 Mg m^{-3}), *Acacia nilotica* (1.29 Mg m^{-3}), and *Prosopis juliflora* (1.32 Mg m^{-3}). A minimum reduction was recorded under *Azadirachta indica* (1.48 Mg m^{-3}) over the initial value of 1.57 Mg m^{-3}. The bulk density of the surface soil (0–75 mm) under the control remained unchanged, whereas under the 75–150 mm soil layer, it was slightly improved (Tab. 8.18). Soil porosity under the ten-year-old plantation in the 0–75 mm soil layer increased from 40.7% to 54.3%. However, under the control plot, the soil porosity was almost unchanged. The highest soil porosity in the 0–75 mm soil layer was recorded under *Casuarina equisetifolia* (54.3%) and the minimum under *Azadirachta indica* (44.1%). The maximum moisture content in the 0–75 mm soil layer was under *Casuarina equisetifolia*, followed by *Terminalia arjuna*, *Acacia nilotica*, and *Prosopis juliflora* and minimum under *Eucalyptus teriticornis*. There was a significant improvement in the infiltration rate under the tree plantation over the control and initial values. The highest infiltration rate after ten years of tree plantation was recorded under *Prosopis juliflora*, followed by *Casuarina equisetifolia*, *Pongamia pinnata*, *Pithecellobium dulce*, *Acacia nilotica*, *Azadirachta indica*, *Terminalia arjuna*, *Prosopis alba*, *Eucalyptus tereticornis*, and *Cassia siamea*.

The biological properties of the soil are largely affected by the microorganisms' status in the soil and nutrients held by these organisms. The seasonal dynamics

Tab. 8.18 Ameliorative effects of different tree species on the physical properties of soil ten years after plantation.

Tree species	Bulk density (Mg m^{-3})		Soil porosity (%)		Cumulative infiltration rate (mm d^{-1})
	0–75 mm	75–150 mm	0–75 mm	75–150 mm	
Terminalia arjuna	1.47	1.52	44.5	42.6	21.20
Azadirachta indica	1.48	1.56	44.1	41.1	21.70
Prosopis juliflora	1.32	1.46	50.2	44.9	26.30
Pongamia pinnata	1.36	1.57	48.6	40.7	24.30
Casuarina equisetifolia	1.21	1.42	54.3	46.4	25.80
Prosopis alba	1.37	1.61	48.3	39.2	20.00
Acacia nilotica	1.29	1.58	51.3	40.4	21.90
Eucalyptus tereticornis	1.38	1.51	48.0	43.0	19.70
Pithecellobium dulce	1.25	1.58	52.8	40.4	23.10
Cassia siamea	1.46	1.48	45.0	44.1	15.80
Natural fallow	1.50	1.57	43.4	40.7	11.80
Initial	1.57	1.60	40.7	39.6	2.10
LSD ($P \leqslant 0.05$)	0.08	0.11	3.26	0.76	6.34

Source: [55].

of microbial C, N, and P measured in three tropical forest soils ranged from 466–662, 48–72, and 21–30 μg^{-1}, respectively, being the greatest in the dry season and the lowest in the wet season [65]. Soil microorganisms are the most active fractions of soil organic matter and therefore play a central role in the flow of plant nutrients in ecosystems. They constitute a transformation matrix for organic materials in the soil and act as labile reservoirs for plant available N and P [66].

Some of the studies have characterized the microflora of saline-sodic soils [67–69] where microbial populations were measured at different salinity and alkalinity levels in the soil [70]. They observed that the number of microorganisms decreased with the increase in EC and pH. Excessive amounts of salt present in the soil adversely affected the soil microbial populations and their activities. The dehydrogenase activity and microbial biomass carbon (MBC) were determined in the salt-affected soils of northeast India [71]. Corresponding to soil organic matter, microbial biomass-carbon also decreased with soil depth.

Forest growth over 40 years has found to reclaim the soil in many properties. Several soil characteristics were studied comparatively in forested as well as non-forested sodic soils of the surrounding areas to observe the degrees of reclamation in the degraded sodic soils. MBC, Microbial biomass nitrogen (MBN), and Microbial biomass phosphorus (MBP) decreased significantly from the surface to a depth of 45 cm (Tab. 8.19). This decrease was about 90% in MBC and 65% in MBP from the surface soil. The mean MBC from 0–45 cm deep was 131 $\mu g \ g^{-1}$ in forested soil, which was approximately three times greater than that in non-forested sodic soils. MBC, MBN, and MBP varied significantly between forested and barren sodic soils. MBC, MBN, and MBP were at the maximum in the summer season and minimum in the winter/rainy season, because in the winter, microbial activity becomes slow due to low temperatures.

Tab. 8.19 Biological properties of forested (F) and non-forested (C) sodic soils (μg g^{-1}).

Character	State	Depth (cm)			Mean	LSD$_{0.5}$
		0–15 mean ± SD	15–30 mean ± SD	30–45 mean ± SD		
MBC	F	285.33±87.66	55.0±33.15	33.33±13.57	124.55±44.79	15.0
	C	89.33±6.65	32.0±9.16	19.66±4.61	46.99±2.27	
MBN	F	53.16±3.09	19.93±5.96	10.2±0.75	27.43±2.60	4.37
	C	14.33±3.76	8.26±0.11	4.96±0.77	9.18±1.94	
MBP	F	25.76±7.0	15.53±4.31	10.66±2.24	17.31±2.38	7.0
	C	9.7±3.81	5.53±0.76	4.13±1.17	6.45±1.65	

Source: [72].

8.9   Conclusions

To sustain the production of food and fuel, each portion of land needs to be best utilized corresponding to ecology and land use capabilities. This means that salt-affected degraded lands should either be reclaimed for agricultural purposes or put under afforestation. Rehabilitation of salt-affected soils for crop production is, in fact, proceeding at a snail's pace because salt-affected soils either are owned by resource-poor marginal farmers or belong to a village community or government agencies. The current gap between the supply and demand of food, fuel, fodder, and timber is likely to worsen in the near future as a consequence of the continuing degradation of lands and reduced per capita land availability. Harnessing the productivity potential of salt-affected soils through the plantation of multipurpose tree species has the potential to put them under use for biomass and bioenergy production and also to improve the productivity of degraded lands. Various research efforts have been made to identify the tree species that are suitable for afforestation and biomass and bioenergy production in salt-affected soils. Besides providing biomass in terms of fuel, fodder, and timber, afforestation will also lead to the bioamelioration of salt-affected soils. Afforestation of salt-affected soils will not only help ecological and environmental considerations but also be useful for energy conservation and relieving pressure on traditionally cultivated lands.

References

[1] Brown S, Lugo AE. Rehabilitation of tropical lands: a key to sustaining development. Restoration Ecology 1994; 2 (1): 97–111.
[2] Sharma RC, Rao BRM, Saxena RK. Salt affected soils in India—Current assessment. In: Advances in Sodic Land Reclamation, Proceedings; International Conference on Sustainable Management of Sodic Lands, Lucknow, India, 2004; 1–26.
[3] Yadav JSP. Improvement of saline alkali soils through biological methods. Indian Forester 1975; 101(7): 385–395.
[4] Yadav JSP. Salt affected soils and their afforestation. Indian Forester 1980; 106: 259–272.
[5] Chaturvedi AN, Behl HM. 1996. Biomass production trials on sodic site. The Indian Forester 1996; 122: 439–455.
[6] Singh B. Rehabilitation of alkaline wastelands on the Gangetic alluvial plains of Uttar Pradesh, India through afforestation. Land Degradation and Rehabilitation 1989; 1: 305–310.

[7] Hansen EA, Baker JB. Biomass and nutrient removal in short rotation intensively cultured plantations. In: Proceedings Symposium on Impact of Intensive Harvesting on Forest Nutrient Cycling. State University of New York. Syracuse, New York, 1979; 130–151.

[8] Brady NC. The Nature and Properties of Soils. Macmillan Publishers, New York 1990.

[9] Lugo AE, Wang D, Borman FH. A comparative analysis of biomass production in five tropical tree species. Forest Ecology and Management 1990; 31: 153–166.

[10] Singh GB, Gill HS. Ameliorating effect of tree species on characteristics of sodic soils at Karnal. Indian Journal of Agricultural Science 1992; 62: 142–146.

[11] Garg VK, Jain RK. Influence of fuel wood trees on sodic soils. Canadian Journal of Forestry Research 1992; 22: 729–735.

[12] Tripathi KP, Singh B. The role of revegetation for rehabilitation of sodic soils in semi-arid subtropical forest. Restoration Ecology 2005; 13 (1): 29–38.

[13] Bhargava GP, Sharma RC, Pal DK, Abrol IP. A case study of the distribution and formation of salt affected soils in Haryana state. In: Proceedings of International Symposium on Salt affected Soils, CSSRI Karnal, 1980; 83–91.

[14] Sidhu PS, Sehgal JL. National symposium on land and water management in Indus Basin. Ecological Society, Punjab Agricultural Unioversity, Ludhiana. 1978; 276–286.

[15] Bhargava GP, Kumar R. Characteristics, extent and genesis of sodic soils of the Indo-Gangetic alluvial plain. In: Sharma DK, Rathore RS Nayak AK, Mishra VK (Eds.). Sustainable Management of Sodic lands. Central Soil Salinity Research Institute, Regional Research Station, Lucknow 2011; 28–42.

[16] Szabolcs I. Review of research on salt affected soils. Natural Resources Research 1979; 15: 137 (UNESCO, Paris).

[17] Bhumbla DR, Abrol IP. Saline and sodic soils. In: Soil and Rice Symposium, IRRI,Los Banos, Laguna, Philippines, 20–23 September, 1979; 719–738.

[18] Tomar OS, Gupta RK, Dagar JC. Afforestation techniques and evaluation of different tree species for waterlogged saline soils in semiarid tropics. Arid Soil Research and Rehabilitation 1998; 12: 301–316.

[19] Sharma RC, Singh R, Singh YP, Singh GB. Sodic soils of Shivri experimental farm; site characteristics, reclamability and use potential for different land uses. Central Soil Salinity Research Institute, Karnal, 2006; 36.

[20] Gupta GN, Prasad KG, Mohan S, Subramaniam V, Manivachakam P. Effect of alkalinity on survival and growth of tree seedlings. Journal of Indian Society of Soil Science 1988; 36: 537–542.

[21] Singh K, Yadav JSP. Growth response and cationic uptake of *Eucalyptus* hybrid at varying levels of soil salinity and sodicity. India Forester 1985; 111: 1123–1135.

[22] Yadav JSP, Singh K. Response of Casuarina equisetifolia to soil salinity and sodicity. Journal of Indian Society of Coastal Agriculture Research 1986; 4: 1–8.

[23] Gupta GN, Mohan S, Prasad KG. Salt tolerance of selected tree species. Journal of Tropical Forest 1987; 3: 217–226.

[24] Bandhopadhyay AK, Dutt SK, Bal AR. Salt tolerance of Casuarina equisetifolia. In: Proceedings of International Symposium on Afforestation of Salt Affected Soils, CSSRI, Karnal, 1987; 2: 17–32.

[25] Singh YP, Minhas PS, Tomar OS, Gupta RK, Raj K. Water use and salt tolerance of Shisham (*Dalbergia sissoo*) during establishment stages. Arid Soil Research and Rehabilitation 1996; 10: 379–390.

[26] Singh UN, Bhatt DN, Yadav JSP. Growth and biomass production of certain forest species as influenced by varying pH levels. In: Proceedings of International Symposium on Afforestation of Salt Affected Soils, CSSRI, Karnal, 1987; 2: 51–62.

[27] Dagar JC, Tomar OS. Utilization of salt affected soils and poor quality waters for sustainable bio-saline agriculture in arid and semiarid regions of India. 12th ISCO Conference Beijing 2002.

[28] Singh YP, Singh GB, Sharma DK. Biomass and bio-energy production of ten multipurpose tree species planted in sodic soils of Indo-Gangetic plains. Journal of Forestry Research 2010; 21 (1): 63–70.

[29] Dagar JC, Singh NT. Agroforestry options in reclamation of problem soils. In: Thampan PK (Ed.). Trees and Tree farming. Peekay Tree Crops Development Corporation, Cochin, India, 1994; 65–103.

[30] Gupta RK, Tomar OS, Minhas PS. Managing salt affected soils and waters for afforestation. CSSRI, Karnal, India 1995; Bull.7/95, 23.

[31] Chhabra R, Abrol IP. Effect of amendments and nutrients on the performance of selected tree species in sodic soils. Report CSSRI, Karnal, 1986; 23–25.

[32] Gill HS, Abrol IP. Salt affected soils, their afforestation and its ameliorating influence. International Tree Crop Journal 1991; 6: 261–274.

[33] Singh GB, Abrol IP, Cheema SS. Effect of gypsum application on mesquite (*Prosopis juliflora*) and soil properties in an abandoned sodic soils. Forest Ecology Management 1990; 29: 1–14.

[34] Singh GB, Singh NT, Tomar OS. Agroforestry in salt affected soils CSSRI, Karnal, 1993; 65, Research Bulletin 17.

[35] Singh, GB, Abrol IP, Cheema SS. Effect of management practices on mesquite (*Prosopis chilensis*) in highly alkaline soils. Indian Journal of Agricultural Science 1989; 59: 1–7.

[36] Shah SA. Letter to the editor. Indian Forester 1957; 83: 472.

[37] Tomar OS, Minhas PS, Gupta RK. Potentialities of afforestation of waterlogged saline soils. In: Singh P, Pathak PS, and Roy MM (Eds.). Agroforestry Systems for Degraded Lands. Oxford and IBH publishing Co.Pvt Ltd. New Delhi, 1994; 1: 11–120.

[38] Leather JW. Reclamation of reh or usar lands. Agriculture Ledger 1987; 4: 129–138.

[39] Sodic Land Reclamation Committee. U.P. Report 1938–1939. Superintendent, Printing and Stationary, Allahabad, 118.

[40] Khan WAW, Yadav JSP. Characterization and afforestation problems of saline alkali soils. Indian Forester 1962; 83: 259–271.

[41] Pande GC. Afforestation of Usar Lands. In: Proceedings of 11th All India Silviculture Conference. FRI, Dehradun, India, 1967.

[42] Yadav JSP, Singh K. Tolerance of certain forest species to varying degree of salinity and alkalinity. Indian Forester 1970; 96: 587–599.

[43] Yadav JSP, Bhumbla DR, Sharma OP. Performance of certain forest species on saline sodic soils. In: Proceedings of International Symposium on New Development in the Field of Salt Affected Soils of Sub-commission on Salt Affected Soils. International Society of soil Science held at Cairon 1975; 683–690.

[44] Ghosh RC. Handbook of Afforestation Techniques. Controller of Publications, New Delhi, 1977.

[45] Sandhu SS, Abrol IP. Growth responses of *Eucalyptus teriticornis* and *Acacia nilotica* to selected cultural treatments in a highly sodic soil. Indian Journal of Agricultural Sciences 1981; 51: 437–443.

[46] Gill HS, Abrol IP. Salt affected soils and their amelioration through afforestation. In: Amelioration of Soil by Trees 1986; 43–53.

[47] Grewal SS, Abrol IP, Singh OP. Agroforestry on alkali soils. Effect of some management practices on initial growth, biomass production and chemical composition of selected tree species. Agroforestry Systems 1986; 4: 221–32.

[48] Gill HS, Abrol IP, Gupta RK. Afforestation of salt affected soils. In: Technologies for Wasteland Development. ICAR, New Delhi, 1990, 355–380.

[49] Singh, GB, Abrol IP, Cheema SS. Effect of irrigation on Prosopis juliflora and soil properties of an alkali soil. International Tree Crops Journal 1990; 6: 81–99.

[50] Tomar OS, Kumar M, Gupta RK, Minhas PS. Raising nursery of Ramakanti Kikar (*Acacia nilotica* var.*cupressiformis*) with saline water. Indian Forester 1996; 123: 148–152.

[51] Singh GB, Singh NT. Mesquite for the revegetation of salt lands. CSSRI, Karnal, 1993; Bull. No.18, 24.

[52] Singh B, Tripathi KP, Jain RK, Behl HM. Fine root biomass and tree species effects on potential N mineralization in afforested sodic soils. Plant and Soil 2000; 219: 81–89.

[53] Singh GB, Singh H, Bhojvaid PP. Sodic soils amelioration by tree plantations for wheat and oat production. Land Degradation and Rehabilitation 1998; 9: 453–462.

[54] Dagar JC, Sharma HB, Shukla YK. Raised and sunken bed technique for agroforestry on alkali soils of North India. Land Degradation and Development 2001; 12: 107–118.

[55] Singh YP, Singh GB, Sharma DK. Ameliorative effect of multipurpose tree species grown on sodic soils of Indo-Gangetic alluvial plains of India .Arid Land Research and Management 2011; 25: 55–74.

[56] Singh GB, Gill HS. Raising trees in alkali soils. Wasteland News 1990; 6: 15–18.

[57] Mishra A, Sharma SD, Khan GH. Rehabilitation of degraded sodic lands during a decade of Dalbergia sissoo plantation in Sultanpur district of Uttar Pradesh, India. Land Degradation and Development 2002; 13: 375–386.

[58] Fisher RF. Amelioration of soil by trees. In. Lacate DS, Weetman GF and Powers RF (Eds.). Sustained productivity of Forest Soils. U.B.C. Faculty of Forestry, University of British Columbia, Vancouuver.1990; 290–300.

[59] Sanchez PA, Palm CA, Davey CB, Szott LT, Russell CE. Tree crops as soil improver in the humid tropics. In Cannel M and Jackson JE (Eds.). Attributes of Trees As Crop Plants. Institute of Terrestrial Ecology, Huntington, 1985; 327–358.

[60] Brown S, Lugo AE. Tropical secondary forests. Journal of Tropical Ecology 1990; 6: 1–32.

[61] Prinsely RT, Swift MJ. Amelioration of soils by trees; a review of current concepts and practices. Commonwealth Science Council, London, U.K., 1986; 181.

[62] Mao DM, Min YW, Yu L, Martens R, Insam H. Effect of afforestation in Nigeria I. Effect of agrisilviculture on soil chemical properties. Soil Biology Biochemistry 1992; 24: 865–872.

[63] Kaur B, Gupta SR, Singh GB. Bioamelioration of Sodic soils by silvipastoral systems in north-western India. Agroforestry Systems 2001; 54: 13–20.

[64] Fisher RF. Amelioration of degraded rain forest soils by plantations of native trees. Soil Science Society of American Journals 1995; 59: 544–549.

[65] Raghubanshi AS. Dynamics of soil biomass C, N and P in a dry tropical forest in India. Biology and Fertility of Soils 1991; 12: 55–59.

[66] Jenkinson DS, Ladd JN. Microbial biomass in soil: Measurement and turnover. In: Paul EA, Ladd JN, Dekker M (Eds.). Soil Biochemistry. New York, 1981; 5: 415–471.

[67] James N. Plate counts of bacteria and fungi in a saline soil. Canadian Journal of Microbiology 1959; 5: 432

[68] Rankov. Inst. Pochvoznan Agrotekh 'Pushkarov' 1962; 4: 239.

[69] Rankov V. The salinization of the soil and the development of the nitrogen fixation microorganisms. Agrochimica 1964; 8: 330.

[70] Gupta BR, Bajpai BP. Some microbiological studies in salt affected soils I-pattern of soil microbial population as affected by salinity and alkalinity. Journal of Indian Sodic Soil Society 1974; 22: 176–180.

[71] Batra L, Kumar A, Manna MC, Chhabra R. Microbiological and chemical amelioration of alkali soils by growing karnal grass and gypsum application. Experimental Agriculture 1997; 22: 389–397.

[72] Singh B, Goel LV. Soil amelioration through afforestation. Restoration of Degraded Land to Functioning Forest Ecosystem, 2012; 114–145.

Chapter 9

Bioenergy and Prospects for Phytoremediation

Dimitriou Ioannis

9.1 Introduction

Phytoremediation is the use of trees and other plants such as grasses and aquatic plants to remove, destroy, or sequester hazardous substances from the environment. This chapter reviews the use of biomass production systems cultivated not only for energy but also for cleaning and improving soil quality in terms of reducing hazardous compounds, with due reference to cleaning municipal and industrial wastewater by fertigation and treating and utilizing the wastewater. The most common tree species being used in phytoremediation systems producing biomass for energy are not hyperaccumulators of metals or other hazardous compounds. They are preferred in commercial phytoremediation projects due to their fast and heavy growth and also because of the fact that agronomic practices for their easy management and good growth performance already exist.

Appropriate species for use in different kinds of phytoremediation systems must have, besides heavy and fast growth habits, some of the following characteristics for high phytoremediation performance: high evapotranspiration (ET) ability, high nutrient use efficiency, tolerance to high heavy metal concentrations in the soil, tolerance to anoxic conditions in the roots, and the ability to uptake hazardous compounds. The above traits enable growth under unfavorable environments, but such systems should primarily be seen as a biomass production system. For this, productive soils should be preferred to achieve high growth, preferably in large-scale plantations. In many cases, however, moderately contaminated soils are available for the cultivation of bioenergy systems and other contaminants can already exist in agricultural soils. Therefore, this chapter will mostly focus on the implications for large-scale short rotation forestry (SRF) and short-rotation coppice (SRC) bioenergy systems, mainly with willows and poplars, since these are the main species used in Europe and America for the production of dedicated biomass for energy and which have shown a reported ability for the abovementioned phytoremediation-promoting characteristics. Results obtained in the laboratory will also be used to estimate soil ecological effects but to a smaller extent than large-scale fields.

Poplars and willows have been used for several different types of phytoremediation for soil improvement. This is based on the function of the plants against hazardous compounds, *e.g.*, phytoextraction (ability to accumulate large quantities in the aboveground parts removed by harvest), rhizofiltration (absorption onto plant roots removed from aqueous waste streams), phytotransformation (degradation or metabolization in plant parts), phytovolatilization (volatilized into the air from plant biomass), phytostimulation (degradation in the soil due to secreted plant enzymes or by plant stimulation of microbial biodegradative activity), phytostabilization (when immobilization in the soil occurs by plant exudates), and phytomining (extract of large amounts of metals from soils [1]). Willows and poplars were preferred in commercial phytoremediation projects due to their fast and high growth and the fact that agronomic practices for SRC accommodated easy management and good growth performance already exist, despite the fact that they are not natural hyperaccumulators of metals or other hazardous compounds. However, plants of these species have been reported to evapotranspire high amounts of water [2–3] and to tolerate high heavy metal concentrations in the soil [4–5]. Furthermore, willows have been found to be tolerant to anoxic conditions [6].

9.2 Bioenergy Systems for Soil Phytoremediation

In the text that follows, a description of related research will be presented that concerns the idea behind the concept of phytoextraction of heavy metals and organic compounds from the soil when SRC and SRF are cultivated for energy purposes. Related research concerning the issues-to-be will be considered when the implementation of such systems occurs. Strategies that need to be considered for the optimization of such systems will be discussed.

9.2.1 Phytoextraction of Heavy Metals

Extensive research related to phytoextraction of heavy metals by willow and poplar, *i.e.*, their ability to accumulate large quantities in the aboveground parts

removed after harvest, has been conducted. In the present review, it is phytoextraction that is our interest compared to other phytoremediation means because
the main aim is to clean the soil of the SRC fields from heavy metals after regular
harvests of aboveground biomass. Willows have been reported from the early
stages of their commercial bioenergy use to take up large amounts of cadmium
(Cd) [7–8]. Initially, the focus was on Cd uptake by willows, followed by research on the uptake by willow of other metals such as copper (Cu), lead (Pb),
zinc (Zn), chromium (Cr), and nickel (Ni) [9–10]. Metal uptake by poplars was
studied at the later stages after poplar gained constant interest as an alternative species to willow for biomass production for energy purposes [11–12]. The
phytoremediation potentials of willows and poplars, while not being hyperaccumulators of hazardous metals, have been reported to be high, as indicated by
the high accumulation of metals in the plant biomass [13–14].

Before getting into more details about the implications of the reported results, it is worth mentioning that substantial related research was conducted
in controlled laboratory conditions where individual willow and poplar plants
were grown in contaminated soils [15–16] or in hydroponic systems [17]. Very
promising results for uptake of certain metals in willow and poplar plant parts
were reported from those experiments and speculations for the great potential
of cleaning contaminated soils with willow and poplar were expressed. Although
results from pot trials have been validated in some cases in field situations [18],
concerns about the difference in conditions between controlled small-scale experiments (often artificially mixed heavily contaminated soils and favorable plant
growth) and large-scale field situations (often non-uniform and moderate contamination and lower plant growth) have been raised [19]. Such extrapolations
from the laboratory to the field should, however, be drawn cautiously and generalizations for implications under the field conditions should be avoided.

Many studies have proposed the use of a range of chelating agents such as
EDTA, EDDS, oxalic and citric acids, etc. to increase the positive metal uptake
rates by willow and poplar plants [20–21]. Despite the positive results for induced
phytoextraction indicated in the previously mentioned work, chelating agents
have been reported to cause toxicity symptoms to the plants. Leaching of metals
and negative impacts on soil biota have been reported, questioning the potential
future use of chelate-assisted phytoextraction [22]. This, combined with the
high costs involved in the application of chelating agents in large fields, makes
the extensive use of chelating agents highly uncertain in commercial willow and
poplar SRF in the future. Therefore, such issues are of limited interest for serious
discussions in this chapter.

Great variations in the metal uptake abilities of willows and poplars have been
reported in different SRC fields. This probably depends on different contamination levels for field use. Vandecasteele et al. [23] suggested that Cd uptake in
aboveground plant parts tends to increase with increasing Cd in the soil. This is

also reported in other studies with elevated metal concentrations where willows and poplars absorbed significant amounts of heavy metals in aboveground tissues [24], compared to less-contaminated soils [25]. Moreover, the differences in the uptake patterns of willow and poplar species and clones have been reported [26]. Therefore, for the effective use of SRC on clean soils, much attention should be paid to the selection of the clone in relation to the contamination source and level at the site. Nevertheless, it will probably be unlikely to find predictable uptake patterns for all metals for the accumulation in aboveground biomass. Only genotypes preferring more mobile elements such as Cd and Zn can be selected for a specific site. The mobility and plant availability of metals in the soil might also be responsible for the great differences in uptake patterns. For example, Eriksson and Ledin [27] indicated that plant available Cd concentrations in the soil were reduced in a willow SRC field, but the higher uptakes of different metals in willow shoots were not found when plant available fractions differed due to pH changes in a field willow experiment. In all, it seems that for cleaning soils contaminated by a certain compound, a "site-specific" approach with pre-testing of several clones to identify the best performing ones for further use in large-scale instances should be performed in advance, although difficulties due to the heterogeneity of the localization of the pollution are to be expected [28].

The aforementioned doubts raise the question as to which soils can be satisfactorily remediated by the phytoextraction of heavy metals with willow and poplar SRF for energy purposes and what strategies should be followed for the best remediation combined with the best economic value. Although willow and poplar have shown better phytoextraction efficiency than other species used [29], recent studies suggest that short-term remediation is not to be expected in heavily contaminated soils such as mine spoils or heavily contaminated industrial sites due to the unrealistically long time periods needed [30]. Also, such sites might be polluted in deep layers, which cannot be cleaned with poplars or willows that are appropriate for rather shallow contamination because most of their active roots are concentrated near the soil surface [31]. However, moderately contaminated soils with metal concentrations just above the metal threshold criteria offer great potential for cleaning the soil from metals. Such moderately contaminated land that can be considered appropriate for SRC and SRF for energy purposes can be agricultural land that has elevated metal amounts not only due to extensive P fertilization but also due to continuous sludge applications. Berndes et al. [32] calculated that 100 times more Cd would be removed by willow SRC than the harvest of straw in Sweden if SRC will be grown in arable land with elevated Cd concentrations due to phosphate fertilization. These amounts would compensate for the atmospheric deposition each year and would drastically reduce the amount of Cd in arable soils in Sweden. This would also give economic incentives for farmers to be compensated for reducing Cd in the soil (i.e., 10% of the total revenue). Lewandowski *et al.* [33] made similar calculations suggesting that phytoextraction with willow cultivation for a certain period can allow the future use of moderately contaminated fields for more profitable food production, thus increasing farmers' income.

A current extensively used way to increase bioenergy farmers' income in certain counties when SRC is cultivated is the application of sewage sludge [34]. Sludge contains not only P and N that are used as fertilizer to SRC but also heavy metals that can accumulate in the soil when applied with sludge for a number of years. Therefore, an increase in the biomass of SRC combined with increased metal uptake would result in a balance between metal input with sludge application and metal output with SRC harvest. Based on the field results, Dimitriou [35] calculated that the amounts of metals applied with sludge and after the uptake in SRC stems was within legal limits for such practices. Furthermore, if Cd in the soil would continue to reduce as in the initial years of the experiment, a 26% reduction of the total Cd in the upper soil layer could be expected in 25 years. Significant reductions of the levels of Cu and Zn were also calculated. Similar results after sewage sludge applications were reported by Lazdina *et al.* [36] who also found increased metal concentrations in willow shoots compared to the control by 4%–8%. This indicated the potential for SRC fields to receive sewage sludge in consecutive years without drastically affecting soil quality. Several willow clones were grown in historically sewage sludge-amended fields to test the effects of long-term sewage sludge applications [37]. Results underline the potential for using willow to reduce metal amounts, but indicate great differences among clones in the uptakes of different metals at the same site.

Different patterns of metal concentrations were found in either the bark, wood, or leaves in comparison to the shoots. Cd and Zn concentrations were much higher in the leaves than in the shoots [38]. It has been suggested that leaf harvest would significantly reduce the soil concentrations of these elements in SRC fields [39]. Vandecasteele *et al.* [40] suggested that Cd and Zn were accumulated in the aboveground willow parts compared to the other metals accumulated in the roots. However, others suggest that most of the metals are concentrated in the roots and small amounts are accumulated in the aboveground biomass [41]. Therefore, there have been suggestions to remove both the leaves and roots of SRC if a maximum soil cleaning effect is projected. However, keeping in mind that SRC is considered more appropriate for moderately contaminated and simultaneously productive soils, this option should not be considered as appropriate. Species or clones that have the highest biomass growth and potential ability to store more metals in the shoots at a certain site should be preferred for commercial SRC fields.

9.2.2 SRCs and Rhizodegradation of Organic Pollution

Besides the positive effects of SRC remediating soils from heavy metals, willow and poplar SRCs have been reported to remediate a series of organic compounds, such as chlorinated solvents, explosives, petroleum hydrocarbons, cyanides, pesticides, and others [for chlorinated solvents: 42–46; for explosives: 47–49; for petroleum hydrocarbons: 50–51; for ethanol-blended gasoline: 51; for pesticides: 52–55].

Soils polluted with such compounds are usually characterized as heavily polluted and are therefore not considered fit for the production of agricultural crops. The plant roots degrade the different compounds in the soil and, in most cases, these are not absorbed into the harvested parts, as is the case with heavy metals. Although the focus of this chapter is on willow and poplar SRF systems producing biomass for energy in productive soils, which are not heavily polluted with organic compounds, poplar and willow show the ability to treat some compounds of interest in agriculture, such as pesticides, and their ability to remediate such compounds needs to be examined more closely.

9.3 Bioenergy Systems for Water Phytoremediation

In the next part of this chapter, a description of the concept behind using SRC and SRF plantations as bioenergy systems that treat and utilize municipal and industrial wastewaters, related research on issues to be considered when implementation of such systems occur as well as descriptions of selected phytoremediation systems in Sweden as more concrete examples on how this concept works will be presented. For this, indicative examples with municipal wastewater and landfill leachate (industrial wastewater) will be given to cover different cases of SRC and SRF wastewater phytoremediation systems.

9.3.1 Phytoremediation Systems with Municipal Wastewater

Municipal wastewater contains N and P and is, in most cases, a well-balanced nutrient solution that can be used for fertilizing plants. However, for sanitary reasons, it is only suitable for use on non-food and non-fodder crops, such as SRC and SRF, for bioenergy purposes. Large SRC plantations equipped with drip or sprinkler irrigation systems were established during the 1990s in Sweden, adjacent to wastewater treatment plants, to improve the efficiency of N treatment while producing biomass irrigated with wastewater. This was based on observations for low N leaching from early established willow SRC fields for energy. Bergström and Johansson [56] measured very low N concentrations (less than 1 mg N L^{-1}) in the groundwater of an intensively fertilized willow SRC field in southern Sweden. Measurements of N in the surface groundwater at the same field for a period of eight years, with average annual application rates of 112 kg N ha, showed that N concentrations remained below 1 mg N L^{-1} for the whole period, except during the year of establishment [57]. These results were in agreement with those of Mortensen et $al.$ [58] who measured close-to-zero N concentrations in drainage water from Danish SRC fields, except for the establishment year. The maximum N concentrations in the drainage water for that year were up to 100 mg N L^{-1} for the plots fertilized with 75 kg N, but were high even for control plots that did not receive any N (maximum ca. 60 mg N L^{-1}).

Therefore, for SRC plantations treated with wastewater, it was assumed that if leaching was not occurring, with a biomass production of 10 tonnes of dry matter per hectare and the N concentration in the willow shoots 0.5%, then 50 kg of N per hectare would be removed from the field at harvest each year. Differences in N leaching to the groundwater from SRC compared to the reported common N leaching figures from arable crops are rather high and they could be attributed, in some cases, to the lower input of fertilizer applied to SRC compared to "normal" fertilization rates for arable crops. To examine if SRC is equally good in N leaching performance under the situation with high N fertilization along with wastewater irrigation, differently planned experiments were carried out. Concentrations of N in the drainage water below 5 mg N L^{-1} were recorded in an experimental willow SRC field in Northern Ireland, where ca. 200 kg N ha yr^{-1} was applied [59]. Moreover, Sugiura et $al.$ [60] applied much higher amounts (ca. 300 kg N ha yr^{-1}) and N concentrations in the drainage water at different depths were between 5–10 mg L^{-1}. This figure is rather low considering the high application rate and in comparison with the findings for other arable crops. The above findings suggest that, in general, leaching of N from SRC in comparison to arable crops is significantly lower and a shift from arable crops to SRC indicates an improvement in the groundwater quality and, consequently, in the surface water quality in certain areas, even when N fertilization exceeds the recommended levels for good agriculture practices.

A practical example of such systems is established in Enköping, a town of about 20,000 inhabitants in central Sweden (Fig. 9.1). The N-rich wastewater from the dewatering of sludge, which was formerly treated in the wastewater plant, is distributed to an adjacent 75 ha willow plantation during the growing season. This water contains approximately 800 mg of N per liter and accounts for about 25% of the total N treated in the wastewater treatment plant. The water is pumped into lined storage ponds during the winter and is used for irrigating a

Fig. 9.1 A 75 ha willow phytoremediation system at Enköping, Sweden. Foreground: wastewater treatment plant; middle: ponds for winter storage of wastewater; background: willow fields irrigated by wastewater from sludge.

Credit: Pär Aronsson, SLU.

willow SRC during the summer (May to September). The system was designed so that conventionally treated wastewater could be added and mixed with nutrient-rich wastewater to promote plant growth. The willows are irrigated for about 120 days annually. The system treats about 10 tonnes of N and 0.2 tonnes of P per year in an irrigation volume of 200,000 m^3 of wastewater, of which 20,000 m^3 is water derived from the dewatering of sludge after sedimentation and centrifugation. Irrigation ceases automatically on rainy days. Irrigation rates reach a daily mean value of about 2.5 mm during the growing season [61]. Possible environmental hazards associated with such applications, *e.g.*, nutrient leaching is monitored, and the results, so far, indicate minimal risks after wastewater application [62]. The biomass produced in this system in the form of willow chips, after the harvest of 75 ha field, is sold in the local district heat and power plant, which uses only biomass as its fuel.

9.3.2 Phytoremediation Systems with Landfill Leachate

Landfill leachate, water that has percolated through landfills, is usually treated together with municipal wastewater in wastewater treatment plants. This is generally costly for the landfill operator and involves high energy consumption because the leachate must be transferred away from the site for treatment.

Therefore, landfill operators are becoming interested in alternative solutions for the on-site treatment of leachate. One method is to aerate it and then use it to irrigate willow SRC, either on restored parts of the landfills or on adjacent arable fields. The aim is to promote plant growth and minimize the potentially negative effects of the usually high ionic strength of landfill leachate with chloride concentrations, often in the order of 1,000 mg L^{-1}. The low establishment costs compared with conventional on-site engineered systems are considered as the main advantages of this method. A willow SRC plantation established on a restored cover of the landfill decreases leachate formation by means of high ET and a near-to-zero net discharge of landfill leachate can be achieved by recycling this wastewater into a SRF willow or poplar plantation, even in the humid climatic conditions of northern Europe. At the same time, hazardous compounds in the leachate (*e.g.*, ammonium and a range of persistent and potentially toxic organic substances) are taken up by the SRF plants or are retained in the soil-plant system. High concentrations of ammonium ions in the water can be an environmental hazard. However, if it is carefully monitored, ammonium can also be considered a source of N for the SRF plants.

Differences in leachate composition from various landfills under different soil and climatic conditions as well as differences in the uptake of chemicals by different clonal materials need to be considered in the design and management of leachate treatment systems involving the irrigation of SRC or SRF. During the 1990s, several systems were established in Sweden for treating landfill leachate by irrigation of willow SRC, established either on restored parts of landfills or on adjacent arable fields. Similar systems have been tested in the UK, USA, Poland, and elsewhere and scientific studies on the treatment efficiency of such systems show promising results with variable efficiency [63–66]. Studies from the UK have reported toxicity symptoms in willow plants irrigated by landfill leachate, probably due to its high ionic strength [67]. In Swedish treatment systems using SRC, plant die-back has been reported, possibly due to leachate irrigation [68]. It has been suggested that nutrient imbalances could also be the reason for such die-back, but this has not been scientifically verified. In a greenhouse pot experiment [69], willow plants were found to be at least equally sensitive to sodium as to chloride on a molar basis. At quite moderate concentrations, clear negative effects were observed on the plants at concentrations of 200 mg L^{-1} for sodium and 600 mg L^{-1} for chloride.

There are currently about 20 sites in Sweden where landfill leachate is used to irrigate willow SRC phytoremediation systems in sprinkler or drip irrigation systems. For example, at Högbytorp in central Sweden, a system operated by the Ragnsells Avfallsbehandling AB company, stores and aerates the landfill leachate in ponds and then pumps it into a 5 ha willow SRF field, which is irrigated daily during the growing season with approximately 2–3 mm of wastewater. Research results from field experiments conducted at that site testing the biological treatment efficiency of the system indicate that the aboveground plant growth was not significantly affected by irrigation with landfill leachate compared to no irrigation or irrigation with tap water [70]. Moreover, the concentrations of N in groundwater increased as a result of irrigation with landfill leachate. This clearly showed varying irrigation water concentrations and loads over time. For a satisfactory retention of N, P, and heavy metals, case-specific tests with the existing wastewater available under the specific local soil and climatic conditions seeking the optimum application loads are essential. As an example, the relative retention of total N was found to be more or less linear in Aronsson et al. [70], even at loads exceeding 2,000 kg total N. However, such application rates should not be used since the plant requirements are much lower and leaching can be significant in actual terms.

References

[1] Glass D. US. and International Markets for Phytoremediation. D. Glass Associates, Inc., Needham, Massachusetts, USA. 1999.

[2] Persson G, Lindroth A. Simulating evaporation from short-rotation forest : variations within and between seasons. J Hydrol 1994; 156: 21–46.

[3] Bungart R, Hüttl RF. Growth dynamics and biomass accumulation of 8-year-old hybrid poplar clones in a short-rotation plantation on a clayey-sandy mining substrate with respect to plant nutrition and water budget. Europ J Forest Res 2004; 123: 105–115.

[4] Hammer D, Kayser A, Keller C. Phytoextraction of Cd and Zn with *Salix viminalis* in field trials. Soil Use and Manage 2003; 19: 187–192.

[5] Laureysens I, Blust R, De Temmerman L, Lemmens C, Ceulemans R. Clonal variation in heavy metal accumulation and biomass production in a poplar coppice culture: I. Seasonal variation in leaf, wood and bark concentrations. Environ Poll 2004; 131: 485–94.

[6] Jackson MB, Attwood PA. Roots of willow (*Salix viminalis* L.) show marked tolerance to oxygen shortage in flooded soils and in solution culture. Plant Soil 1996; 187: 37–45.

[7] Perttu KL. Sludge, wastewater, leakage water, ash : a resource for energy forestry. Rapport/Avdelningen för Skoglig Intensivodling. Institutionen för Ekologi och Miljövård, Sveriges Lantbruksuniversitet 1992; 47: 7–19. (in Swedish)

[8] Riddell-Black DM. Heavy metal uptake by fast growing willow species. Rapport/Avdelningen för Skoglig Intensivodling, Institutionen för Ekologi och Miljövard, Sveriges Lantbruksuniversitet 1994; 50: 133–144.

[9] Granel T, Robinson B, Mills T, Clothier B, Green S, Fung L. Cadmium accumulation by willow clones used for soil conservation, stock fodder, and phytoremediation. Austr J Soil Res 2002; 40: 1331–1337.

[10] Dos Santos Utmazian MN, Wenzel WW. Cadmium and zinc accumulation in willow and poplar species grown on polluted soils. J Plant Nutr Soil Sci 2007; 170: 265–272.

[11] Sebastiani L, Scebba F, Tognetti R. Heavy metal accumulation and growth responses in poplar clones Eridano (*Populus deltoides* × *maximowiczii*) and I-214 (*P.* × *euramericana*) exposed to industrial waste. Environ Exp Bot 2004; 52: 79–88.

[12] Licht LA, Isebrands JG. Linking phytoremediated pollutant removal to biomass economic opportunities. Biomass Bioenergy 2005; 28: 203–218.

[13] Aronsson P, Perttu, K. Willow vegetation filters for wastewater treatment and soil remediation combined with biomass production. Forestry Chronicle 2001; 77(2): 293–299.

[14] Rockwood DL, Naidu CV, Carter DR, *et al.* Short-rotation woody crops and phytoremediation : opportunities for agroforestry? Agroforestry Syst 2004; 61: 51–63.

[15] Landberg T, Greger M. Interclonal variation of heavy metal interactions in *Salix viminalis*. Environ Tox Chem 2002; 21: 2669–2674.

[16] Wieshammer G, Unterbrunner R, Garcia TB, Zivkovic MF, Puschenreiter M, Wenzel WW. Phytoextraction of Cd and Zn from agricultural soils by *Salix* spp. and intercropping of *Salix caprea* and *Arabidopsis halleri*. Plant Soil 2007; 298: 255–264.

[17] Kuzovkina YA, Knee M, Quigley MF. Cadmium and copper uptake and translocation in five willow (*Salix* L.) species. Intern J Phytorem 2004; 6: 269–287.

[18] Robinson BH, Mills TM, Petit D, Fung LE, Green SR, Clothier BE. Natural and induced cadmium-accumulation in poplar and willow: implications for phytoremediation. Plant Soil 2000; 227: 301–306.

[19] Dickinson N, Baker A, Doronila A, Laidlaw S, Reeves R. Phytoremediation of inorganics: realism and synergies. Intern J Phytorem 2009; 11: 97–114.

[20] Schmidt MWI, Schwark L, Wiesenberg GLB. How relevant is recalcitrance for the stabilization of organic matter in soils? J Plant Nutr Soil Sci 2008; 171: 91–110.

[21] Komarek M, Tlustos P, Szakova J, Chrastny V .The use of poplar during a two-year induced phytoextraction of metals from contaminated agricultural soils. Environ Pollut 2008; 151: 27–38.

[22] Evangelou MWH, Ebel M, Schaeffer A. Chelate assisted phytoextraction of heavy metals from soil: effect, mechanism, toxicity, and fate of chelating agents. Chemosphere 2007; 68: 989–1003.

[23] Vandecasteele B, De Vos B, Tack FMG. Cadmium and zinc uptake by volunteer willow species and elder rooting in polluted dredged sediment disposal sites. Sci Total Environ 2002; 299: 191–205.

[24] Unterbrunner R, Puschenreiter M, Sommer P, *et al.* Heavy metal accumulation in trees growing on contaminated sites in Central Europe. Environ Poll 2007; 148: 107–114.

[25] Klang-Westin E, Eriksson J. Potential of *Salix* as phytoextractor for Cd on moderately contaminated soils. Plant Soil 2003; 249: 127–137.

[26] Meers E, Vandecasteele B, Ruttens A, Vangronsveld J, Tack FMG. Potential of five willow species (*Salix* spp.) for phytoextraction of heavy metals. Environ Exp Bot 2007; 60: 57–68.

[27] Eriksson J, Ledin S. Changes in phytoavailability and concentration of cadmium in soil following long term *Salix* cropping. Water Air and Soil Pollution 1999; 114: 171–184.

[28] Keller C, Hammer D, Kayser A, Richner W, Brodbeck M, Sennhauser M. Root development and heavy metal phytoextraction efficiency: comparison of different plant species in the field. Plant Soil 2003; 249: 67–81.

[29] Rosselli W, Keller C, Boschi K. Phytoextraction capacity of trees growing on a metal contaminated soil. Plant Soil 2003; 256: 265–272.

[30] Dickinson NM, Pulford ID. Cadmium phytoextraction using short-rotation coppice *Salix*: the evidence trail. Environ Intern 2005; 31: 609–613.

[31] Rytter RM, Hansson AC. Seasonal amount, growth and depth distribution of fine roots in an irrigated and fertilized *Salix viminalis* L. plantation. Biomass and Bioenergy 1996; 11: 129–137.

[32] Berndes G, Fredrikson F, Borjesson P. Cadmium accumulation and *Salix*-based phytoextraction on arable land in Sweden. Agriculture Ecosystems and Environment 2004; 103: 207–223.

[33] Lewandowski I, Schmidt U, Londo M, Faaij A. The economic value of the phytoremediation function—Assessed by the example of cadmium remediation by willow (*Salix* ssp.). Agricultural Systems 2006; 89: 68–89.

[34] Dimitriou I, Rosenqvist H. Sewage sludge and wastewater fertilisation of Short Rotation Coppice (SRC) for increased bioenergy production—biological and economic potential. Biomass and Bioenergy 2011; 35(2): 835–842.

[35] Dimitriou I. Performance and sustainability of short-rotation energy crops treated with municipal and industrial residues. Doctoral diss. Dept. of Short Rotation Forestry, SLU. Acta Universitatis agriculturae Sueciae vol. 2005; 44.

[36] Lazdina D, Lazdins A, Karins Z, Kaposts V. Effect of sewage sludge fertilization in short-rotation willow plantations. Journal of Environmental Engineering and Landscape Management 2007; 15: 105–111.

[37] Pulford ID, Riddell-Black D, Stewart C.Heavy metal uptake by willow clones from sewage sludge-treated soil: the potential for phytoremediation. Intern J Phytorem 2002; 4: 59–72.

[38] Dimitriou I, Eriksson J, Adler A, Aronsson P, Verwijst I. Fate of heavy metals after application of sewage sludge and wood-ash mixtures to short-rotation willow coppice. Environ Poll 2006; 142: 160–169.

[39] Puschenreiter M, Wenzel W.Short rotation forestry and phytoextraction—a good combination? Centralblatt für das gesamte Forstwesen 2007; 124: 189–200.

[40] Vandecasteele B, Meers E, Vervaeke P, De Vos B, Quataert P, Tack FMG. Growth and trace metal accumulation of two *Salix* clones on sediment-derived soils with increasing contamination levels. Chemosphere 2005; 58: 995–1002.

[41] Landberg T, Greger M. Differences in uptake and tolerance to heavy metals in *Salix* from unpolluted and polluted areas. Applied Geochemistry 1996; 11(1–2): 175–180.

[42] Gordon M, Choe N, Duffy J, et al. Phytoremediation of trichloroethylene with hybrid poplars. Phytoremediation of Soil and Water Contaminants 1997; 664: 177–185.

[43] Shang TQ, Doty SL, Wilson AM, Howald WN, Gordon MP. Trichloroethylene oxidative metabolism in plants: the trichloroethanol pathway. Phytochemistry 2001; 58: 1055–1065.

[44] Xiang S, Chen YiTai, Rao LB, Duan HP. Phytotoxicity, uptake and degradation of 2,4-dichlorophenol in *Salix integra*. China Environmental Science 2008; 28 (10): 921–926.

[45] Larsen M, Burken J, Machackova J, Karlson UG, Trapp S. Using tree core samples to monitor natural attenuation and plume distribution after a PCE spill. Environmental Science and Technology 2008; 42(5): 1711–1717.

[46] Mills T, Arnold B, Sivakumaran S, et al. Phytoremediation and long-term site management of soil contaminated with pentachlorophenol (PCP) and heavy metals. Journal of Environmental Management 2008; 79: 232–241.

[47] Brentner LB, Mukherji ST, Merchie KM, Yoon JM, Schnoor JL, Van Aken B. Expression of glutathione S-transferases in poplar trees (*Populus trichocarpa*) exposed to 2,4,6-trinitrotoluene (TNT). Chemosphere 2008; 73(5): 657–662.

[48] Van Aken, B, Yoon JM, Just CL, Schnoor JL. Metabolism and mineralization of hexahydro-1,3,5-trinitro-1,3,5-triazine inside poplar tissues (*Populus deltoides* × *nigra* DN-34). Environmental Science and Technology 2004; 38: 4572–4579.

[49] Yoon JM, Oh BT, Just CL. Schnoor JL. Uptake and leaching of octahydro-1,3,5,7-tetranitro-1,3,5,7-tetrazocine by hybrid poplar trees. Environmental Science and Technology 2002; 36: 4649–4655.

[50] Palmroth MRT, Pichtel J, Puhakka JA. Phytoremediation of subarctic soil contaminated with diesel fuel. Bioresource Technology 2002; 84: 221–228.

[51] Corseuil HX, Moreno FN. Phytoremediation potential of willow trees for aquifers contaminated with ethanol-blended gasoline. Water Research 2001; 35: 3013–3117.

[52] Skaates SV, Ramaswami A, Anderson LG.Transport and fate of dieldrin in poplar and willow trees analyzed by SPME. Chemosphere 2005; 61 (1): 85–91.

[53] Aitchison EW, Kelley SL, Alvarez PJJ, Schnoor JL. Phytoremediation of 1,4-dioxane by hybrid poplar trees. Water Environment Research 2000; 72: 313–321.

[54] Burken JG, Schnoor JL. Uptake and metabolism of atrazine by poplar trees. Environmental Science and Technology 1997; 31: 1399–1406.

[55] Predieri S, Figaj J, Rachwal L, Gatti E, Rapparini F. Selection of woody species with enhanced uptake capacity: the case-study of Niedwiady resort pollution by pesticides stored in bunkers. Minerva Biotecnologica 2001; 13: 111–116.

[56] Bergström L, Johansson R. Influence of fertilized short-rotation forest plantations on nitrogen concentrations in groundwater. Soil Use Manage 1992; 8: 36–40.

[57] Aronsson P. Nitrogen Retention in Vegetation Filters in Short-Rotation Willow Coppice. Acta Universitatis Agriculturae Sueciae, Silvestria, 2000; 16.

[58] Mortensen J, Nielsen KH, Jørgensen U. Nitrate leaching during establishment of willow (*Salix viminalis*) on two soil types and at two fertilization levels. Biomass Bioenergy 1998; 15(6): 457–466.

[59] Werner A, McCracken A. The use of Short Rotation Coppice poplar and willow for the bioremediation of sewage effluent. Aspects of Applied Biology, Biomass and Energy Crops III 2008; 90: 317–324.

[60] Sugiura A, Tyrrel S, Seymour I. Growth and water use of *Salix viminalis*, *Populus trichocarpa* and *Eucalyptus gunnii* field trial plantation irrigated with secondary treated effluent. Aspects of Applied Biology, Biomass and Energy Crops III 2008; 90: 119–126.

[61] Dimitriou I, Aronsson P. Willows for energy and phytoremediation in Sweden. Unasylva 2005: 221 (56): 46–50.

[62] Dimitriou I, Aronsson P. Wastewater and sewage sludge application to willows and poplars grown in lysimeters—Plant response and treatment efficiency. Biomass and Bioenergy 2011; 35(1): 161–170.

[63] Bialowiec A, Wojnowska-Baryla I, Agopsowicz M. The efficiency of evapotranspiration of landfill leachate in the soil-plant system with willow *Salix amygdalina* L. Ecological Engineering 2007; 30(4): 356–361.

[64] Alker GR. Phytoremediation of nutrient rich wastewaters and leachates using *Salix*. Ph.D thesis. Department of Civil and Environmental Engineering, Imperial College of Science, Technology and Medicine, University of London, 1999.

[65] Dimitriou I, Aronsson P, Weih M. Stress tolerance of five willow clones after irrigation with different amounts of landfill leachate. Bioresource Technology 2006; 97 (1): 150–157.

[66] Watzinger A, Reichenauer TG, Blum WEH, Gerzabek MH. Leachate production and gas composition in a revegetated landfill cover under landfill leachate irrigation. In: Christersen T, Cossu R and Stegmann R (Eds.). Ninth International Waste Management and Landfill Symposium, Proceedings. Sardinia, Italy, 2003; 390–391.

[67] Stephens W, Tyrrel SF, Tiberghien JE. Irrigating short rotation coppice with landfill leachate: constraints to productivity due to chloride. Bioresource Technology 2000; 75 (3): 227–229.

[68] Dimitriou I, Aronsson P, Weih M. Evaluation of leaf length and fluctuating asymmetry as stress indicators after irrigation of willow with landfill leachate. In: Christersen T, Cossu R and Stegmann R (Eds.). Ninth International Waste Management and Landfill Symposium, Proceedings. Sardinia, Italy 2003: 594–595.

[69] Dimitriou I, Aronsson P. Landfill leachate treatment on Short-Rotation Willow Coppice: resource instead of waste? In: Landfill Research Focus. Nova Publishers, New York, USA, 2007: 263–291.

[70] Aronsson P, Dahlin T, Dimitriou I. Treatment of landfill leachate by irrigation of willow coppice—Plant response and treatment efficiency. Environmental Pollution 2010; 158: 795–804.

Part V
Life Cycle Assessment Principles

Chapter 10

Eight Principles of Uncertainty for Life Cycle Assessment of Biofuel Systems

Adam J. Liska

10.1 Introduction: Regulatory LCA

New environmental regulations in the USA and Europe require a reduction of greenhouse gas (GHG) emissions from transportation fuels as a component of climate change mitigation policy. The US Energy Independence and Security Act of 2007 (EISA) requires GHG emission reductions from the life cycles of biofuels compared to gasoline, by 20% for ethanol from maize grain (maize-ethanol), 60% for cellulosic ethanol, and 50% for other advanced biofuels. To determine these reductions, the US Environmental Protection Agency (EPA) employs life cycle assessment (LCA) methods which were not used previously in national environmental regulations. These regulations, entitled the "Renewable Fuel Standard 2" (RFS2), build on concurrent state efforts by the California Air Resources Board (CARB) under the Low Carbon Fuel Standard (LCFS). These regulations can affect billions of dollars in financial incentives and market access for the existing biofuel industry and they will determine how new feedstocks for biofuels are developed in the future.

Over roughly the last twenty years, LCA has been applied to biofuel production systems for determining GHG emissions and energy efficiency, but these evolving methods have been inconsistent [1–3]. These methods are used to estimate direct emissions from the life cycle from crop production to finished fuels, while also considering upstream emissions such as from fertilizer production. Contrary to these relatively simple analyses, the assessments currently developed under state and federal law are generally far more complex by including global modeling. The use of global models has been encouraged by findings that indirect effects from biofuel production, which are international in scope, lead to additional GHG emissions that were not previously recognized. Emissions related to indirect land use change (ILUC) from biofuel production are now quantified under RFS2 and LCFS legislation [4]. However, accounting for one indirect emission further necessitates the evaluation of other indirect changes in global emissions [5–6], which has led to the immense complexity now seen in federal LCA regulations.

Fundamentally, LCA integrates diverse data sources associated with an industrial process to: (i) quantify environmental impacts as continuous variables (*e.g.*, GHG emission rates) and (ii) guide improvements in efficiency. The related field of risk assessment attempts to estimate the probability of discrete events that are not easily predicted, such as the timing of system failures [7]. In LCA, the most probable performance of a specific type of system operating in the recent past, currently, or in the recent future is estimated based on measurements of patterns and frequencies in industry (*e.g.*, parameter values). By assembling a set of frequencies describing the system (although with incomplete information), LCA is based on a probability theory that states that the frequencies of future events will be approximated by past frequencies, given enough replicated observations under similar conditions [8].

The models employed in LCA are regulatory tools to archive knowledge, interpret and predict the links between industrial activities and outcomes of interest, communicate findings, and explore uncertainty and shortcomings in understanding [9].

10.2 Eight Principles of Uncertainty for LCA of Biofuel Systems

This chapter proposes eight principles of uncertainty for LCA of biofuels that will help to minimize errors in estimating direct and indirect emissions when designing and implementing regulatory LCA methods. Two main types of uncertainty arise in LCA models and other regulatory settings: (i) parameter uncertainty arises due to spatial and temporal variability in the numerical value of a parameter and a lack of information concerning this variability and its actual value at any one point (principles 1, 3–6, 8) and (ii) model uncertainty arises due to the incoherence between the structure of a model and the system under investigation and includes uncertain system boundaries (principles 1–2, 6, 8).

The eight proposed principles below were developed because of how they relate to one another:
1. biofuel systems are highly variable and complex;
2. invariable LCA methods to assess this complexity do not exist;

3. information deficiencies are extensive in assessing this complexity;

4. analysis of localized systems can reduce some variability and uncertainty;

5. sensitive factors are often uncertain and undermine the accuracy of LCAs by orders of magnitude;

6. expanding LCA system boundaries to a global level tends to increase uncertainty and restricts the accuracy of using LCA for predicting system performance;

7. clear presentation of data in LCA can ensure that biases are limited; and

8. reference systems are just as complex and uncertain.

These principles were developed based on previous research and new analyses presented here.

10.3   Principle 1: Biofuel Production Is a Complex System of Systems

The biofuel production process may best be conceptualized and characterized as a system of systems (SoS). The emerging discipline of SoS engineering is defined by the International Council on Systems Engineering in a manner appropriate for describing biofuels:

"System of systems applies to a system of interest whose system elements are themselves systems; typically, these entail large-scale inter-disciplinary problems with multiple, heterogeneous, distributed systems" [10].

Five interdependent subsystems in fuel supply chains have been identified, which are similar for both biofuels and fossil fuels. These subsystems include: (i) feedstock production, (ii) feedstock logistics, (iii) feedstock-to-fuel conversion, (iv) fuel distribution, and (v) fuel end-use. In the case of maize-ethanol, for example, these components would comprise, respectively: (i) a cropping system (e.g., rain-fed maize-soybean rotation), (ii) a grain harvesting, transportation, and storage system, (iii) a biorefinery and associated regime for co-product processing and use (e.g., feeding of distiller grains to livestock), (iv) a rail or potentially a pipeline, distribution network to fuel blenders, and gas stations, and (v) use of ethanol as either a 10% blend with gasoline (E10) in most cars or use of an 85% ethanol blend (E85) in flex fuel vehicles. Within each of these five systems, technical, spatial, and temporal variabilities add to the uncertainty in defining its performance.

The field of industrial ecology seeks to characterize the environmental impacts of the life cycles of production systems, with the goal of improving system performance. Among the methods in industrial ecology, LCA is recognized for the analysis of a SoS [11]. In LCA, the complexities in the subsystems investigated must be greatly simplified and reduced to one or a few parameters (e.g., efficiencies) that best characterize the performances of the supply chain components. Using these efficiencies, LCA models are built to generate transparent emission inventories from a complex SoS (Fig. 10.1).

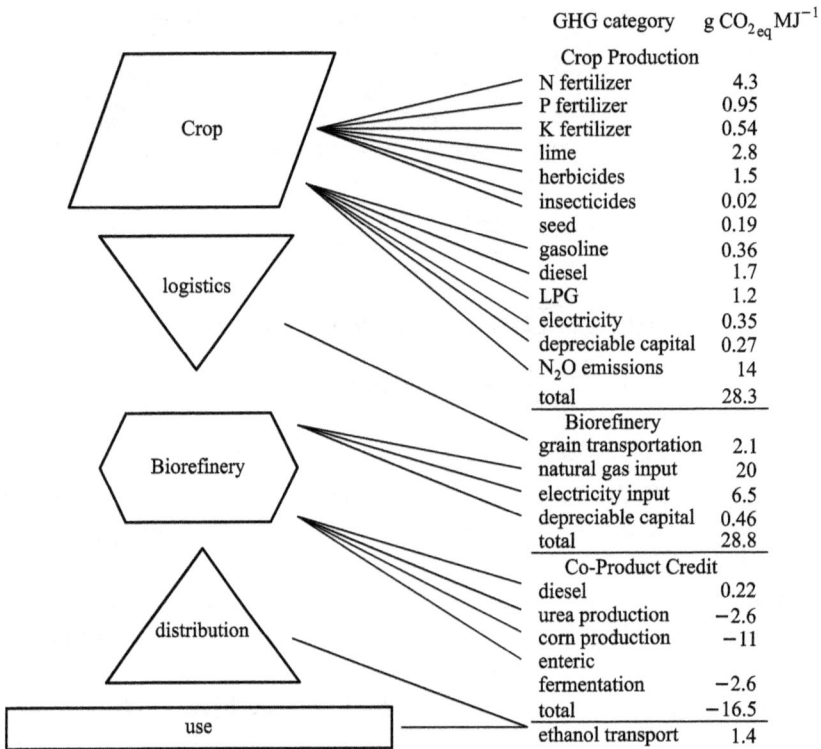

GHG category	g CO_{2eq} MJ^{-1}
Crop Production	
N fertilizer	4.3
P fertilizer	0.95
K fertilizer	0.54
lime	2.8
herbicides	1.5
insecticides	0.02
seed	0.19
gasoline	0.36
diesel	1.7
LPG	1.2
electricity	0.35
depreciable capital	0.27
N_2O emissions	14
total	28.3
Biorefinery	
grain transportation	2.1
natural gas input	20
electricity input	6.5
depreciable capital	0.46
total	28.8
Co-Product Credit	
diesel	0.22
urea production	−2.6
corn production	−11
enteric fermentation	−2.6
total	−16.5
ethanol transport	1.4

Fig. 10.1 A system of systems to GHG emission inventory for maize-ethanol. Inventory categories and data from [12].

When applying LCA models to biofuels, the feedstock employed generally determines many of the key characteristics of the biofuel SoS. For example, the production of ethanol from either grain or non-grain biomass requires the use of different harvests and logistical practices, biorefinery infrastructure and conversion methods, and co-product types [13]; ethanol and biodiesel also have different fuel distribution and use systems. Feedstock types also largely determine average gross bioenergy yields, which can greatly differentiate biofuel systems, and determine the profitability and adoption of these systems (Fig. 10.2). Energy yield data is also essential for characterizing the thermodynamic efficiency of

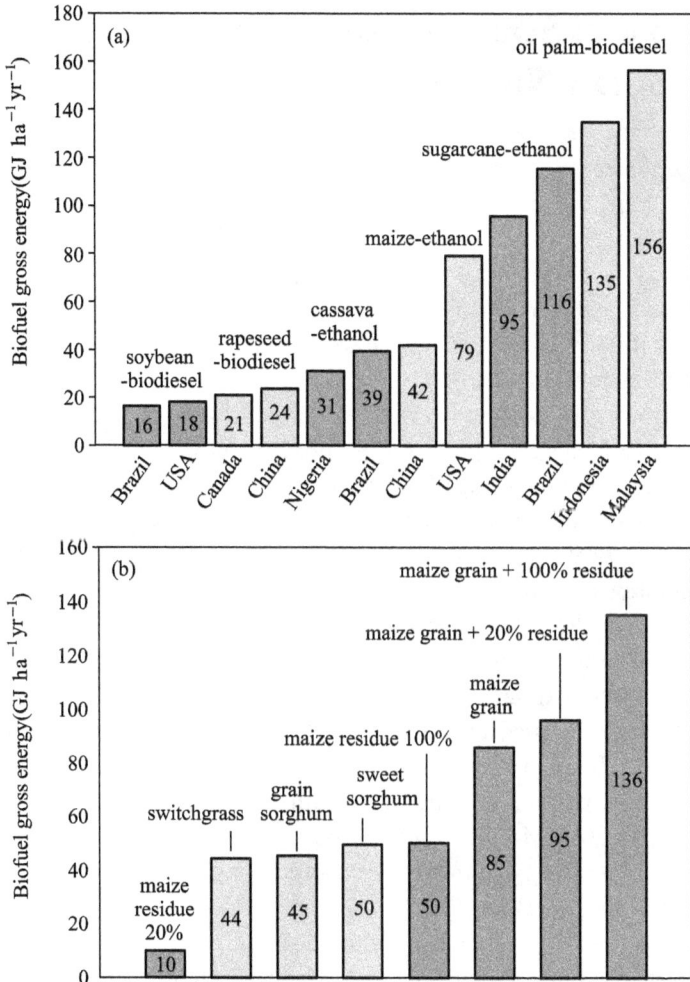

Fig. 10.2 Biofuel gross energy yields from average ethanol and biodiesel production systems using food crops globally (a) and from ethanol in Nebraska from various crops (b), based on reported average crop yields and field studies in Nebraska; co-products are not included.

Source: data from [1] (a) and [15] (b).

the life cycle (*e.g.*, net energy efficiency) [14]. For example, soybean biodiesel produced in Brazil and the USA has about a ten-fold lower gross energy yield than biodiesel from oil palm in Malaysia, which can determine land use efficiency. Even looking within the state of Nebraska in the USA, theoretical gross energy yields for an integrated system producing cellulosic ethanol from maize residue and ethanol from grain can have about a two-fold higher yield compared to average ethanol yields from sorghum and switchgrass.

10.4 Principle 2: Standardized LCA Methods for Biofuels Do Not Exist

The International Organization for Standardization provides general recommendations for the LCA for any metric or production system analyzed. They suggest, among other things, that "LCA is an iterative technique. Therefore, the scope of the study may need to be modified while the study is being conducted as additional information is collected" [16]. Generic standards that appropriately commend the continued improvement of LCA have been inadequate for defining consistent LCA practices for biofuels.

Controversy has historically surrounded the assessment of the net energy balance (*i.e.*, energy outputs/energy inputs) of the production of maize-ethanol. Most of the past studies of the life cycle of maize-ethanol have used LCA models with roughly 300 to 400 parameters, mainly composed of a combination of input parameters (*e.g.*, application intensities or efficiencies) and emission factors (*e.g.*, GHG emission intensities for primarily CO_2, N_2O, and CH_4). In 2006, the Energy Resources Group's Biofuel Analysis MetaModel (EBAMM) was used to estimate the most appropriate values for key system parameters within consistent boundaries of six major studies [17]. The study found a 20% positive net energy return over energy invested and a 13% GHG emission reduction compared to gasoline. GHG, Regulated Emissions, and Energy use in Transportation (GREET) model from the US Argonne National Laboratory was one of the models analyzed in the EBAMM study. The GREET model has received the most development out of all life cycle models for biofuels and it now serves as a component model for both CARB and EPA regulatory LCA methods.

Unlike measuring the mass of molecules, the methods employed in the LCAs of biofuels are not absolute but are dependent on relative system boundaries in addition to uncertain parameter values. The Biofuel Energy System Simulator (BESS) model was developed based on the EBAMM model but used new survey statistics for biorefinery energy efficiency and found that the life cycle of ethanol from maize was substantially more efficient than previously estimated. In Liska *et al.* (2009), maize-ethanol was found to reduce GHG emissions compared to gasoline by 51% on average for natural gas-powered biorefineries (based on direct emissions), which made up 90% of the USA's ethanol industry in 2008 [12, 15]. Some of the data employed in that analysis was found to be less representative of the biofuel systems in question (*e.g.*, electricity GHG intensity) and some parameter values were changed [18–20]. Based on the suggested changes and new co-product analyses, an updated analysis of maize-ethanol was found to reduce GHG emissions by $46.5 \pm 2.3\%$ compared to gasoline, corresponding to an intensity of 52.2 ± 2.8 g of C dioxide equivalent per megajoule (g CO_{2eq} MJ^{-1}) of energy in the fuel [20]. Using a modified GREET model, the CARB currently finds the same class of biorefineries to have an intensity of between 60.1 to 68.4 g $CO_{2_{eq}}$ MJ^{-1} [21] and the EPA estimates that maize-ethanol will have an intensity of roughly 43 g $CO_{2_{eq}}$ MJ^{-1} in 2022 [22], not including indirect land use change emissions.

These examples highlight some of the difficulties of having no standard LCA methods for biofuels, which is a reflection of the few guidelines that are specific enough to generate consistent quantitative measurements describing these systems. Seemingly small changes in system boundaries in these models can markedly change LCA results. Comparing denatured ethanol (which contains a low level of gasoline) and oxygenated gasoline (which contains a low level of ethanol), as done by the CARB, instead of a comparison of pure ethanol and pure petroleum, increases the GHG intensity of maize-ethanol by roughly 3 to 7 g $CO_{2_{eq}}$ MJ^{-1} (or 3% to 7%) and biases against the use of ethanol [19–20].

Today, the models employed by regulators are the nearest methods to being defined as standards, as they determine economic incentives such as market access and subsidies. Yet, difficulties exist as these immense, multi-faceted models now estimate global changes and likely do not accurately predict actual system performance. There appears to be a USA consensus in the use of g $CO_{2_{eq}}$ MJ^{-1} as the standard GHG emission metric because it is being adopted in regulations [1], although European observers prefer g $CO_{2_{eq}}$ km^{-1}, despite variable fuel efficiency (km MJ^{-1}), with different vehicle types [23]. To successfully assess the absolute results from any LCA, the corresponding regulatory policy or relative frame of reference must be identified.

10.5   Principle 3: Empirical Data Are Scarce for Most Aspects of Biofuels

To define a biofuel SoS using LCA, each subsystem must be sufficiently characterized and particularly those that contribute the most to GHG emissions (see Section 10.6–10.8). In the case of maize-ethanol produced in the USA, the US Department of Agriculture (USDA) provides recent data on crop yields at the county and state levels, updated annually, and fertilizer rates at the state level are updated every few years. To determine GHG emissions from these inputs, standard emission factors are available from the EPA and the Intergovernmental Panel on Climate Change (IPCC).

More limited data are available for most other parameters. First, as of 2010, the last released USDA survey data on fossil fuel use for the USA's maize production was from 2001; the average energy use during that time is suspected to have decreased due to the use of more no-till practices [12]. Biorefinery efficiencies have been based on limited recent surveys, often representing less than a quarter of the industry capacity (see Section 7.1) [12] and upstream emissions, such as from N fertilizer production, are not well-characterized for specific suppliers. Ecosystem emissions such as from N_2O and soil organic carbon (SOC) loss to CO_2 have only been measured in limited studies (see Sections 7.2 and 7.3). In general, the most accurate average (*i.e.*, expected value) for any variable will incorporate data from the full range of the probability distribution of observable values, instead of the use of clearly limited data that biases the analysis and misrepresents the systems [8, 24].

To overcome these data deficiencies, regulators and other federal agencies are conducting more thorough and frequent surveys (the US National Agricultural Library has recently initiated the development of a LCA database, http://www.lcacommons.gov [25]). The current LCA approach taken by the EPA, however, uses more industry averages and less data specifically for regulated facilities because such data collection was explicitly stated to be too burdensome. There are clearly declining marginal returns on investment for the collection of additional data; yet, regulators and those regulated must weigh the costs and benefits of increased investments.

10.6   Principle 4: Local Biofuel LCAs Reduce Uncertainty and Errors

Accuracy in LCA is achieved from the "bottom-up" based on measurements for individual system parameters, and it cannot be verified from the "top-down" using GHG emissions measurements of the entire SoS. Where system boundaries are fixed, uncertainty in LCA primarily originates from an information deficiency, and in general, more information is used to reduce this uncertainty [26]. The analysis of an individual biofuel production system can incorporate more easily accessible and well-defined information compared to an analysis of a whole industry containing many biorefineries; most LCAs combine a set of frequencies from different aspects of the system measured at different places and times.

Crop production contributes approximately 50% of positive life cycle GHG emissions from maize-ethanol and the use of state values for local refineries would reduce errors in estimating cropping emissions because of the variability between states [12] (Fig. 10.1). Use of the 12-state Midwest average GHG intensity of 263 kg CO_{2eq} per Mg of grain corresponds to a roughly 48% GHG reduction compared to gasoline; yet, the use of individual state values produces a range of GHG reductions from 40% to 56% [12]. Use of the industry average provides a more favorable assessment to underperforming states and does not recognize the higher efficiencies of other states (*e.g.*, Iowa; Fig. 10.3). Differences in emission intensities by state are primarily due to declining crop yields from north to south due to higher plant respiration and lower soil carbon levels in the south, which requires higher rates of N fertilizer to achieve the desired crop yields due to less indigenous N in the soil [12].

The uncertainty in defining the shape of the probability distribution function for a variable is, in general, thought to be a major source of model uncertainty [27]. The distribution of biorefineries relative to agricultural emissions is one example of a non-normal (*e.g.*, lognormal or Weibull) distribution, although it is for a calculated metric and it does not arise from a single measurement. Biorefineries tend to be built where grain yields and nutrient use efficiencies are highest, thus maximizing profitability and establishment in states with lower GHG emissions per unit of crop yield (Fig. 10.3).

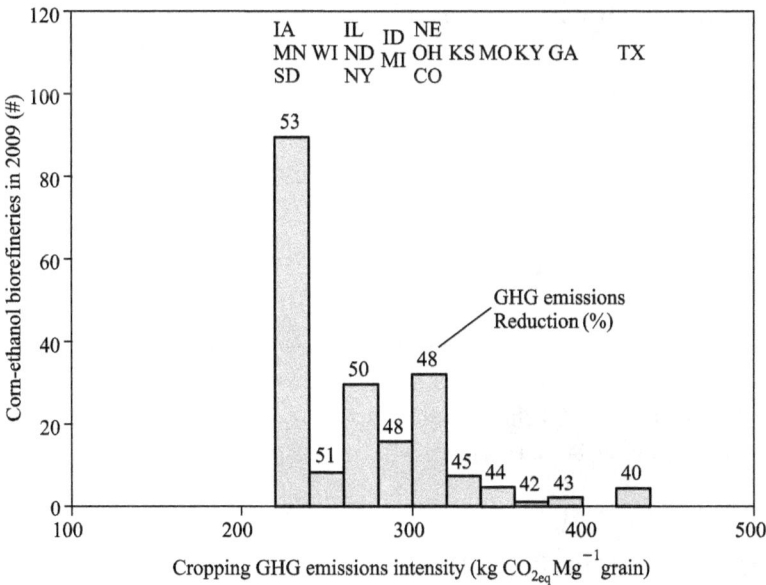

Fig. 10.3 Distribution of ethanol biorefineries having specific crop GHG emissions per unit of grain produced.
Source: based on data from [12].

Another source of regional variability is associated with co-product production and use (allocation of emissions among co-products in LCA is another major issue for determining emissions that are related to model structure; related to Section 10.4 [28]). Dry mill biorefineries generally produce dry, modified, or wet distiller grains with solubles (DGS), which can be variable from year to year. Beef cattle substitute more GHG-intense maize grain in their diets with DGS, compared to dairy cattle and swine, which substituted relatively less grain and more soybean meal in their diets [20]. Co-product credits (emission off-sets, Fig. 10.1) based on variable substitution efficiencies have been found to range from 12 g CO_{2eq} MJ^{-1} for dry DGS fed to dairy and swine to 18 g CO_{2eq} MJ^{-1} for wet DGS fed to beef cattle (Midwest average at 15.2). Using recent industry statistics, the natural gas efficiency of dry mill biorefineries ranged between 8.33 MJ L^{-1} of ethanol when producing all dry distiller grains to 4.91 MJ L^{-1} when producing all wet distiller grains. Combining variable natural gas use due to co-product processing with variable co-product credits resulted in GHG emission reductions at 43%–55%, compared to gasoline for Midwest average maize-ethanol, corresponding to 56–44 g CO_{2eq} MJ^{-1} [20].

The above variabilities in cropping systems and co-product feeding is not currently recognized by federal or state regulators when assessing individual facilities. Significant variability in these systems necessitates that state-level agricultural GHG assessments be performed to ensure accuracy for regulating GHG emissions from individual biorefineries, instead of taking broad averages across USA agriculture. Use of state averages could reduce the errors associated with estimates by more than 20%, in some cases.

10.7   Principle 5: Sensitive Parameters Cause Order of Magnitude Changes

In addition to the relatively minor variability presented above, common variability in the value of sensitive parameters can lead to order of magnitude changes in GHG emissions estimates. Of the three examples below, natural gas efficiency is the least sensitive, but it is more sensitive than the parameters above.

10.7.1   Biorefinery Natural Gas Efficiency

Natural gas use per unit of ethanol produced at the biorefinery appears to be the parameter by which normal variations lead to the largest differences in GHG emission intensities of the maize-ethanol life cycle. Using a 2001 survey of wet and dry mills, the EBAMM model employed biorefinery thermal energy input values for natural gas and coal at 13.9 MJ per liter of ethanol, in total [17]. From 2001 to 2006, the capacity of the USA's ethanol industry grew by roughly threefold and, by 2008, 90% of the installed biorefinery capacity was dry mills and 89% of the capacity was powered by natural gas [15]. Based on multiple independent surveys from 2006, the efficiency of new natural gas dry mills was found to be roughly 7.7 MJ of natural gas per liter of ethanol produced, on average [12] (corresponding well with a much larger industry survey in 2008 [29]), thereby reducing thermal energy requirements at the biorefinery from 67% of the life cycle energy inputs to 56% of the inputs from 2001 to 2006. By substituting the 2006 efficiency value for the previous 2001 efficiency (for wet and dry mills) in the EBAMM model (thus, from 13.9 to 7.7 MJ L^{-1}), maize-ethanol is found to reduce the life cycle GHG emission compared to gasoline by 55% (corresponding to 42 g $CO_{2_{eq}}$ MJ^{-1}), compared to the previous updated finding of a 13% reduction [17, 19]. This example shows a greater than four-fold difference in GHG emission reductions and clearly demonstrates the sensitivity of this single parameter and the need for accuracy in its definition.

10.7.2   Agricultural N_2O Emissions

Nitrous oxide (N_2O) is a potent GHG with a global warming potential that is 298 times CO_2 on a mass basis and is produced by agroecosystems via the denitrification of nitrate in soils and water [30]. In maize production, direct and indirect N_2O emissions from synthetic N applications are estimated to be roughly 36% of cropping GHG emissions based on default emission factors from the IPCC [15]. Additional N_2O emissions from crop biomass and manure constitute another 13% of emissions, making N_2O alone nearly 50% of cropping GHG emissions in maize systems, based on the IPCC values. The IPCC default values are used in national GHG emission inventories and represent a broad international consensus based on available studies.

Yet, N_2O emissions are highly variable due to soil moisture and temperature differences and field measurements are costly and limited. When not calibrated with direct measurement data, six models were recently shown to predict N_2O emissions with a range nearly six-fold from 3.8 to 21 kg N ha^{-1} yr^{-1}, suggesting that the use of N_2O emission models without measurement data is "quite uncertain at this time" [31].

A recent analysis from Crutzen et al. [32] suggested that N_2O emissions downstream from field N application could lead to higher total emission rates than predicted by the IPCC. Whereas, the IPCC suggests that 1.33% of N application in maize systems is converted to N_2O on average (from direct and indirect losses) [15], Crutzen et al. [32] controversially propose that N_2O emissions are 3%–6% of N applied due to additional background N_2O emissions produced downstream. Inclusion of these variable N_2O rates leads to dramatically different results in the life cycles of biofuels. At 1.5% of N converted, roughly 15 g $CO_{2_{eq}}$ MJ^{-1} is added due to N_2O in the maize-ethanol life cycle [33]. At 5% N conversion, 41 to 56 g $CO_{2_{eq}}$ MJ^{-1} is added to the life cycle from N_2O emissions, thus changing

GHG emission reductions of this biofuel relative to gasoline from roughly 40% to zero. Further research is needed to better quantify actual direct and indirect N_2O emissions and this will be an important factor for all crop-based biofuels.

10.7.3 Soil Organic Carbon Dynamics and CO_2 Emissions

Cropping systems associated with biofuel production can have a range of impacts on soil quality. Yet, for the LCA of biofuel production, three examples of ethanol production systems show that changes in SOC are perhaps the most critical factors in determining net GHG emissions.

Many studies have assumed that ethanol from residue leads to a biofuel system with the potential for large GHG reductions compared to gasoline (*e.g.*, 84%–106%) [34]. In producing cellulosic ethanol from maize residue, the impact of residue removal on SOC loss and its impact on life cycle emissions is limited in recent scientific literature. Recent summaries of field research have found that crop residue removal generally tends to reduce SOC levels [35–36]. If SOC is lost due to oxidation to CO_2 based on a broadly accepted understanding of soil processes [37] (assuming soil erosion is also limited), then a simple calculation can determine the GHG impact of this loss. Removing 25% of maize residue could reduce SOC by roughly 0.3 Mg C per hectare per year, which would add roughly 88 g CO_{2eq} MJ^{-1} to other production emissions in this system; similar results are found at the 100% removal level (Tab. 10.1). Inclusion of this emission from SOC cancels out nearly all of the GHG benefits of this system, reducing emission reductions from roughly 90% to roughly 0–30%. These results challenge the prevailing understanding of soil processes in the LCA of this system.

Rates of SOC losses from maize residue removal were recently applied to sweet sorghum, a similar C4 crop, in a scenario in which all residue was removed [38]. By incorporating estimated SOC loss into the life cycle emission inventory, ethanol from sweet sorghum was found to be roughly 10%–20% more GHG-intense when compared to gasoline (Tab. 10.1). Alternatively, when all residue was assumed to be left on the field, assuming no net SOC change, ethanol from sweet sorghum reduced GHG emissions compared to gasoline by 50% [38]. Thus, if not managed properly, SOC loss has previously been shown to be able to possibly negate all GHG benefits.

In a third example, C sequestration (transfer of atmospheric CO_2 to SOC) is a key variable for dedicated energy crops. Sequestration reduces net life cycle emissions in switchgrass by more than 70%, which has led to estimates that this system will reduce GHG emissions by up to 94% compared to gasoline [39]. However, limited measurements of SOC dynamics under harvested switchgrass and energy crops lead to the current uncertainty in determining accurate sequestration rates.

Tab. 10.1 Net CO_2 emissions from SOC in the life cycle of ethanol from sweet sorghum (sugar only), maize residue, and switchgrass (latter two as cellulosic).

Biofuel system	SOC loss	Energy yield	SOC adder[a]	Production emissions[b]	Life cycle total	GHG reduction[c]
	Mg C hm^{-2}	GJ hm^{-2}		g CO_{2eq} MJ^{-1}		%
maize residue, 25%[d]	0	13	0	10	10	89
maize residue, 25%[d,e]	0.30	13	88	10	98	−3
maize residue, 100%[d,e]	0.80	50	58	10	68	28
sweet sorghum[f]	0	50	0	46	46	52
sweet sorghum[e,f]	0.80	50	59	46	105	−11
switchgrass[g]	−0.27	60	−16	22	6	94

a. The SOC adder (g CO_{2eq} MJ^{-1}) is determined by multiplying net SOC dynamics per hectare by 44/12 to convert to grams of CO_2, then dividing by energy yield in ethanol, and correcting for units. b. Production emissions from residue use are approximations. c. Reduction of GHG emissions is compared to gasoline estimated at 95 g CO_{2eq} MJ^{-1}. d. Energy yields from residue removal [15]. e. SOC loss from residue removal [36]. f. Sweet sorghum yields [38]. g. Switchgrass data [39]. These calculations are consistent with calculations using more complex models [4, 12].

10.7.4   Setting an Uncertainty Standard for Biofuel LCA

The examples above are essential for understanding some of the main sources of uncertainty in the LCA of biofuels and should be considered when making decisions about setting acceptable uncertainty limits. Stochastic quantitative Monte Carlo methods can be used for integrating known parameter variabilities for a range of variables to accurately estimate an expected value of a population of systems [40–42]. Unfortunately, complete distributions for most parameters are unavailable and data on the most sensitive parameters are often neglected (*i.e.*, N_2O and CO_2 from SOC).

It seems appropriate that regulators should now establish an acceptable threshold for parameter uncertainty when characterizing GHG emissions from the direct life cycle. For example, parameter variability may lead to emission results that are less than ±5%–15% of the mean value (provided by the regulator) and may be subsequently neglected, while measured variability that likely leads to actual GHG emissions being outside of that range must be incorporated into LCA methods to minimize bias.

10.8   Principle 6: Indirect Emissions Are Numerous and Highly Uncertain

10.8.1   Indirect Land Use Change

The assignments of GHG emissions from various sources related to biofuel production follow two general approaches in LCA, so-called attributional and consequential approaches [6]. Attributional LCA is an approach in which emissions are quantified from components of the fuel production life cycle and allocation procedures are used when more than one product is produced by the system [28]. Alternatively, consequential LCA attempts to identify the total marginal changes in any and all direct and indirect emissions that would occur as a consequence of some change in the output of the fuel. The consequential approach is thus more exhaustive and relevant in evaluating the consequences of new policies.

Global conversions of forests and grasslands to agriculture have contributed roughly one fifth of the global anthropogenic GHG emissions in the 1990s and roughly one third since 1750 [5]. Yet, the estimation of ILUC and its associated GHG emissions have been highly controversial. In spite of this controversy, ILUC estimates are included in state and federal LCAs. Resulting ILUC from biofuel production is based on the notion that the global agricultural economy is in an equilibrium, where production equals consumption. In response to a new biofuel industry, global agricultural markets need to meet the new demands in addition to the existing demands for food and feed. Because agricultural yields are slowly increasing, rapid growth in biofuel production must be sustained by increasing the size of the existing global agricultural land base or by less consumption from existing consumers. Regardless of existing trends in deforestation, there is assumed to be an additional marginal incentive to convert forested land to agriculture from the development of new biofuel industries. This incentive is in the form of an increased price that is transmitted through international agricultural markets from the source of demand (*e.g.*, maize in the USA for ethanol) to distant agricultural markets and associated deforestation (*e.g.*, soybean expansion in Brazil) [43]. Because these models estimate the most likely marginal change in land conversion based on a multivariate analysis, the impacts of biofuels cannot be directly verified by measurements [44]. Based on this understanding, deforestation rates can be observed to be declining but these rates would have declined even faster, or even reversed, without biofuels.

Global ILUC is quantitatively estimated by taking recent trends in agricultural productivity, agricultural supply and demand, commodity prices, trade substitutions, international land conversion rates, and emission models to predict an uncertain future [5–6, 45]. Global econometric models were developed to analyze the impacts of specific policies on agricultural markets, but are now also used to estimate ILUC; *e.g.*, the Food and Agricultural Policy Research Institute (FAPRI) model [4] and Global Trade Analysis Project (GTAP) [46]. Because ILUC projections are expected to occur in a probable future, projections become more inaccurate with the time horizon as new variability accumulates. This is due to many unforeseen changes in global crop production and policies that may change the incentives governing land conversion around the world, such as global climate accords that could dramatically slow deforestation in the foreseeable future and reduce projections of ILUC emission rates. Because most models project ILUC over 20–30 years into the future, it is very likely they will predict absolute land conversion with a high degree of error.

Despite these uncertainties, the mean value for the most recent estimates is between 14 and 30 g CO_{2eq} MJ^{-1} for maize-ethanol production and these values have been tending to get smaller with further analysis (Fig. 10.4), with at least one exception [47]. The ILUC emissions analysis by Hertel *et al.* [46] provides a wide distribution of probable ILUC rates. They also state that there is further uncertainty and these estimates should be interpreted as "order of magnitude" in accuracy [46]. Emissions from ILUC in Brazil from sugarcane-ethanol are equally uncertain [48], but tend to be getting larger with more analyses (Fig. 10.4).

Hertel *et al.* [46] reported a mean value of 27 g CO_{2eq} MJ^{-1} with a coefficient of variation of 0.46; two standard deviations (SD) are shown; a minimum was reported at 444 g in total over 30 years or 14.8 g CO_{2eq} MJ^{-1} for the non-normal distribution (Fig. 10.4a). This was reported as a corrected value for total marginal ethanol liters over 30 years (1127 billion liters: increasing from 0 to 50.1 b. liter from 2001 to 2015, then constant at 50.1 until 2030; see Hertel *et al.* (C) in Fig. 10.4a.). The upper value of Lapola *et al.* [48] reports total marginal emissions divided by total marginal liters of ethanol over 30 years in Brazil (746 billion liters; from 2003 to 2020, increasing from 0 b. liter to 35.53, then constant at 35.53 until 2032) and the lower value is the total marginal emissions divided by the total ethanol industry production over 30 years (1181 b. liters); Lapola *et al.* also includes direct land use change (DLUC) emissions. Mostly FAPRI or GTAP models were used (Fig. 10.4).

10.8.2 Multiple Indirect Effects and Global Economic Forecasting

Deforestation is not the only indirect change in GHG emissions from the global agricultural economy due to biofuel production. A multitude of GHG emission sources and sinks are indirectly affected including emissions of CH_4 and N_2O from livestock, CH_4 emissions from rice, soil C dynamics from changing cropping patterns, and reclamation of dry and degraded land, among potential others

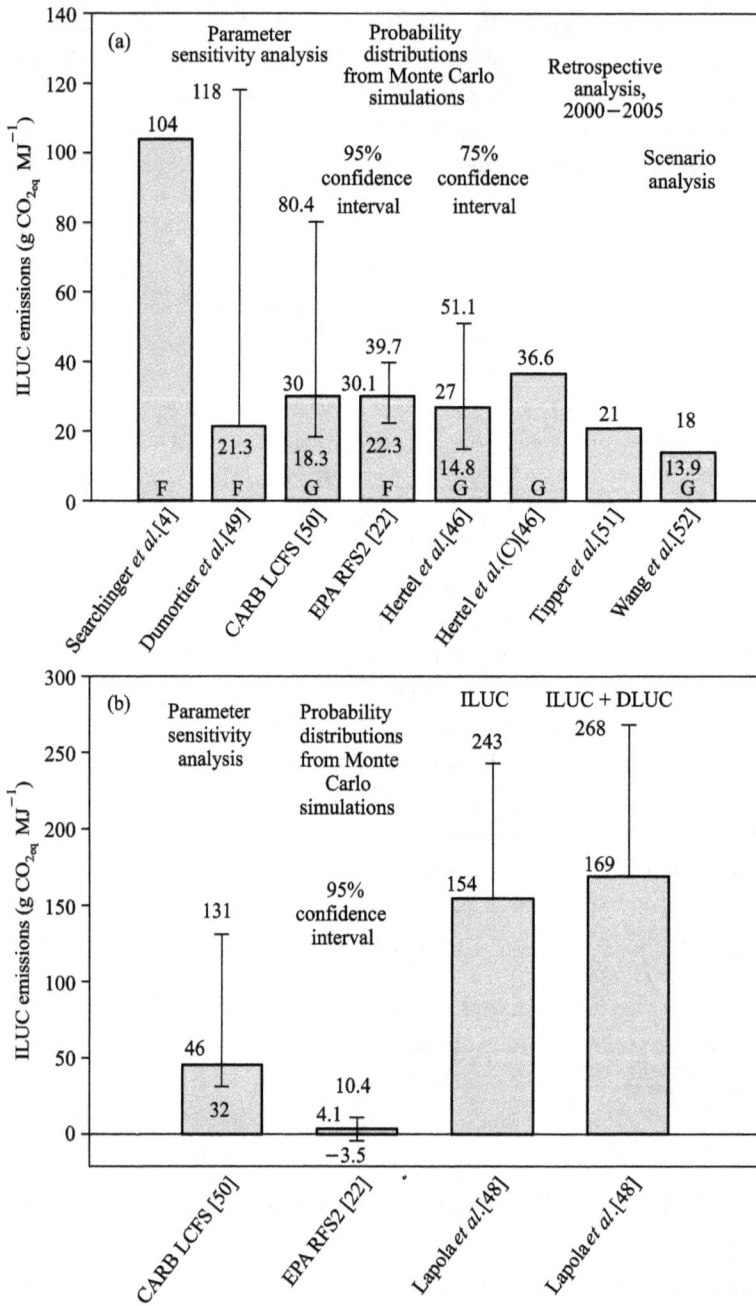

Fig. 10.4 Estimated means and uncertainties of projected ILUC GHG emission rates due to maize-ethanol production over a 30-year period (a) and sugarcane-ethanol over a 30-year period (b).

[5–6]. Recent estimates showed that projected declines in livestock from rising grain prices from biofuel production could offset nearly 50% of the positive emissions from ILUC [5].

In the EPA's RFS2 LCA methodology, multiple changes in direct and indirect GHG emissions from the USA and global agricultural economies due to ethanol production are quantified using at least eight highly complex models, incorporating tens of thousands of parameters [6, 22]. Because the EPA's approach attempts to characterize so many diverse and disparate systems, namely, global changes in agricultural and ecosystem GHG emissions over roughly 20 years, this method is likely associated with a large degree of error (e.g., 30,000 emission factors are used to estimate emissions from land conversion alone, as one of two data sets included in the EPA's partial error analysis, leading to a 95% confidence interval that is ±28% of the mean, Fig. 10.4a). It is clear why no similar LCA approaches are found in the scientific literature: the uncertainty is too large and the probability is too low for accurately predicting the future global economy and land use over a period of 20 years. In terms of complexity, the next closest LCA (but much simpler than the EPA's) estimated the ILUC emissions due to biofuels by combining a LCA model (GREET) and a global econometric model (FAPRI [4]), which has been subject to prolonged controversy [5–6, 44–53].

When projecting global agricultural changes and ILUC, the magnitude of changes due to marginal price signals are determined by trade and agricultural markets, among other issues, that depend on global economic and financial conditions [53]. A recent survey of economic predictions (data that are included in GTAP and FAPRI) emphasizes how quantitative estimates are often associated with large errors (e.g., the Black-Scholes equation [54]). Recurring financial crises undermine the accuracies of predictions made using economic models, among other issues. For example, in the period from 1800 (or independence) to 2008, 79% of countries in Europe and the Americas, on average, have experienced a sovereign default or debt rescheduling every 33.5 years (based on 39 countries); the author's calculations are based on [55]. In addition, 98% of countries globally, on average, have also experienced a banking crisis every 37.6 years (based on 66 countries from 1945, or independence, to 2008).

These trends suggest that economic failures regularly set countries off of trend line growth rates and undermine predictions of economic conditions. If scientists are not good at predicting economic contexts, they are also not good at predicting relative changes in these uncertain futures. For example, the Soviet Union economic growth projections made in 1990 would have hypothetically estimated large biogenic C losses due to projected economic growth. However, in the real course of events, the Soviet Union experienced an economic and political collapse and the 1990s witnessed a vast accumulation of C in the region [56–57] and economic conditions likely would not have transmitted price signals as effectively. With the debt crises in the USA and Europe, it may be possible that economic models are currently overestimating growth and ILUC in the future, which suggests that these projections should include statistical estimates of regular economic failures, thus more accurately representing a more probable future, to a degree possible.

These circumstances raise important questions: What is an acceptable level of uncertainty when predicting multiple global indirect effects? What precedents should regulators recognize when establishing an acceptable uncertainty threshold? What is the most efficient use of resources in producing multi-sector LCAs?

In comparison with the EPA methods, global integrated models (combining social, economic, demographic, and environmental variables) have been developed and used over the past 40 years. Though, the accuracy of these results has been perhaps one of the greatest scientific controversies of the 20th century, dating back to Thomas Malthus [58–60]. Global models used in *The Limits to Growth* studies (1972, 1992, 2004) and emission scenarios for the IPCC, however, present a limited number of scenarios for the future and, unlike those of the EPA, they do not attempt to provide a single point estimate but provide a range of potentially equal probable results for consideration and explore the sensitivity of the relationships between multiple factors [61]. Despite the uncertainty and controversy surrounding these studies (and ILUC), the Millennium Ecosystem Assessment (2005) has generally confirmed the ominous projections for the environment from *The Limits to Growth* (1972) [59, 61]. This research strongly suggests that modeling efforts may not produce accurate point estimates, but the general relationships elucidated in modeling studies can provide valuable insight for managing our resources and precautionary actions taken today are likely to reduce the risk of more environmental degradation [5].

10.9   Principle 7: Transparency Is Essential for Regulatory LCA

The intent to quantify all significant indirect emissions leads to a contradiction concerning transparency—the ability to see all of the methods and data used. The EPA has sought to ensure a high standard of transparency and has recognized that a lack of transparency may conceal biases in results. In 2001, the USA's government-wide guidelines for information quality were established. The associated guidelines state: "The more important benefit of transparency is that the public will be able to assess how much an agency's analytic result hinges on the specific analytic choices made by the agency. Concreteness about analytic choices allows, for example, the implications of alternative technical choices to be readily assessed" [62].

To evaluate the technical choices made in use of data for LCA, ISO standards specify the need for qualifying information to supplement data used. The ISO standard specifies these requirements: "The data quality requirements should address: time-related coverage; geographical coverage; technology coverage; precision, completeness and representativeness of the data; consistency and reproducibility of the methods used throughout the LCA; sources of the data and their representativeness; uncertainty of the information" [16].

By using many tens of thousands of parameters, the RFS2 LCA is not likely to be 100% transparent. From the author's discussions with the model developers, in some cases, the data are not available to inspect and are not supported by references and different models provide conflicting results [45].

For implementing regulations with acceptable uncertainty limits, all observers are likely to agree that regulatory LCA methods should only be as complex as can be practically and transparently reviewed and supported by accurate data. Evaluating all global indirect effects in one LCA is excessively complex, particularly for contentious EPA regulation. It is also clear that if sufficient transparency and accuracy are not achieved, indirect effects should be considered to be excluded from regulations, merely because they greatly expand the number of variables employed, magnify the uncertainty involved, and lead to more and more arbitrary results, with corresponding severe penalties for the regulated parties; final EPA values today appear to be politically negotiated results. Where great complexity is to be characterized (such as the global ripple effects of biofuel production across all sectors), proportional analytical resources should be employed to adequately acquire the needed data, explore the uncertainty, and determine the limitations of the methods employed. This has been done to some degree in the RFS2, but greater analysis is needed. Analysis of likely but uncertain indirect effects may be more appropriately investigated during policy analysis before passage and implementation of environmental regulations and a more limited analysis provided by conventional LCA methods may be more appropriate for regulating the performances of supply chains.

Yet, after the passage of the EISA, regulators still desire to approximately know the actual GHG emission impacts from this policy. Thus, an apparent contradiction inherent in LCA methods emerges: use less-certain and less transparent methods but include comprehensive estimates of global changes in GHG emissions due to biofuel production that incorporate a precautionary approach (the EPA's approach) or use a more certain and more transparent assessment of biofuel supply chains that can be adequately monitored and regulated, while excluding uncertain global GHG consequences from biofuel production, but providing more reliable predictions of system performance (the conventional LCA approach used in most studies). This appears to be the core question related to ILUC and LCA, but it also appears to be largely unanalyzed.

10.10 Principle 8: Fossil Fuel Reference Systems Are Diverse and Uncertain

In 2008, 580 of the 651 largest oil fields globally (contributing >60% of the global production) were reported to have passed their peak production rates and are now producing an average of about 5%–6% less oil each year [63]. Because of growing global demand and limited oil reserves, unconventional sources of petroleum are being developed. These unconventional forms of petroleum (heavy oil, oil sands, oil shale, natural gas-to-liquids, and coal-to-liquids) are generally more costly to produce. However, as the price of oil has risen to over $90 per barrel in March, 2013, these alternative petroleum sources are now profitable to produce, but they are also more energy- and GHG-intensive to produce (Fig. 10.5). These unconventional sources are becoming a greater fraction of the feedstock for gasoline, as the lighter crudes are depleted.

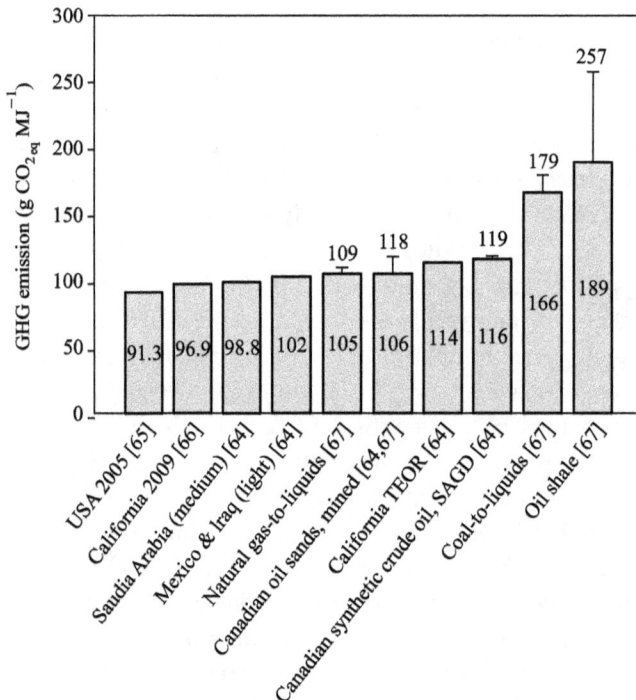

Fig. 10.5 Estimates of GHG emissions from gasoline reference systems (various sources). SAGD= steam-assisted gravity and drainage; TEOR = thermally enhanced oil recovery.

Canadian oil sands (tar sands) are an important example. Over the next 20 years, with only considering growth from oil sands production, Canada was projected by the IEA to have the second greatest oil production growth globally behind Saudi Arabia [63]. By 2020, based on Canadian industry projections, it was previously projected that oil sands could contribute as much as 20% of the USA's gasoline supply, up from a current fraction of 7% [5].

A recent analysis of the GHG intensities of gasoline blendstock from a range of countries around the world found that many countries are above what was estimated as the USA's average in 2005 at 91.3 g $CO_{2_{eq}}$ MJ^{-1} (Fig. 10.5). Because of new diverse sources of oil being developed, it has been suggested that each gasoline producer should also be assessed in the life cycle GHG-intensity of its gasoline blendstock. This would put all fuels from every producer side-by-side for comparison using LCA: this would be ideal for effective fuel policy to reduce GHG emissions [68]. Where this cannot be done, at least the baseline for the average petroleum fuel should be assessed on an annual or biannual basis because of significant trends toward increased GHG-intense fuel sources.

In addition to changing direct production emissions, indirect emissions from petroleum fuels have received little attention. Indirect emissions from deforestation in Canada from oil sands petroleum is one example [69]. Military security associated with the acquisition of petroleum has also been estimated to contribute an additional ~20% to the baseline life cycle emissions of gasoline from the Middle East, potentially offsetting ILUC emissions due to biofuels [6, 70].

10.11   Conclusions

The above principles and examples should be considered when designing accurate and transparent LCA methods for researchers, corporations, and regulators. Of the issues addressed here, the emissions of N_2O, CO_2 from SOC and assessing multiple indirect emissions appear to be the current dominant challenges for reducing the uncertainty of biofuel LCA. These principles should be further developed to minimize uncertainty and the use of arbitrary methods for the LCA of biofuel, particularly in regulatory settings.

References

[1] Liska AJ, Cassman KG. Towards standardization of life-cycle metrics for biofuels: greenhouse gas emissions mitigation and net energy yield. J Biobased Materials Bioenergy 2008; 2: 187–203.

[2] Gnansounou E, Dauriat A, Villegas J, Panichelli L. Life cycle assessment of bio-fuels: energy and greenhouse gas balances. Biores Tech 2009; 100: 4919–4930.

[3] Menichetti E, Otto M. Energy balance and greenhouse gas emissions of biofuels from a life-cycle perspective. In: Howarth RW, Bringezu S (Eds.). Biofuels: Environmental Consequences and Interactions with Changing Land Use. Ithaca, NY, US: Cornell University, 2009; 81–109.

[4] Searchinger T, Heimlich R, Houghton RA, et al. Use of US croplands for biofuels increases greenhouse gases through emissions from land-use change. Science 2008; 319: 1238–1240.

[5] Liska AJ, Perrin RK. Indirect land use emissions in the life cycle of biofuels: regulations vs. science. Biofuel Bioprod Biorefin 2009; 3: 318–328.

[6] Sanchez ST, Woods J, Akhurst MA, et al. Accounting for indirect land use change in the life cycle assessment of biofuel supply chains. J Royal Society Inter 2012; 9: 1105–1119.

[7] Aven T. Quantitative Risk Assessment: The Scientific Platform. NY, USA: Cambridge University Press, 2011.

[8] Feller W. An Introduction to Probability Theory and its Applications, Volume I, 3rd edition. NY, USA: John Wiley and Sons, 1968.

[9] National Research Council. Models in Environmental Regulatory Decision Making. Washington DC, USA: National Academies Press, 2007.

[10] Duffy M, Garrett B, Riley C, Sandor D. Future transportation fuel system of systems. In: Jamshidi M (Ed.). System of Systems Engineering: Innovations for the 21st Century. Hoboken, NJ, USA: John Wiley and Sons, 2009; 409–442.

[11] Graedel T, Allenby B. Industrial Ecology and Sustainable Engineering. Upper Saddle River, NJ, USA: Prentice-Hall, 2010.

[12] Liska AJ, Yang HS, Bremer VR, et al. Improvements in life cycle energy efficiency and greenhouse gas emissions of corn-ethanol. J Indust Ecol 2009; 13: 58–74.

[13] Gupta RB, Demirbas A. Gasoline, Diesel, and Ethanol Biofuels from Grasses and Plants. NY, USA: Cambridge University Press, 2010.

[14] Bakshi BR, Gutowski TG, Sekulić DP. Thermodynamics and the Destruction of Resources. NY, USA: Cambridge University Press, 2011.

[15] Liska AJ, Perrin RK. Energy and climate implications for agricultural nutrient use efficiency. In: Clay DE, Shanahan JF (Eds.). GIS Applications in Agriculture—Volume Two: Nutrient Management for Energy Efficiency. Baco Raton, FL, USA: CRC Press 2011; 1–17.

[16] International Organization for Standardization. ISO 14040: Environmental Management—Life cycle Assessment—Principles and Framework. Geneva, Switzerland: ISO 1997.

[17] Farrell AE, Plevin RJ, Turner BT, Jones AD, O'Hare M, Kammen DM. Ethanol can contribute to energy and environmental goals. Science 2006; 311: 506–508.

[18] Plevin, R. Modeling corn ethanol and climate: a critical comparison of the BESS and GREET models. J Indust Ecol 2009; 13: 495–507.

[19] Liska AJ, Cassman KG. Response to Plevin: implications for life cycle emissions regulations. J Indust Ecol 2009; 13: 508–513.

[20] Bremer VR, Liska AJ, Klopfenstein TJ, *et al.* Emissions savings in the corn-ethanol life cycle from feeding co-products to livestock. J Environ Qual 2010; 39: 1–11.

[21] Carbon Intensity Lookup Table for Gasoline and Fuels that Substitute for Gasoline. Sacramento, CA, USA: California Air Resources Board, 2009 (Accessed August 14, 2012, at http://www.arb.ca.gov/fuels/lcfs/121409lcfs_lutables.pdf).

[22] Renewable Fuel Standard Program (RFS2) Regulatory Impact Analysis. Washington, DC: USA Environmental Protection Agency, 2010 (Accessed August 14, 2012, http://www.epa.gov/otaq/renewablefuels/420r10006.pdf).

[23] Cherubini F, Bird ND, Cowie A, Jungmeier G, Schlamadinger B, Woess-Gallasch S. Energy- and greenhouse gas-based LCA of biofuel and bioenergy systems: key Issues, ranges and recommendations. Resources Conserv Recycl 2009; 53: 434–447.

[24] Heath GA, Mann MK. Background and reflections on the Life Cycle Assessment Harmonization Project. J Industrial Ecology 2012; 16: S8–S11.

[25] Cooper JS, Noon M, Kahn E. Parameterization in life cycle assessment inventory data: review of current use and the representation of uncertainty. Int J Life Cycle Assess 2012; 17: 689–695.

[26] Klir GJ. Uncertainty and Information: Foundations of Generalized Information Theory. Hoboken, NJ, USA: John Wiley and Sons, 2006.

[27] Drosg M. Dealing with Uncertainties: A Guide to Error Analysis, second edition. Berlin, Germany: Springer, 2009.

[28] Kim S, Dale BE. Allocation procedure in ethanol production system from corn grain I. System expansion. Int J Life Cycle Assess 2002; 7: 237–243.

[29] Mueller S. 2008 National dry mill corn ethanol survey. Biotechnology Letters 2010; 32: 1261–1264.

[30] Eggleston HS, Buendia L, Miwa K, Ngara T, Tanabe K. 2006 IPCC Guidelines for National Greenhouse Gas Inventories. (IPCC National Greenhouse Gas Inventories Programme). Hayama, Japan: IGES, 2006.

[31] David MB, Del Grosso SJ, Hu X, *et al.* A modeling denitrification in a tile-drained, corn and soybean agroecosystem of Illinois, USA. Biogeochemistry 2009; 93: 7–30.

[32] Crutzen PJ, Mosier AR, Smith KA, Winiwarter W. N_2O Release from agro-biofuel production negates global warming reduction by replacing fossil fuels. Atmos Chem Phys Discuss 2007; 7: 11191–11205.

[33] Mosier AR, Crutzen PJ, Smith KA, Winiwarter W. Nitrous oxide's impact on net greenhouse gas savings from biofuels: life-cycle analysis comparison. Intl J Biotech 2009; 11: 60–74.

[34] Sheehan J, Aden A, Paustian K, *et al.* Energy and environmental aspects of using corn stover for fuel ethanol. J Indust Ecol 2004; 7: 117–146.

[35] Wilhelm WW, Johnson JMF, Karlen DL, Lightle DT. Corn stover to sustain soil organic carbon further constrains biomass supply. Agronomy J 2007; 99: 1665–1667.

[36] Anderson-Teixeira KJ, Davis SC, Masters MD, DeLucia EH. Changes in soil organic carbon under biofuel crops. GCB Bioenergy 2009; 1: 75–96.

[37] Kutsch WL, Bahn M, Heinemeyer A. Soil Carbon Dynamics: An Integrated Methodology. NY, USA: Cambridge University Press, 2010.

[38] Wortmann CS, Liska AJ, Ferguson RB, Klein RN, Lyon DJ, Dweikat I. Dryland performance of sweet sorghum and grain crops for biofuel in Nebraska. Agronomy J 2010; 102: 319–326.

[39] Schmer MR, Vogel KP, Mitchell RB, Perrin RK. Net energy of cellulosic ethanol from switchgrass. Proc Nat Acad Sci USA 2008; 105: 464–469.

[40] Lloyd SM, Ries R. Characterizing, propagating, and analyzing uncertainty in life-cycle assessment. J Indust Ecol 2007; 11: 161–179.

[41] Spatari S, MacLean HL. Characterizing model uncertainties in the life cycle of lignocellulose-based ethanol fuels. Environ Sci and Technol 2010; 44: 8773–8780.

[42] Mullins KA, Griffin WM, Matthews HS. Policy implications of uncertainty in modeled life-cycle greenhouse gas emissions of biofuels. Environ Sci and Technol 2011; 45: 132–138.

[43] Naylor RL, Liska AJ, Burke MB, et al. The ripple effect: biofuels, food security, and the environment. Environment 2007; 49: 30–43.

[44] O'Hare M, Delucchi M, Edwards R, et al. Comment on "Indirect Land Use Change for Biofuels: Testing Predictions and Improving Analytical Methodologies" by Kim and Dale: statistical reliability and the definition of the indirect land use change (iLUC) Issue. Biomass Bioenergy 2011; 35: 4485–4487.

[45] National Research Council. Renewable Fuel Standard: Potential Economic and Environmental Effects of U.S. Biofuel Policy. Washington DC, USA: National Academies Press, 2011.

[46] Hertel TW, Golub A, Jones AD, O'Hare M, Plevin RJ, Kammen DM. Global commodity trade analysis identifies significant land-use change and greenhouse gas emissions linked to U. S. corn ethanol production. BioScience 2010; 60: 223–231.

[47] Plevin RJ, O'Hare M, Jones AD, Torn MS, Gibbs HK. Greenhouse gas emissions from biofuels' indirect land use change are uncertain but may be much greater than previously estimated. Environ Sci and Tech 2010; 44: 8015–8021.

[48] Lapola DM, Schaldach R, Alcamo J, et al. Indirect land-use changes can overcome carbon savings from biofuels in Brazil. Proc Natl Acad Sci USA 2010; 107: 3388–3393.

[49] Dumortier J, Hayes DJ, Carriquiry M, et al. Sensitivity of Carbon Emission Estimates from Indirect Land-Use Change. Working Paper 09-WP 493. Ames, IA, USA; Center for Agricultural and Rural Development at Iowa State University, 2009.

[50] Proposed Regulation to Implement the Low Carbon Fuel Standard. Volume II Appendices. Sacramento, CA, USA: California Air Resources Board, 2009 (Accessed August 14, 2012, at http://www.arb.ca.gov/fuels/lcfs/030409lcfs_isor_vol2.pdf).

[51] Tipper R, Hutchison C, Brander M. A Practical Approach for Policies to Address GHG Emissions from Indirect Land Us Change Associated with Biofuels. Technical Paper-TP-080212-A. Edinburgh, UK: Ecometrica and Green Energy 2010.

[52] Wang MQ, Han J, Haq Z, Tyner WE, Wu M, Elgowainy A. Energy and greenhouse gas emission effects of corn and cellulosic ethanol with technology improvements and land use changes. Biomass Bioenergy 2011; 35: 1885–1896.

[53] Keeney R, Hertel TW. The indirect land use impacts of United States biofuel policies: the importance of acreage, yield, and bilateral trade responses. American Journal of Agricultural Economics 2009; 91: 895–909.

[54] Taleb NN. The Black Swan: The Impact of the Highly Improbable. NY, USA: Random House 2007.

[55] Reinhardt CM, Rogoff K. This Time is Different: Eight Centuries of Financial Folly. Princeton, NJ, USA: Princeton University Press, 2009.

[56] Vuichard N, Ciais P, Belelli L, Smith P, Valentini R. Carbon sequestration due to the abandonment of agriculture in the former USSR since 1990. Global Biogeochem Cycles 2008; 22: GB4018.

[57] Henebry GM. Global change: carbon in idle croplands. Nature 2009; 457: 1089–1090.

[58] Constanza R, Leemans R, Boumans RMJ, Gaddis E. Integrated global models. In: Constanza R, Graumlich LJ, Steffen W (Eds.). Sustainability or Collapse? An Integrated History and Future of People on Earth. Cambridge, MA, USA: MIT Press, 2007: 417–445.

[59] Meadows DL. Evaluating past forecasts: reflections on one critique of the limits to growth. In: Constanza R, Graumlich LJ, Steffen W (Eds.). Sustainability or Collapse? An Integrated History and Future of People on Earth. Cambridge, MA, USA: MIT Press, 2007: 399–415.

[60] Lomborg B. Environmental alarmism, then and now. Foreign Affairs 2012; July/August: 24–40.

[61] Meadows DH, Randers J, Meadows D. Limits to Growth: The 30-year Update. White River Junction, VT, USA: Chelsea Green 2004.

[62] Executive Office of the President. Guidelines for Ensuring and Maximizing the Quality, Objectivity, Utility, and Integrity of Information Disseminated by Federal Agencies; Republication. Washington, DC: U.S. Office of Management and Budget 2002.

[63] International Energy Agency. World Energy Outlook 2008. Paris, France: OECD/IEA 2008.

[64] Jacobs Consultancy and Life Cycle Associates. Life Cycle Assessment Comparison of North American and Imported Crudes. Calgary, AB, Canada: Alberta Energy Research Institute, 2009.

[65] National Energy Technology Laboratory. Development of Baseline Data and Analysis of Life Cycle Greenhouse Gas Emissions of Petroleum-Based Fuels. Washington DC, USA: US Department of Energy, 2009.

[66] Detailed CA-GREET pathway for California Reformulated Gasoline Blendstock for Oxygenate Blending (CARBOB) from Average Crude Refined in California. Version: 2.0. Sacramento, CA, USA: California Air Resources Board, 2009 (Accessed August 14, 2012, at http://www.arb.ca.gov/fuels/lcfs/011209lcfs_carbob.pdf).

[67] Brandt AR, Farrell AE. Scraping the bottom of the barrel: greenhouse gas emission consequences of a transition to low-quality and synthetic petroleum resources. Climate Change 2007; 84: 241–263.

[68] DeCicco JM. Biofuels and carbon management. Climatic Change 2012; 111: 627–640.

[69] Rooney RC, Bayley SE, Schindler DW. Oil Sands mining and reclamation cause massive loss of peatland and stored carbon. Proc Nat Acad Sci USA 2012; 109: 4933–4937.

[70] Liska AJ, Perrin RK. Securing foreign oil: a case for including military operations in the climate change impact of fuels. Environment 2010; 52: 9–22.

Chapter 11

Energy and GHG Emission Assessments of Biodiesel Production in Mato Grosso, Brazil

Antonella Baglivi, Giulia Fiorese, Giorgio Guariso, and Clara Uggè

11.1   Introduction

The strong increase in worldwide biodiesel production, which rose at a double-digit yearly rate from 1.7 million tons produced in 2003 to 17.3 in 2010 [1], has been preceded by a similar increase of interest in research activities. The number of scientific papers catalogued by Google Scholar as having "biodiesel" in the title rose from 129 in 2000 to 868 in 2005 and to 2,330 in 2010.

Such an evident evolution was however not always seen as a positive development both in theory and in practice. Some authors (see Pimentel and Patzek [2]) calculated that the production of biodiesel from soybean has a negative energy budget. Furthermore, the Food and Agriculture Organization (FAO) has expressed concerns about the negative effects on food availability and prices due to the expansion of biodiesel production. Indeed, the 2007 "tortilla riots" in Mexico were sparked after a 400% increase in the local price of corn and are an impressive demonstration of the effect that extensive biodiesel development may have.

Given these concerns on the increased use of biodiesel and the many others that may arise (see [3]), the development of biofuels should be carefully planned by taking into account the socio-economic and environmental features of the specific geographical location. In fact, even if biofuels are tradable goods (see, for example, the commercial routes from Brazil to the USA in Bauen *et al.* [4]), a sustainable use of biomass should be realized on a local scale within comprehensive plans that consider all of the main impacts and externalities, from carbon flows to crop diversity. This requires a vision that goes beyond the sole energy balance or the production costs involving, for instance, the siting of energy conversion facilities close enough to where feedstocks are available. The production of biofuels, themselves, rather than heat, is an example of the many facets of the problem: the direct combustion of biomass produces much more energy than its transformation into a liquid fuel; however, clearly, liquid fuels can serve uses, such as transport, that the direct supply of heat cannot.

Many countries, including Argentina, Brazil, China, Europe and the USA, are pushing forward large plans to include at least 10% of biofuels in car fuel. As a consequence, extended territories have been cultivated with crops suitable for this energy conversion. For example, *Jathropa curcas* [5] or *Pistacia chinensis* [6] are becoming common crops for the production of biodiesel. These crops do not directly impact the food market (as it happens with soybeans, for example), but may still compete with food crops for arable land and water.

Brazil is certainly a country that offers many opportunities to the agro-energy sector. It has large availability of land suitable for cultivation, abundant water and sunlight, and low labor costs as well as high agricultural technology. According to the FAO [7], ethanol produced from sugarcane in Brazil is the most competitive biofuel in the world, being the only one able to compete with fossil fuels without incentives. With respect to biodiesel, in 2004, the Brazilian government launched the National Program for the Production and Use of Biodiesel (PNPB) with the aim of ensuring sustainable economic production of biofuels, also taking into consideration social inclusion and regional development. The main step of the PNPB action was the introduction of Law 11097 in 2005. It provided for the optional use of 2% biodiesel in the car fuel mix (the so-called B2) until early 2008 and made it mandatory starting from that year. Then, starting from January 2010, the required mixture increased up to 5% (B5), as in the CNPE Resolution No. 6 of September, 2009.

As already mentioned, the objectives of the PNPB were aimed not only at reducing oil dependency and increasing export opportunities, but also at decreasing greenhouse gas (GHG) emissions, promoting renewable energies, and social development. Issues related to the social aspects are particularly important in Brazil. For instance, the 2006 Agricultural Census confirmed the uneven structure of Brazilian agriculture: extensive properties occupy 75.7% of cropland, even though they represent only 15.6% of the number of farms, resulting in a Gini index of 0.87. In particular, the average area of a family farm is 18.4 ha, while that of business farms is 309.2 ha [8]. On the contrary, family agriculture employs about 3/4 of its workers in the countryside, with a density per hectare close to ten times higher than industrial farms. In 2006, family agriculture was producing 34% of countryside revenues, with a productivity of 677 reais/ha, almost double with respect to industrial cultivation. This clearly means that, while family agriculture exploits land in a more effective way, the revenues per worker are much lower.

Within the PNPB, the Brazilian government decided to promote regional development and to increase employment through a tax reduction for transformation plants that produce biodiesel from seeds obtained by "family" agriculture. According to the definition in the 2006 Brazilian Agroenergy Plan of the Ministry of Agriculture, this can be defined based on few specific characteristics: (i) the management of the productive unit and the investments made in it are carried out by individuals who have blood or marriage ties; (ii) most of the work is equally provided by the members of the family; (iii) the ownership of the means of production (though not always of the land) belongs to the family and it is transferred within the family in case of death or retirement of the heads of the productive unit; (iv) the surface of the land detained is up to four fiscal units (a fiscal unit goes from 15 to 90 ha in Mato Grosso). Biodiesel producers participating in the PNPB may receive a "Selo Combustivel Social" (SCS—Social Fuel Seal), which entitles them to priority in selling their products. In exchange, biodiesel producers are required to purchase a certain percentage (e.g., 15% in Mato Grosso) of raw materials from family farmers and to assist farmers in the selection of oil crops according to the different soil and climate characteristics and other technical issues (Official Gazette, July 19, 2009).

Taking into account this energy and institutional context, this chapter illustrates a method to define a biofuel exploitation plan based on two phases. The first phase consists of the local analysis of land and climate features in order to understand which types of crops can be successfully grown. This is performed on a local scale using Information Geographic System (GIS) data and software [9–10]. Once this step is accomplished, the area can be divided into a number of cells (or parcels) that are small enough to be considered as having the same soil and climate characteristics. This subdivision may follow some administrative scheme or may simply be a regular grid in a large, uniform area. Clearly, a finer subdivision allows a very detailed analysis; however, on the other hand, it may slow down the subsequent phase. As usual, a compromise between detail and speed must be found. The second phase consists of the formulation and solution of a mathematical programming problem, normally of the mixed integer type [11–12], and its solution. The problem explicitly considers an economy-related objective such as the maximization of the net energy produced (the potential net energy from biofuels minus the energy necessary to grow and transport feedstock), but also allows the determination of environment-related indicators such as the impact on crop diversity or GHG balance. The optimal solution provides the land to be cultivated in each parcel with each crop, the best location for plants, and the energy used in all the process stages under various assumptions about external conditions. Other aspects, such as social impacts on "family" agriculture, will also be taken into account and discussed in this chapter.

It is important to underline that while the approach presented is very general, it must be tailored to the specific situation: the results strongly depend on the local reality under consideration and can hardly be extended to other cases. For instance, in the current study, some transformation plants were already built and some other issues, such as the extension of the area under consideration, were already fixed at the local level.

11.2   Study Area

The area considered in this work lies within a region with very specific climatic and ecological characteristics, called "Cerrado", and occupies a small portion of southeastern Mato Grosso (part of the Midwest region of Brazil). Its surface (Fig. 11.1) is 15,031 km^2 and includes the municipalities of Rondonopolis (3,566 km^2), Pedra Preta (3,590 km^2), Alto Garças (3,134 km^2), and Alto Araguaia (4,741 km^2) [13]. The total population is about 237,000 with Rondonopolis representing more than 80% [14].

Temperatures and rainfall are those typical of the Cerrado: the average temperature is 25°C, with a maximum of 40°C in the summer. The dry season begins in April and continues until September. The coldest months are June and July, with temperatures ranging between 10°C and 20°C. The wettest months are November, December, and January (with an average precipitation ranging from 400 to 640 mm in the three months) [15].

The soil is generally poor and therefore considered not very fertile. It has an acidic pH, varying from 4.3 to 6.2, and has a high aluminum content and low availability of nutrients such as phosphorus, calcium, magnesium, potassium, organic matter, zinc, and clay. The prevalent soils in the area are of the "latosols" and "podzols" types [16]. The elevation is between 175 m and 897 m and slopes are rather small: 98.65% of the area has less than 20° [17].

Animal farming is the most relevant activity in the study area (it takes up 62% of the area or 9,394 km^2) as in the entire state. Specifically, 14% of Brazilian bovines are in fact in farms in Mato Grosso [18]. However, only about 3% of the bovines of Mato Grosso are located in the four municipalities considered in this study. Animal farming is prominent in the north and the northeast of the region [18]. Farming of medium (e.g., swine) and small (e.g., poultry) animals is much less common; the two activities in Mato Grosso contribute only 4% each to the Brazilian animal count.

The state of Mato Grosso has water resources and a climate that are favorable to the development of agricultural activities. In fact, agriculture has developed incredibly over the past few years, thanks to research carried out at EMBRAPA (Empresa Brasileira de Pesquisa Agropecuária, i.e., Brazilian Agricultural Research Corporation), which has, for example, improved the soil quality allowing extensive cultivations [19].

Indeed, agricultural and animal farming contributed 22% of the gross domestic product (GDP) of Mato Grosso in 2010 [20]. This was R$59.6 billion (exchange rate was about R$1=0.6 USD), corresponding to 1.6% of the national GDP and to a GDP per capita of about R$20,000.

In the study area, the cultivation of soybean was predominant in 2011, exceeding 60% of the cultivated area [21]. Other common crops were cotton (18% of the surface), corn (13%), sorghum (3%) and sugarcane, bananas, coconuts, and few others in much smaller percentages. There are, however, important changes in time: the area occupied by soybean, for instance, was 17% larger in 2005 and 10% lower in 2008 with respect to the data in 2011.

This approximately reflects the general situation of the agricultural sector of Mato Grosso: soybean is, in fact, the most common crop in the state, with 6.5 million ha grown in 2011. In the same year, corn occupied 1.9 million ha, cotton 0.7 million ha, and sugar cane 0.23 million ha. According to the most recent statistics [21], Mato Grosso alone represents 24.9% of the production of cereals and leguminous and oil crops of all of Brazil.

The tropical savannah of the Cerrado is rich in plant and animal biodiversity and is currently threatened by the progressive expansion of these extensive soybean plantations. Further growth of the agricultural areas should thus be avoided [22].

The state of Mato Grosso is one of the Brazilian leaders for biodiesel production with 14 active plants and a production of 504.4 million liters in 2007 [23]. The biodiesel production process is based on transesterification, a reaction process in which vegetable oil reacts with an alcohol (methanol in 82% of Brazilian plants) to produce esters (biodiesel) and glycerol, a byproduct that can also be commercialized [24]. Soybean is the most important feedstock for the production of biodiesel in Brazil. If we consider only oilseeds, 95% of the biodiesel produced in 2011 (70,436 thousand tons) was derived from soybeans, followed by cotton seeds (4.6%), peanuts (0.6%), sunflower (0.2%), and castor bean and rapeseed (0.1% each) [25].

Specifically, in the studied area, there are currently two main oil extraction facilities (Fig. 11.1): one is located in the municipality of Rondonopolis, with a capacity of 245.5×10^3 m^3 yr^{-1}, and the other is located on the border between the town of Alto Araguaia and the neighboring state of Goias, with a capacity of 235.3×10^3 m^3 yr^{-1} [26]. There are five other smaller plants located near Rondonopolis. For the sake of simplicity, we incorporated these minor plants with the largest one and assumed a new aggregate capacity of 258.1×10^3 m^3 yr^{-1} [26]. The transformation of oil into biodiesel takes place in the same plants. The transformation capacity of the region is thus divided between the two opposite sides of the area. We have assumed a 2% oil loss during the extraction process and we have disregarded the potential energy production from soybean meal, castor cake, and other byproducts.

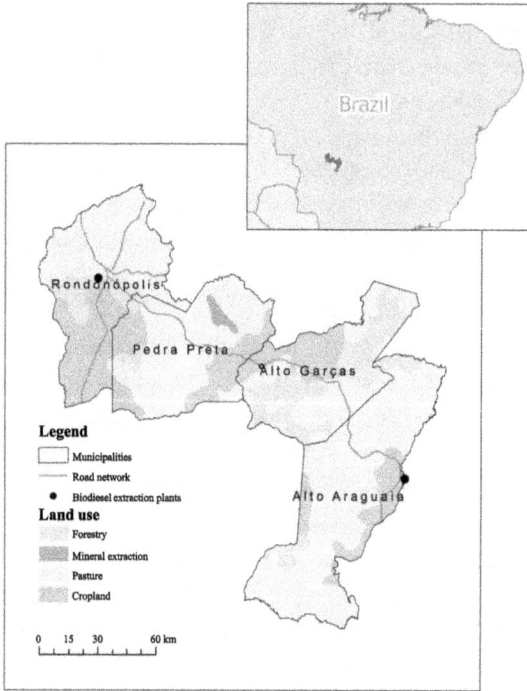

Fig. 11.1 The study area at the southern border of Mato Grosso, Brazil.

11.3   Methods

The method that we applied in order to define the efficient production of biodiesel can be summarized in the following four-step procedure [9]:

1. Selection of potential oilseed crops: the climatic characteristics of the area are analyzed and compared with the phytology and climatic requirements of a set of potentially interesting oilseed crops.

2. Identification of the area suitable for oilseed crops: only areas with the altitude, slope, and soil type that is suitable for the cultivation of the selected crops are included in the study. Additionally, maps of adaptability for each oleaginous crop are created, which quantify the degree of adaptability of the crops to the local characteristics of the territory.

3. Identification of the area available for cultivation: not all of the areas suitable for agriculture, which have been identified in step 2, can be considered available for energy production. In fact, social and political restrictions must be taken into account. Current land uses are therefore analyzed and assumptions are made with respect to the amount of land that can be converted to oilseed crops.

4. Optimal allocation of crops and of conversion plants: an optimization problem is formulated and solved in order to determine all of the details of the development plan. In particular, the solution of the problem determines which crop should be cultivated in each portion of suitable and available land, the location of the conversion plants and the allocation of the oilseed crops to each plant.

Despite the final solution that such a procedure should point out is the optimal use of both land and plants, it is important to look at these types of studies in the correct perspective. When planning for a large area, such as the one at hand, and for a horizon of some years, the results obtained by any mathematical procedure are only useful for highlighting the most relevant trade-offs, the effect of some scenario variable, and the plausible environmental consequences of the decisions being considered. In no way should such results be viewed as the actions to be straightforwardly adopted. The reason for this is simple: it is evident that real life is always in a transient state and each local situation is specific. The introduction of new crops is an adaptive process: farmers understand how to manage them in steps and thus the land productivity changes in time. The local market adapts more or less rapidly to the new products and byproducts but is influenced by the almost unpredictable fluctuations of national and international markets. The change in crops modifies the carbon balance in the soil and the time to reach a new equilibrium with the new aboveground biomass may be dozens of years [27]. The natural yearly climatic evolution may differ substantially and for a prolonged period from the average conditions assumed in any planning study and such average conditions may never be replicated in the future under climate changes.

A planning model, like the one that will be illustrated later, must always be seen as a decision support tool—a technique of reasoning over the problem in a systematic way and a means for understanding the consequences of given assumptions. Certainly, a lot of factors that may appear relevant to some stake-holders are disregarded, others are simplified, and others prove to be irrelevant. The (limited) scope of the study must thus remain clear from the beginning.

11.3.1 Crop Selection

Preceding studies in Brazil have already identified the most interesting oil crops for each region. In particular, those considered more suitable for Mato Grosso [5] are the following: soybeans and cotton (which are already cultivated in the area) and castor, groundnut, sunflower, and palm (which are not yet widely cultivated).

Each of these crops has special needs in terms of soil quality, elevation, slope, sunlight, temperature, and precipitation patterns. Crops may also have different growing cycles, intended as the timeframe from sowing to harvesting (or to explant). All of the crops above have cycles of a few months, with the sole exception of the palm tree, which is a perennial crop that can last for some 20 years.

Requests of the various crops, as far as ambient temperature is concerned, are summarized in Table 11.1. They are subdivided into two different ranges. The first range constitutes the optimal yearly excursion for growth, *i.e.*, the range that, with all other climatic variables at their best values, gives the maximum yield in terms of oilseeds. The second range is the acceptable one and represents temperature variations that are still tolerable for plant growth, even with a reduced yield.

Tab. 11.1 Optimal and acceptable temperature ranges for oil crops [28].

Crop	Optimal range (°C)		Acceptable range (°C)		Cycle length (months)
	Min	Max	Min	Max	
Soybean	20	30	10	40	6
Sunflower	8	34	−5	35	4
Cotton	18	30	14	38	5
Castor	20	35	18	40	5
Peanut	25	35	10	40	4
Palm tree	25	27	17	40	–

Each crop also requires a certain amount of precipitation during the growing cycle to provide the maximum yield. Optimal values for crops under consideration are presented in Table 11.2. In principle, a reduced precipitation input may be compensated by irrigation, but this option can be reasonably considered only where irrigation systems are already in place and are regularly managed. This means that in practice, these are only suitable for much smaller areas within highly developed agricultural systems. Even disregarding the important impact

Tab. 11.2 Optimal precipitation range for oil crops [28].

Crop	Precipitation (mm/cycle)
Soybean	450–800
Sunflower	500–700
Cotton	700–1300
Castor	600–700
Peanut	450–700
Palm tree	1800–2000

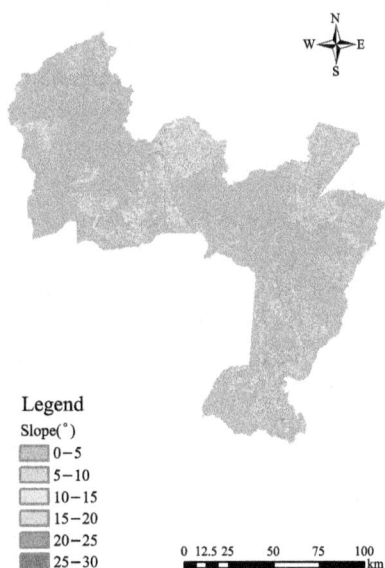

Fig. 11.2 Range of slopes present in the territory [18].

that this option may have on the water balance, it is out of the question for the large, sparsely populated region under consideration.

As an example of the thematic maps produced in this phase, a discretized range of slopes is presented in the area as shown in Figure 11.2.

Crops considered in this study for family farming are limited to castor and an intercropping of castor and peanut. These crops best lend themselves to a more artisanal cultivation as they do not require heavy agricultural mechanization. Castor, in particular, has been indicated by the PNPB as the most advisable choice due to its adaptability to various climatic and soil conditions. Furthermore, the intercropping of castor with peanuts exploits the nitrogen stored by the latter to avoid almost completely the use of chemical fertilizers [29]. Castor is also the crop with the minimum environmental impact under several viewpoints, according to Padula *et al.* [30].

11.3.2 Identification of the Area Suitable for Cultivation

Climatic and soil characteristics that determine the suitability of a certain area may be defined by qualitative and quantitative attributes. Qualitative features, such as pedology, normally correspond to a yes/no condition. If the feature is favorable, the crop may be cultivated, otherwise it cannot. Quantitative features may represent different suitability levels and thus a difference in crop yield. These quantitative features are normally combined to obtain a production index (PI), which is often a linear, or piece-wise linear, function of the most relevant soil and climatic parameters. When a set of yield values under different conditions is available, the coefficients of the assumed function can be obtained by linear regression [31–32]. Such a procedure is acceptable for limited areas and known management conditions. In cases like the one at hand, such an approach has limited usefulness. First, data on actual yields of the region is not available. Second, reliable estimates require long sets of data to filter natural climatic variations. Third, and possibly more important, yields "naturally" change in time because of the evolution of management technologies and farming expertise.

Whatever the selected approach, a number of resulting thematic maps can be computed to determine, for instance, where all of the optimal growing conditions are contemporarily met for each crop type. These are simply the results of an intersection operation of the thematic maps, which separately represent such conditions. For instance, while the land does not have high elevations or slopes, areas with slopes greater than 20° and altitude above 750 m a.s.l. were excluded in order to facilitate the use of agricultural machinery. Additionally, the soil type suitable for the cultivation of the species of oil crops considered is "latosols" and "podzols". Therefore the remaining areas were excluded from the allocation process.

The area suitable for oil crops is presented in Figure 11.3a as an example of the results obtained.

For suboptimal conditions, the number of possible combinations of thematic maps can become extremely high and have uncertain meanings. It is thus better to combine all maps in a way that it is already representative of the final result that we aim to obtain. So, if the purpose of the plan is to maximize the energy output, the map combination can already be oriented to that purpose. Given the

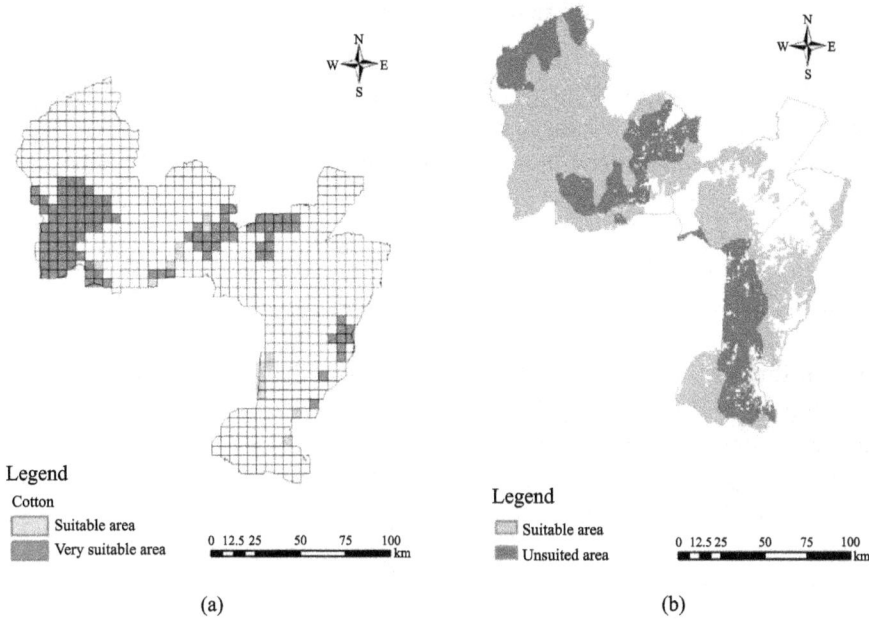

Legend

Cotton

Suitable area

Very suitable area

(a)

Legend

Suitable area

Unsuited area

(b)

Fig. 11.3 Area suitable for all oil crops (a) and its discretization for cotton cultivation (b).

limited amount of data available regarding the land productivity of the region and the speculative nature of this work, a simplifying assumption has been used. The quantitative features have been transformed into qualitative ones by assigning a value of "very suitable" to all areas where a given feature has the best values, a value of "suitable" to areas with admissible values, and a value of "unsuited" to all other areas, for each specific crop. Then, to be conservative, the superposition of features has been done using the lowest value, *i.e.*, if one of the features was "suitable" but all of the others were "very suitable", an overall "suitability" has been assumed. If any of the features were classified as "unsuited", the area was classified in the same way. As an example, palm trees were excluded from the study because all of the territory is "unsuited" as far as precipitation is concerned. In fact, the average annual precipitation in the study area is 900 mm [16], while the water requirement of palm trees is 1,800–2,000 mm yr^{-1}.

Finally, as suggested in Fischer *et al.* [10], a "suitable" condition was assumed to represent a loss of productivity of 20%; *i.e.*, the gross energy yield of these areas was assumed to be 80% of the average achievable in the most suitable conditions.

11.3.3 Settings and Constraints Specific for the Case Study

Physical factors, such as climate and soil, must then be combined with another long list of factors that represent the scenario within which planning decisions will be taken. One such constraint has already been mentioned: at least 15% of the raw material must be purchased by producers with the SCS from family farms. Other constraints are easy to determine, such as the presence of protected areas and natural parks or the necessity of not interfering with urban spaces.

Some others are a bit fuzzier and depend on the way in which we want to formulate the final optimal planning problem. For instance, consistent with some concerns already outlined in this study, we have considered only the possibility of modifying current non-food agricultural land without attempting to change the land use in other areas presently devoted to pastures or the mining industry or to reduce the present forest cover (see Fig. 11.1). This strongly limits the space of possible biodiesel plans, but avoids the conflicts for land with food crops and non-agricultural activities, possibly easing the adoption of something similar to the solutions proposed here. Entering the full debate on land use changes would necessitate a much wider and deep discussion involving such topics as the general economy of the country as well as the reduction of other ecosystem services.

The area that was finally considered as suitable for the selected crops and available for their cultivation amounts to 262,500 ha, *i.e.*, the total area was planted with soybeans and cotton in 2011.

To allow the formulation of the optimal planning problem, the last operation to be performed is the subdivision of the suitable and available areas into a number of discrete units that can be considered as homogeneous under all aspects. The region of Cerrado is large and with few variations, which indicates that relatively large cells can be assumed. In particular, square cells with sizes equal to 2,500 ha each has been adopted and resulted in a total of 105 cells available to the allocation of the crops (see example for cotton in Fig. 11.3b). This rather coarse definition, sufficient for this initial framing of the problem, would probably need to be refined when one of the following plans will be discussed and possibly adopted.

Given this discretization, we assume family farming to be evenly dispersed on the territory (most current studies, such as da Silva César and Otávio Batalha [34], indeed report that this dispersion is one of the main difficulties for effectively involving family farmers into the development plans), *i.e.*, we reserve a suitable percentage of each cell to family farming in such a way that the overall production satisfies the percentage of supply required by law.

As for oil extraction and the subsequent conversion into biodiesel, we assumed that the process is similar for all of the considered oilseeds. Furthermore, since some of the considered oilseeds are not currently used in Mato Grosso (nor in

Brazil), the assumption was almost unavoidable. We thus adopted the same energy requirement of 2.0125 kcal for processing 1 kg of oil, as indicated for sunflower and soybean by Gazzoni *et al.* [35].

11.3.4   Problem Formulation

The overall crop allocation problem can now be formulated as a mathematical programming problem aiming to maximize the net energy from the system and evaluate some measure of the environmental and social impacts. The decision variables z_{ijsf} are the fractions of biomass cultivated in cell i, with crop s and method f ($f = 1$ extensive, $f = 2$ family) and hauled to plant j for processing. Currently, we are considering 105 cells times 5 possible crops times 2 existing plants, which makes a total of 1,050 decision variables plus the fraction of surface to be reserved for family farming in all cells.

The maximization of the net energy output of the system can be written as follows [36]:

$$\max_{\{z\}} J_e = \sum_{i=1}^{N_c} \sum_{j=1}^{N_p} \sum_{s=1}^{N_s} \sum_{f=1}^{2} A_f \left[E_b w_s p_s s_{is} z_{ijsf} - e_{tr} p_s s_{is} d_{ij} z_{ijsf} - (e_s^{agr} + e_s^{pr} p_s s_{is} z_{ijsf}) \right] \quad (11.1)$$

The first term represents the energy output (at the conversion plants), the second is the energy spent to transport the feedstock to the plants, and the third is the energy employed for the cultivation and conversion processes. More precisely:

N_c, N_p, and N_s represent, respectively, the number of cells, plants, and crops considered;

E_b is the energy content of a unit mass of biodiesel (set to 38 MJ kg^{-1} biodiesel);

p_s is the optimal crop productivity (*i.e.*, obtainable in the best conditions) in terms of the mass of seeds per unit area;

s_{is} represents the suitability of cell i for crop s: as explained before, it is equal to 1 for "very suitable" conditions or to 0.8 for "suitable" ones;

w_s is the amount of biodiesel (kg) that can be extracted from a unit weight of seeds type s; it takes into account both the oil content of the seeds and the efficiency of the extraction operations;

A_f is the land surface (ha) available in each cell for extensive ($f = 1$) or family farming ($f = 2$), the sum of which is constant since we have assumed a regular grid.

As for the second term:

e_{tr} is the energy necessary to transport a unit of biomass over a unit distance, assumed to be 0.5 MJ km^{-1} kg^{-1} following Bovolenta [37];

d_{ij} is the distance between cell i and plant j. Such distance cannot follow the road network since there is only one main road crossing the region, practically linking the plant at the western border with that on the eastern side. The distance has thus been computed using GIS tools in the following way: first, the geometric distance from the cell center to the existing roads has been computed, and second, the distance along the road has been added. In the future, more roads may be available, but this way of proceeding is again one of the assumptions of this initial planning study.

Finally, the last term represents the energy costs of biodiesel production. They are composed by two factors, the first being the energy necessary for the cultivation itself and the second is the transformation of seeds into biodiesel, so:

e_s^{agr} is the energy of all agricultural operations to grow a unit area of crop s;

e_s^{pr} is the energy needed to process a unit weight of seeds from crop s to extract biodiesel.

All of the numerical values used in this study are summarized in Table 11.3.

Tab. 11.3 Productivity, oil content, and energy requirements for the cultivation of selected oil crops

Oil crop	Yield[e] (kg oilseeds ha^{-1})	Share of oil in seeds[e] (%)	Cultivation energy (GJ ha^{-1})
Soybean	2,985	19	11.03[a]
Cotton	1,800	25	56.22[b]
Sunflower	1,700	40	11.67[a]
Castor and peanut	1,139	49	7.10[c]
Castor	1,500	50	10.86[d]

Source: a. [35]; b. [38]; c. [29]; d. [39]; e. [28].

The constraints of the problem are:
The use of land in each cell cannot exceed its availability:

$$\sum_{j=1}^{N_p} \sum_{s=1}^{N_s} \sum_{f=1}^{2} z_{ijsf} \leqslant 1, \ \forall i \tag{11.2}$$

The biomass shipped to each plant cannot exceed the plant capacity C_j:

$$\sum_{i=1}^{N_c} \sum_{s=1}^{N_s} \sum_{f=1}^{2} p_s s_{is} z_{ijsf} \leqslant C_j, \ \forall j \tag{11.3}$$

The seeds produced by family farming ($f = 2$) exceeds the proportion imposed for the Biofuel Social Seal:

$$\alpha \sum_{i=1}^{N_c} \sum_{p=1}^{N_p} \sum_{s=1}^{N_s} p_s s_{is} z_{ijs2} \geqslant 0.15 \sum_{i=1}^{N_c} \sum_{p=1}^{N_p} \sum_{s=1}^{N_s} \sum_{f=1}^{2} p_s s_{is} z_{ijsf} \qquad (11.4)$$

where α is a decision variable representing the ratio of family farm surface in each cell;
The non-negativity of decision variables:

$$\alpha \geqslant 0, z_{ijsf} \geqslant 0, \ \forall i, j, s, f \qquad (11.5)$$

Some specific consideration must be added about cotton. Cotton is already grown in the area under study as a supply for the textile industry. The net balance of the energy production of cottonseed oil is negative because its energy output, derived from oil, is much less than the energy used for the cultivation and the processing of its seeds (Tab. 11.4). The production chain of cotton textile fibers, however, considers the oil seeds as byproducts: this means that the net energy balance should not include the (high) energy spent for the production of seeds, equivalent to 56,222 MJ ha^{-1} [38], and the overall balance of biodiesel production (see again Tab. 11.3) may return a positive value. The positivity and convenience of the production still depend on the energy spent for transportation that, assuming a very conservative approach, is computed as if the cottonseeds should be transported from the fields to the plant, while they would probably already be available at some textile plants.

Tab. 11.4 Energy balance of cotton (energy for transport excluded).

Cotton	Energy flows (GJ ha^{-1})
Cultivation[a]	56.22
Transformation[b]	3.79
Energy output	15.73
Net balance as by-product	11.94

Source: a. [39]; b. [35].

If it is assumed to grow cotton for solely an energy purpose, *i.e.*, for the production of biodiesel, seeds are no longer byproducts but are the raw material of the process. The energy balance in this case would be heavily negative (-40.490 GJ ha^{-1}) and the cultivation of cotton would be considered only to increase crop diversity, but with a strong decrease of the net energy production. Since this situation is clearly unrealistic (there would be many other options to increase biodiversity without resorting to such an energy-intensive crop), in the following cottonseeds will always be considered as "free" byproducts of other activities. This means, in practice, that cotton production will be limited to the current area.

Furthermore, the capacity of the transformation plants C_1 and C_2 require some clarification. The plant in Alto Araguaia is located right on the border with the neighboring state of Goias and some of its capacity must thus be reserved for processing the feedstock coming from outside our study area. It was therefore assumed that the raw material taken from this area of Mato Grosso can be processed up to 187.9×10^3 m^3 yr^{-1}. Additionally, some capacity of both plants is presently producing edible oil. Clearly, full biodiesel production would disrupt the local food market and thus, as a first hypothesis, we have assumed to involve biodiesel production in only half of the plant capacities, leaving the rest available for food production. This is also consistent with the figures in Padula et al. [30], which report an idle capacity of Brazilian biodiesel plants close to 60%.

Each cultivation plan, suggested by the solution of the above problem under different external assumptions, has a number of impacts, some of which can be quantitatively examined.

11.3.4.1 Impact on crop diversity

The impact on crop diversity can be formulated in many different ways. Clearly, the concept of biological diversity implies a richness of environmental components that cannot be simplified in the presence of one or two different crops (see Mendonça [40] for a discussion of the current biodiversity loss in the Cerrado). However, it is important to evaluate such an impact since the solution of the above problem can be to cultivate, wherever possible, only the most energy-efficient crop, i.e., develop only a single, large monoculture with well-known ecological problems. On the contrary, looking at this impact, we can measure whether a certain plan favors more variable land use. This constitutes a small improvement from the ecological viewpoint, but may also be beneficial from the economic side since the local agriculture may not be subject to the price fluctuations of a single product.

To assess the diversification of crops in the area, we use the well-known Shannon index H[41]. This value is calculated for each municipality in order to avoid high uniformity in one part of the territory that can be masked by more diversity in another part. The overall index is then calculated as a weighted sum of those in the different municipalities, as follows:

$$H = \sum_{m=1}^{4} \frac{A_m}{A_T} H_m = \sum_{m=1}^{4} \frac{A_m}{A_T} \left(-\sum_{s=1}^{N_s} q_s^m \log(q_s^m) \right) \tag{11.6}$$

where A_m is the surface available for oil crops in municipality m and A_T the total agricultural area of the region, H_m is the Shannon index computed in each municipality m, and q_s^m is the proportion of available area devoted to crop s in municipality m, which obviously depends on the decision variables z_{ijsf}.

11.3.4.2 Impact on greenhouse gas emissions

We are considering oleaginous crops that are not very common for the production of biodiesel, with the exception of soybeans. For this reason, there is very little (e.g., Nogueira [42]) data available in the scientific literature with regards to the life cycle analyses of all of these crops, particularly with respect to GHG emissions. An additional difficulty is inherent to the life cycle approach itself. In principle, the computation should include all emissions generated by the entire cropping process, so some authors include manpower, while some include emissions due to the production of agricultural machines, and others include the final use of residuals and byproducts. Even on the emissions of a single process component, for instance the production of phosphorous as fertilizer, there are large uncertainties. In an attempt to improve the reliability of this type of estimate, Camargo et al. [43] collected information from a large number of sources in the Farm Energy Analysis Tool (FEAT) model and also allowed each user to add more values. For phosphorous production, they found estimates of the emissions ranging from 0.6 to 1.6 kg $CO_{2_{eq}}$ kg^{-1} and for lime from 0.13 to 0.59 kg $CO_{2_{eq}}$ kg^{-1}.

The data that is comparable across crops include those related to the main agricultural inputs such as fertilizers, chemicals, and diesel usage for farming machines. GHG emissions were therefore estimated only for these agricultural activities. This leads to an underestimation, even though fertilizers (specifically nitrogen and lime) are the greatest contributors to GHG emissions in the agricultural phases. Finally, the amounts of fertilizers, chemicals, and diesel are consistent with the estimate of the energy input as reported in Table 11.5, since the same sources were consulted [29, 35, 38–39]. The emission factors for fertilizers (nitrogen, phosphorus, potash, and lime), chemicals (herbicides and insecticides), and diesel are from Nogueira [42], while that of lime is derived from the FEAT database adding some value from Brazilian studies ([42, 44]). GHG emissions associated to the cultivation of sunflower, soybeans, and castor beans, alone and in association with peanuts, range from 770 to 958 kg $CO_{2_{eq}}$ ha^{-1}. The cultivation of cotton, as for the energy demand, is much higher, up to more than 4 tons of $CO_{2_{eq}}$ ha^{-1}. Gibbs et al. [45] reached a similar conclusion while using a completely different approach.

Tab. 11.5 Emission factors of the selected components of the agricultural phases and crop GHG emissions.

Emission factor[a]		Crop emission (kg CO_{2eq} ha^{-1})				
	Value	Sunflower[b]	Soybean[b]	Castor bean[c]	Cotton[d]	Castor and peanut[e]
Nitrogen	3.97	228.3	0.0	79.4	2580.5	0.0
Phosphorus	1.30	16.9	26.0	58.5	520.0	0.0
Potash	0.71	17.8	14.2	28.4	284.0	0.0
Lime	0.01[a,f,g]	310.0	620.0	0.0	0.0	620.0
Herbicide	25.00	80.3	12.6	296.4	80.0	50.0
Insecticide	29.00	31.0	63.0	80.7	408.9	39.5
Diesel	3.4	182.1	222.6	284.4	183.4	60.7
Total	–	866.3	958.3	827.8	4056.8	770.2

Source: a. [42]; b. [35]; c. [39]; d. [38]; e. [29]; f. [44]; g. [45].

11.3.4.3 Impacts on social inclusion

One of the main objectives of the Brazilian Biodiesel Plan, PNPB, was indeed to favor family farming and protect the land's traditional culture. Measuring how effective a plan is, in this respect, would require data and assumptions about the social aspects that are outside the scope of this study [24].

A number of considerations can, however, be derived from the proposed approach. The requirement of a certain amount of raw materials from family farming to comply with the PNPB specifications results in a certain portion of land (namely, A_2 in Eq. (11.1)) to be reserved for that use.

Additionally, the above optimization problem can be solved with and without constraint (11.4) imposing the percentage of feedstock from family farming. The differences between these situations illustrate all the range of possible options as far as family farming is concerned. Biodiesel producers can in fact operate even without the Social Seal and this may result in the amount of land for family farming being smaller than that implied in the PNPB constraint.

As a result of preserving some family farming, one can easily expect that the efficiency, from the energy viewpoint, will be lower than without it; however, on the other hand, the crop diversity, as measured by the Shannon index, may increase. Family crops may in fact differ from those suggested for the vast industrialized areas.

11.3.5   Other Impacts

Any agricultural plan has a number of other impacts that have not been considered in this study: they range from traffic increases (and consequent air pollution) due to seed and biodiesel transport, to changes in the local labor or agricultural machinery market.

Two aspects appear as particularly relevant—one is the economy and the other is adaptation. The first is as important as it is controversial. Many authors have formulated optimization problems similar to those proposed in this study adopting an economic viewpoint [12], defining prices and costs of any agricultural plans. However, this task remains a challenge. Figure 11.4 (elaborated from IBGE—Instituto Brasileiro de Geografia e Estatìstica—data) shows, for instance, the evolution of soybean prices (net of the inflation rate) in the state of Mato Grosso in the period of 1999–2011.

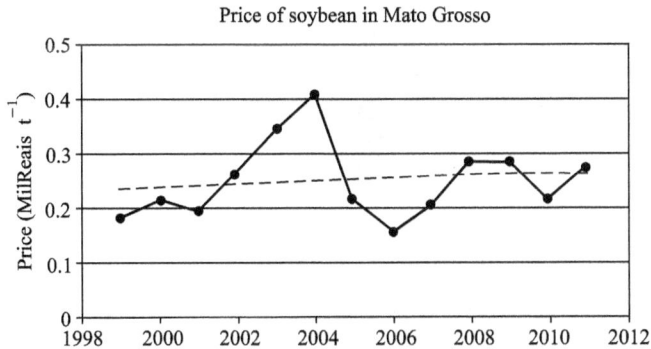

Fig. 11.4 Evolution of soybean prices (Mato Grosso, Brazil).

The evident fluctuations have internal as well as international explanations. Farmers are sensitive to current prices when they take their sowing decisions, so they cultivate a larger surface when the current price is higher. This generates an increased supply in the next year with the effect of reducing the product prices. The opposite is somehow true when the price is low, even if there is a slow, long-term increase in price, as shown by the dashed straight line (Fig. 11.4). Such fluctuations are stronger for crops more affected by international markets. Castor oil, for instance, has more than 700 different uses, ranging from medicines and cosmetics to replacing petroleum in plastics and lubricants and thus its price is influenced by the general evolution of various industrial sectors.

Another puzzling economic point is that biodiesel production in Brazil mainly goes through the so-called methanol route, namely needing methanol for the transesterification process. The value of Brazilian methanol import in US dollars has quadrupled since the inception of the PNPB [24], hindering the positive economic returns expected by the implementation of the plan.

Figure 11.5 (elaborated again from IBGE data) shows the important effect of adaptation and technological improvements. While, obviously, the annual land productivity depends on the specific climatic conditions of the given year, the progressive knowledge and improved cultivation techniques may modify the crop productivity in a significant way. The productivity of soybean in Mato Grosso has grown in the recent past to an almost constant rate, going from about 2,200 kg ha^{-1} at the beginning of the 1990s to 3,100 kg ha^{-1} in 2011. This, once again, makes the results of present and other similar studies just references for political discussions. All of the assumed parameters are in fact due to vary with time.

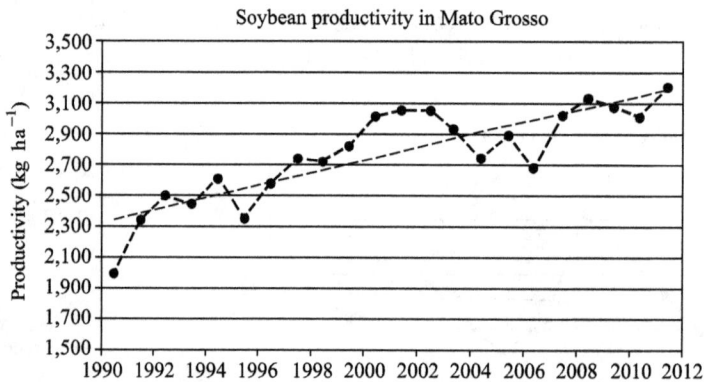

Fig. 11.5 Evolution of soybean productivity (Mato Grosso, Brazil).

11.4 Results

Under the assumptions made above, we define and analyze different alternatives that are created assuming the introduction of new oilseed crops (sunflower, castor, and peanut) in areas currently cultivated with soybeans and cotton, without interfering with other present land uses. As mentioned above, in regards to peanut, we only consider its intercropping with castor as what appears particularly interesting for energy production.

"Current" situation: This alternative assumes to maintain 100% of the area planted to soybean and cotton in 2011 and the area devoted to other oilseeds is then zero. The production of soybean and cotton is allocated according to the maps of adaptability developed in this study in order to maximize efficiency. For the latter reason, this does not correspond precisely to the actual use of agricultural land. However, since only data about the overall production were available without the specific locations of the different crops, this seemed to be a reasonable assumption. Furthermore, it makes the comparison with the other alternatives more consistent since the differences are entirely due to the different crop allocation.

Optimal energy production: The entire area is optimally allocated with the five possible oilseed crops: soybeans, cotton, sunflower, castor, and the intercropping castor-peanut. In particular, with regard to cotton, its seeds are considered as a byproduct (and therefore the agricultural costs are considered "free") until the area allocated does not exceed the area cultivated with cotton in the current situation, for the textile industry. If this amount exceeds the allocated area, additional cotton production is considered for the sole production of biodiesel and it is charged with the related agricultural production costs.

The above two alternatives are particularly significant. The current situation represents a low value of crop diversity and a low value of energy production. It corresponds to about one fourth of the area cultivated with cotton and the rest with soybean. The local soybean productivity specified by the IBGE is indeed lower than in other sites and thus can hardly compensate (from the energy viewpoint) for the energy spent in cultivation, transport, and transformation.

The optimal alternative (that takes into account the development of family farming foreseen in the PNPB) gives a positive (and high) value of net energy production. The area presently cultivated with cotton remains the same, a relatively large area (around 14.5%) is cropped with castor and peanut in family farms, and the rest is cultivated with sunflower.

The maps corresponding to the two alternatives are shown in Figures 11.6 and 11.7, while numerical values are presented in Table 11.6. It is thus impossible to attain a net energy production higher than about two million GJ yr^{-1} under the scenarios previously defined. On the other side, it is possible to increase the Shannon index by 95% from the original 0.120 to 0.403 corresponding to the optimal solution.

Tab. 11.6 Energy balance and GHG emissions.

	Current	Optimal	Family farming
Net energy production (TJ yr^{-1})	674	2,009	176
Biodiesel volume (10^3 m^3 yr^{-1})	154	180	27
Gross energy output (TJ yr^{-1})	4,116	5,774	724
Production energy (TJ yr^{-1})	3,413	3,744	550
Transport energy (TJ yr^{-1})	29	21	2
Shannon index	0.120	0.403	–
Net GHG emissions (t CO_{2eq} yr^{-1})	441.3	392.0	30.3

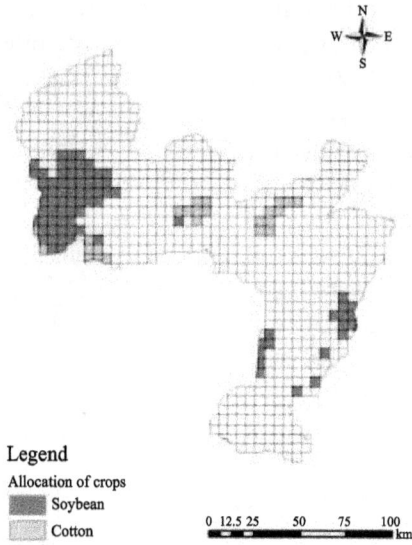

Fig. 11.6 Allocation of crops — the current situation.

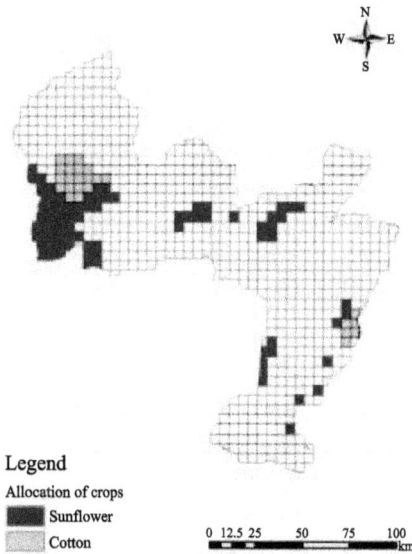

Fig. 11.7 Allocation of crops — the optimal alternative.

Other interesting considerations that emerge from the analysis of Table 11.6 are the very low components due to transport (between 1% and 6% of the net energy production) in contrast to the relevant production costs (they may reach

82% of the gross energy output). In any alternative, the plant capacity is never saturated, which means that the constraints (11.3) are never active. This confirms most current analyses that point out the underutilization of the current Brazilian biodiesel transformation capacity [24].

The optimal plan basically implies the substitution of soybean with sunflower and, since the GHG emissions considered in this study for these two crops are not very different, the overall emission does not change much in the various scenarios. Indeed, this change, together with the increase of family farming with the intercropping of castor and peanuts that avoids some nitrogen fertilizer, reduces the overall GHG emissions on the order of 11%, that, given the uncertainties on this topic previously pointed out, appears as relatively limited.

11.5   Discussion

To deepen the understanding of different components of the problem, another scenario has been analyzed where the construction of a third plant has been considered. From a formal point of view, the problem to be solved remains quite similar to Sections 11.1–11.5, except that a new set of binary decision variables y_i has to be added to represent the existence ($y_i = 1$) or non-existence ($y_i = 0$) of a new plant in cell i. The number of plants N_p obviously becomes equal to 3. The cardinality of this new set of decision variables should be equal to the total number of cells, since this plant may also be built in cells that are not available for oil crops. It is quite unlikely, however, that building a new plant close to the border or to existing plants may be interesting. Only locations close to the barycenter of the biomass production can indeed be useful to improve the situation by reducing transportation activities (we are not considering the additional energy necessary to ship the biodiesel to some other destination external to our domain). In practice, given the scope of this study, it is sufficient to imagine a new plant somehow between existing ones and solve exactly the problem already presented with $N_p = 3$. Even if the selected location is not strictly optimal from a mathematical viewpoint, it will nevertheless point out the main advantages or disadvantages of a solution of this type.

Assuming that a new plant is built halfway between the existing ones, the energy for transport undergoes a reduction of about 33% compared to the case with only two plants, but this represents only 0.3% of the overall energy balance. The construction of a third plant is therefore inconvenient (the energy necessary for building the plant would probably not be repaid in the medium term by that spared thanks to its presence). It is, however, interesting to note that the presence of this third facility changes the crop allocation (see Fig. 11.8). The

cotton and sunflower grown in the municipalities of Alto Araguaia and Rondonopolis (municipalities where there are existing plants) are partly reallocated in
the vicinity of a third plant, on the border between the towns of Alto Garças and
Pedra Preta. This makes the Shannon index equal to 0.375, but does not modify the $CO_{2_{eq}}$ emissions that are only related to the extensions of agricultural
activities.

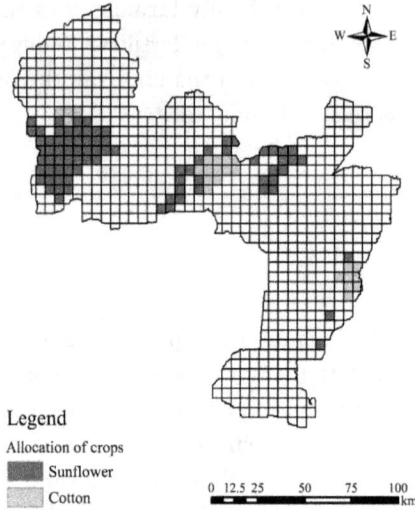

Fig. 11.8 Allocation of crops with a third possible plant.

Another analysis that is worth discussing is the overall impact of family farming. To this purpose, one may solve the above optimization problem by simply
deleting constraint (11.4) on the minimum family production. As expected, the
result of this case shows an increase of energy output and a reduction of input
of the order of some percentage. These combined effects determine a certain
improvement of the net energy balance (+7%). On the other side, this scenario
leads to a consistent reduction of the Shannon index (−46%) that practically
brings it close to the current situation characterized by extended monocultures.
Numerical values for these two cases are reported in Table 11.7.

Finally, it is also interesting to evaluate what could be the effects of adopting fertilizers to improve the productivity of castor-peanut intercropping: the
overall energy output obviously increases by 3%, but the net energy production
decreases by about 1.4%, as already shown by De Albuquerque [29]. In this
situation, given the higher oil productivity, the land required for family farming
to fulfill the national plan constraint is reduced to 12.4% of that available, which
implies a reduction to 0.369 of the Shannon index.

Tab. 11.7 Effects of the construction of an additional plant and of the reduction of family farming.

	Optimal	Additional plant	Difference	No. family farming	Difference
Net energy production (TJ yr^{-1})	2,009	2,016	0.4%	2,148	6.9%
Biodiesel volume (10^3m^3 yr^{-1})	180	180	0%	179	−0.6%
Gross energy output (TJ yr^{-1})	5,774	5,774	0%	5,904	2.2%
Production energy (TJ yr^{-1})	3,744	3,744	0%	3,734	−0.3%
Transport energy (TJ yr^{-1})	21	14	−33.6%	22	3.9%
Shannon index	0.403	0.375	−6.9%	0.217	−46.1%
Net GHG emission (t CO$_{2eq}$ yr^{-1})	392.0	392.0	0%	422.8	7.9%

11.6   Conclusions

This work shows that a mathematical formulation of a biodiesel production plan can help to analyze its multifaceted impacts. The main environmental consequences of such plans may also be analytically evaluated and thus the overall procedure, despite its limitations, may well serve as a basis for further studies and for more informed decisions.

One important limitation of the approach described here is that it disregards time evolution and simply looks at stationary conditions. As already noted, cultivation is an evolutionary process and the change from one traditional crop to another requires time to adapt and learn the methods for the best yield. In quite the same way, the change may modify the metabolism of the soil and release a certain amount of carbon that should be computed in the GHG budget over a number of years.

Another critical aspect is that all of the results presented here evidently depend on the numerical values assigned to the variables involved. The assumed productivity of the crops, for instance, has a strong influence on the optimal energy result and also on the allocation of the various crops. It depends, in turn, on uncertain factors like climate, agricultural practices, precise effects of suboptimal environmental conditions, etc. The values assumed in this study are, as far as possible, those adopted by the local authorities; nevertheless, some of them may appear questionable, such as the low productivity of soybean that has been proven to perform much better in other Brazilian regions.

This notwithstanding, studies like that presented in this chapter may help to clarify some basic questions. For instance, the opportunity emerges for abandoning some first-generation energy crops, such as soybean, to turn toward other options. This is indeed the tendency in many other biodiesel development processes all over the world, where the attention is progressively turning to a better use of byproducts and of other highly productive crops like jatropha (locally known as Pinhão Manso) that should also be probably tested in this area of Mato Grosso.

References

[1] US EIA (Energy Information Administration) International Energy Statistics (Accessed October, 2012, at http://www.eia.gov).

[2] Pimentel D, Patzek TW. Ethanol production using corn, switchgrass, and wood; biodiesel production using soybean and sunflower. Natural Resources Research 2005; 14(1): 65–76.

[3] Nature. Biofuels. Nature 2011, 474 Supplement, S1–S25.

[4] Bauen A, Berndes G, Junginger M, et al. Bioenergy—A sustainable and reliable energy source. A review of status and prospects. IEA Bioenergy, Paris, 2009.

[5] Emerging Markets Online. Biodiesel 2020: A Global Market Survey, 2nd Edition, February, 2008 (Accessed November, 2012, at http://www.emerging-markets.com/biodiesel).

[6] Lu L, Jiang D, Zhuang D, Huang Y. Evaluating the marginal land resources suitable for developing *Pistacia chinensis*—based biodiesel in China. Energies 2012; 5(7): 2165–2177.

[7] FAO. Food and Agriculture Organization of the United Nations. State of Food and Agriculture. Roma, 2008.

[8] IBGE, Instituto Brasileiro de Geografia e Estatística. Censo Agropecuàrio da Agricultura Familiar, 2006 (Accessed October 30, 2012 at: http://www.ibge.gov.br/home/estatistica/economia/agropecuaria/censoagro/default.shtm).

[9] Fiorese G, Guariso G. A GIS-based approach to evaluate biomass potential from energy crops at regional scale. Environmental Modelling & Software 2010; 25(6): 702–711.

[10] Fischer G, Prieler S, van Velthuizen H, Lensink SM, Londo M, de Wit M. Biofuel production potentials in Europe: sustainable use of cultivated land and pastures. Part I: Land productivity potentials. Biomass and Bioenergy 2010; 34(2): 159–172.

[11] Riemke R, de Campos CL, Hamacher S, Oliveira F. Optimization of biodiesel supply chains based on small farmers: a case study in Brazil. Bioresource Technology 2011; 102: 8958–8963.

---

[12] Andersen F, Iturmendi F, Espinosa S, Diaz MS. Optimal design and planning of biodiesel supply chain with land competition. Computers & Chemical Engineering 2012; 47:170–182 (http://dx.doi.org/10.1016/j.compchemeng.2012.06.044).

[13] IBGE, Instituto Brasileiro de Geografia e Estatìstica. Malha Municipal Digital, 2001. (Accessed at December, 2009, at http://www.ibge.gov.br/home/geociencias/default_prod.shtm#TERRIT).

[14] IBGE. Instituto Brasileiro de Geografia e Estatìstica. 2010 Population Census (Accessed October, 2012, at http://www.ibge.gov.br/english/estatistica/populacao/censo2010/default.shtm).

[15] Normais Climatologicas, 1961/1990. (Accessed December, 2009, at http://www.inmet.gov.br/portal/index.php?r=clima/normaisClimatologicas).

[16] EMBRAPA Solos,. Mapa do Solos do Brasil—Escala 1 : 5.000.000, 1981 (Accessed December, 2009, at http://www.cnps.embrapa.br/).

[17] EMBRAPA Brasil Em Relevo, Maps S_21_15_2000 and S_22_15_2000 (Accessed December, 2009, at http://www.relevobr.cnpm.embrapa.br/).

[18] IBGE. Diretoria de Pesquisas, Coordenação de Agropecuária, Pesquisa da Pecuária Municipal 2011, 2011 (Accessed November, 2012, at http://www.ibge.gov.br/home/estatistica/indicadores/agropecuaria/producaoagropecuaria/).

[19] The Economist. The miracle of Cerrado. The Economist Aug. 26th, 2010.

[20] IBGE. Síntese do panorama da economia brasileira, Regional Accounts of Brazil—2010. (Accessed December, 2010, at ftp://ftp.ibge.gov.br/Contas_Regionais/2010/comentarios.pdf).

[21] CONAB—Consolidado e Acompanhamento da Safra 2005/2006, 1° Levantamento, Brasil: Área plantada de grãos (Accessed October, 2012, at http://www.conab.gov.br).

[22] Janssen R, Rutz DD. Sustainability of biofuels in Latin America: risks and opportunities. Energy Policy 2011; 39(10): 5717–5725.

[23] Da Costa ACA, Junior NP, Aranda DAG. The situation of biofuels in Brazil: new generation technologies. Renewable and Sustainable Energy Reviews 2010; 14(9): 3041–3049.

[24] Rathmann R, Szklo A, Schaeffer R. Targets and results of the Brazilian Biodiesel Incentive Program—Has it reached the Promised Land? Applied Energy 2012; 97: 91–100.

[25] PNPB. Produção Oleaginosas—Produção Oleaginosas Brasil. 2012.

[26] ANP (National Agency of Petroleum). Natural Gas and Biofuels (Brazil), Oil, Natural Gas and Biofuels Statistical Yearbook. Agency of Petroleum, Natural Gas and Biofuels—ANP, Rio de Janeiro, 2011.

[27] Gelfand I, Zenone T, Jasrotia P, Chen J, Hamilton SK, Robertson GP. Carbon debt of Conservation Reserve Program (CRP) grasslands converted to bioenergy production. Proceedings of the National Academy of Sciences 2011; 108(33): 13864–13869.

[28] EMBRAPA Algodão, Produtos (Accessed January, 2010, at http://www.cnpa.embrapa.br/).

[29] De Albuquerque FA, Beltrão NEM, de Lima NNC, de Andrade JR, de Melo EBS. Análise energética do consórcio mamona com amendoim. In: III Congresso Brasilero de Mamona. Salvador BA, 2008. Embrapa Algodão.

[30] Padula AD, Santos MS, Ferreira L, Borenstein D. The emergence of the biodiesel industry in Brazil: current figures and future prospects. Energy Policy 2012; 44: 395–405.

[31] Yang J, Hammer RD, Thompson AL, Blanchar RW. Predicting soybean yield in a dry and wet year using a soil productivity index. Plant and Soil 2003; 250(2): 175–182.

[32] Garcia-Paredes JD, Olson KR, Lang JM. Predicting corn and soybean productivity for Illinois soils. Agricultural Systems 2000; 64(3): 151–170.

[33] IBGE Mapas (Accessed December, 2011 at http://mapas.ibge.gov.br/en/).

[34] Da Silva César A, Otávio Batalha M. Biodiesel production from castor oil in Brazil: A difficult reality. Energy Policy 2010; 38(8): 4031–4039.

[35] Gazzoni DL, Felici PH, Coronato RM, Ralish R. Balanço energético das culturas de soja e girassol para produçao de biodiesel. Biomassa & Energia 2005; 2(4): 259–265.

[36] Fiorese G, Gatto M, Guariso G. Utilizzo delle biomasse a scopo energetico: un'applicazione alla Provincia di Cremona. L'Energia Elettrica, Sezione Ricerche 2005; 82: 1–8.

[37] Bovolenta FC. Análise energética comparativa de transporte multimodal da soja. Ciências Agronômicas da Botucatu, Botucatu, Brazil, 2007.

[38] De Albuquerque FA, Beltrão NEM, de Oliveira JMC, et al. Balanço energético de sistemas de produção de algodão no Cerrado do Mato Grosso do Sul. In: VI Congresso Brasilero de Algodão, Uberlândia MG. Embrapa Algodão, 2007. (http://www.cnpa.embrapa.br/produtos/algodao/publicacoes/cba6/trabalhos/index.html).

[39] Chechetto RG, Siqueira R, Gamero CA. Balanço energético para a produção de biodiesel pela cultura da mamona (Ricinus communis L.). Revista Ciência Agronômica 2010; 41(4): 546–553.

[40] Mendonça ML. Monocropping for Agrofuels: The case of Brazil. Development 2011; 54: 98–103.

[41] Shannon CE, Weaver W. The Mathematical Theory of Communication. University of Illinois Press, Urbana 1963; 117.

[42] Nogueira LAH. Does biodiesel make sense? Energy 2011; 36(6): 3659–3666.

[43] Camargo GGT, Ryan MR, Richard TL. The Farm Energy Analysis Tool (FEAT). Department of Agricultural and Biological Engineering, The Pennsylvania State University, September 2011.

[44] De Souza SP, Pacca S, Turra de Ávila M, Borges JLB. Greenhouse gas emissions and energy balance of palm oil biofuel. Renewable Energy November 2010; 35(11): 2552–2561.

[45] Gibbs HK, Johnston M, Foley JA, et al. Carbon payback times for crop-based biofuel expansion in the tropics: the effects of changing yield and technology. Environmental Research Letters 2008; 3(3): 1–10.

Part VI
Global Potential Assessments

Chapter 12

Biomass Potential of Switchgrass and Miscanthus on the USA's Marginal Lands

Varaprasad Bandaru, R. César Izaurralde, and Kaiguang Zhao

12.1   Introduction

Concerns over the depletion of fossil fuels and over climate change that is caused by increased greenhouse gas (GHG) emissions spurred by the use of fossils fuels have motivated strong interest in the development of renewable biofuels globally [1–2]. Corresponding to the Energy Independence Security Act (EISA) of 2007, the United States (US) Congress mandated the annual production of 136 billion liters of renewable fuels by 2022, of which at least 80 billion liters is expected to be obtained from non-grain sources, including 57 billion liters of cellulosic ethanol. To achieve these massive long-term goals sustainably, careful land use planning is necessary. The sustainability of biofuel production mainly depends on two primary factors: (i) choice of cropping system and (ii) type of the land used for biofuels [3–4]. Currently, most biofuels are produced from grain crops growing on croplands, either by replacing other food crops or by crop intensification [5–6]. However, critics have pointed out serious consequences likely to result from these current strategies. For example, the replacement of food crops increases food prices, which is likely to cause cropland expansion elsewhere, leading to indirect land use emissions [7–8]. To avoid this situation, experts have recommended the use of abandoned or non-arable marginal lands for advanced biofuel production, thereby constraining food and feed production to current croplands [9–10]. Even though marginal lands appear to be promising land resources, questions still remain about the availability of marginal lands, their inherent biofuel potential, and possible environmental consequences of their use [11–12].

The definition of marginal lands varies across domains, organizations, regions, and countries based on their management goals. However, biophysical marginality and economic marginality are the two primary variables to be considered when defining marginal lands [13]. Biophysical marginality, which is typically a measure of land productivity for crop cultivation, is widely used in land use planning with the assumption that economic marginality results from biophysical limitations [11]. Primary (productive) agricultural lands are characterized by favorable soil, and landscape and climate features that are suitable for crop cultivation. Conversely, marginal lands are less productive, result in lower economic returns, and may exhibit high vulnerability to environmental risks. For these reasons, marginal lands are typically used in less intensive ways (*e.g.*, haying) [14]. However, because of changes in the demand for land and in policy, some marginal lands have been brought in and out of cultivation depending on needs and opportunities. A good example is the recent land use change in the USA's prairie states that is a result of the increased demand for corn cultivation. Recent studies found that increased corn prices induced conversion of non-arable marginal lands to corn cultivation in the these prairie states [15–16].

Cultivation of non-arable marginal lands for biofuel production may result in negative environmental impacts as consequences of agricultural practices [13, 17–19]; however, the extent of the impact depends on the choice of cropping system [3]. For example, annual crops demand high energy inputs and intensive land management, which cause adverse environmental impacts and also may not be economically viable on marginal lands [4, 11]. In contrast, perennial grasses are tolerant of adverse soil and climatic conditions and they do not require higher energy inputs for cultivation. Therefore, they can be suitable for growing on marginal lands [6, 20–21]. Still, the main uncertainty is whether the USA has sufficient non-arable marginal land to meet long-term biofuel targets. Previous studies, attempted to project marginal lands, required them to meet the EISA target of 57 billion liters of cellulosic ethanol, based on average biomass yields. Based on the average yields of switchgrass (*Panicum virgatum*) in a field scale study, it was estimated that 21 million ha of land would be required to meet the cellulosic ethanol target [20]. Similarly, another study reported that 15.7 and 10.3 million ha of marginal land would be required based on the estimated average biomass values of switchgrass and miscanthus (*Miscanthus × giganteus*), respectively [4]. However, for a precise estimate of land requirements, spatial variability in biomass potentials should be considered since marginal lands vary across the space in terms of their inherent biomass potentials.

Recently, several studies have focused on developing methods for identifying geographical locations of marginal lands. These methods differ in terms of datasets and criteria used in identifying the spatial locations of marginal lands. Some methods are simply based on land quality functions. For example, few studies recognized idle, pasture, or abandoned lands as marginal lands

[22–23]. Another qualitative metric generally used for categorizing agricultural lands is the US Department of Agriculture (USDA) Land Capability Classification (LCC) system. Based on expert knowledge of soil and topographical characteristics, lands are categorized into eight classes (I–VIII) [24]. Land areas in land capability classes IV–VIII are categorized as marginal lands because these lands show the most limitations for cultivation [12–13, 25].

In addition to these qualitative methods, other recent studies have attempted to develop quantitative methods for categorizing marginal lands. A recent study presented a land marginality index based on land quality functions (*e.g.*, soil erodibility, soil tolerance) [11]. Similarly, a few others used different land quality functions to develop multiple criteria for categorizing different types of marginal lands [13, 26]. Important factors that most of the earlier studies did not consider are local climate features. The local climate is a critical factor in determining land suitability for agriculture. In addition, earlier studies did not take historical land use into consideration in their marginal land analyses. Agricultural land use patterns are typically influenced by the physical and environmental attributes of the landscape. For instance, a land use analysis revealed that medium-productivity lands generally remained as grasslands while commodity crops are usually grown on productive lands [27]. Using this rationale, a quantitative methodology was recently developed to classify agricultural lands [4]. This method used crop productivity to represent land suitability for crop cultivation and determined the spatial relationship between land productivity and land use patterns. Based on land use patterns, land productivity threshold levels were identified to classify agricultural lands into productive, marginal, and unproductive lands.

The perennial energy crops that are being extensively studied for biofuel production are switchgrass and miscanthus. While switchgrass, a native American grass, is widely considered a model species for biofuel production in the USA [28–29], European research has focused extensively on miscanthus, which is originally from Japan [30–31]. Both switchgrass and miscanthus are perennial rhizomatous, warm-season C4 grasses that have efficient photosynthetic and nutrient cycling systems. The rhizome systems in perennial grasses allow for the recycling of nutrients between above- and below-ground plant parts, thus maximizing nutrient use efficiency and, due to the C4 photosynthetic systems, these grasses have higher radiation use efficiency than C3 grasses, thus resulting in higher plant biomass [32]. Earlier studies reported that, due to high above- and below-ground biomass potential, not only can these grasses sequester greater amounts of soil carbon, they also have the potential to replace the same amount of fossil energy [9, 20, 33]. Moreover, previous studies found that the low land management and minimum fertilizer requirements of these grasses can benefit the environment by reducing net GHG emissions [34].

Responding to the issues over available marginal lands and their biofuel potentials to meet long-term biofuel targets, this study was initiated with three main objectives: (i) determining the total acreage of available non-arable marginal lands in the USA; (ii) developing meta models for switchgrass and miscanthus using the Environmental Policy Integrated Climate (EPIC) model simulations to predict approximate biomass estimates on a larger scale; and (iii) estimating the total bioenergy potential of switchgrass and miscanthus growing on non-arable marginal lands in the USA.

12.2   Methods

The methods of this study include two subsections. Firstly, we identified non-arable marginal lands using a quantitative methodology developed in a recent study [4]. In the second section, to meet the second objective, we developed empirical models for estimating switchgrass and miscanthus biomasses based on crop productivity and EPIC regional simulations. Further, these empirical models were validated against the results from experimental data. Subsequently, we applied these models to estimate the biomass potentials of switchgrass and miscanthus on the USA's marginal lands. Hereafter, non-arable marginal lands are referred to as marginal lands for conciseness. Detailed methods are described below.

12.2.1   Identification of the USA's Marginal Lands

Two datasets were used to identify the non-crop marginal lands: (i) satellite-based land cover data from the USDA-Cropland Data Layer (CDL) and (ii) USDA National Commodity Crop Productivity Index (NCCPI). The 2009 CDL was used to determine the non-arable lands, including idle lands and grasslands. The NCCPI data, which ranged from 0 to 1, with 0 being the lowest productivity and 1 representing the highest productivity, was used to identify the marginal lands. The land classification method developed recently was adopted to determine the marginal lands [4]. This study found specific land use patterns with respect to the NCCPI and, based on these patterns, the NCCPI threshold values were identified for use in classifying the agricultural lands [4]. Based on their threshold values, the lands were classified as productive (NCCPI >0.47), marginal (0.47 and 0.14), or non-productive (NCCPI <0.14). In this study, we used the same NCCPI range (0.47–0.14) to identify marginal lands.

12.2.2   Processing Land Cover Data

The USDA National Agricultural Statistical Services (NASS) produce annual land cover data for each state based on satellite data [35]. We obtained the 2009 land cover data for each state. The spatial resolution of 2009 land cover data is 56 m, commensurate with the Advanced Wide Field Sensor (AWIFS) onboard the Indian Remote Sensing Satellite RESOURCESAT. The CDL rasters of all of the states were mosaicked to create a USA land cover map, which was then re-projected using Albers Equal-Area Conic Projection. We extracted pixels under idle lands, grasslands, and reclassified as grasslands to represent non-arable lands.

12.2.3   NCCPI

The USDA-NRCS (Natural Resources Conservation Service) has developed a NCCPI model by integrating 30 different soil, climate, and landscape components into the Soil Survey Geographic (SSURGO) database using the fuzzy logic system approach [36]. As NCCPI uses climate and landscape features, along with soil characteristics, it represents inherent land productivity and suitability for crop production more precisely [24]. Soils with high favorable conditions for crop cultivation have higher NCCPI values and soils with adverse conditions have lower NCCPI values. This pattern was evident in earlier study [4]. The NCCPI and SSURGO data for the entire USA were obtained from the USDA-NRCS and the NCCPI values were linked to the SSURGO map unit (MUKEY) in order to create the USA productivity map.

12.2.4   Determination of Marginal Lands

We combined the productivity map with grassland pixels and later extracted pixels that had NCCPI values between 0.14 and 0.43 to determine the marginal lands.

12.2.5   Development of Empirical Models

Simulations using mechanistic models and high-resolution input data sets would provide accurate results; however, its implementation on a large scale is highly challenging due to intensive computation. On the other hand, empirical models can easily be implemented on a large scale but the results may not be as accurate as the mechanistic model simulations. Here, we chose the latter approach so as to develop models that can be easily implemented over large regions. We developed empirical models based on the ordinary least squares (OLS) regression analysis for estimating the biomasses of switchgrass and miscanthus on the USA's marginal lands. We used the NCCPI as a predictor variable to estimate biomass. As mentioned above, the NCCPI represents intrinsic geographical biomass variability effectively as it was estimated by integrating numerous soil, climate, and landscape features, which are determinant factors for crop yields. It was evident in a recent study, which found a close relationship between simulated crop yields and the NCCPI [4]. Moreover, the NCCPI model was developed based on a national interpretation of different productivity features such that the NCCPI values represent land productivity of a wide array of soils across the nation [36]. For example, the maximum NCCPI value for Michigan is limited to 0.87, while in Iowa, the maximum NCCPI value is 0.97. Therefore, the NCCPI is expected to capture interstate as well as intrastate variability in land productivity. The form of the empirical models is:

$$Y = \beta_0 + \beta_1 X + \varepsilon \tag{12.1}$$

where Y is the response variable (Biomass yield), X is predictor variable (NCCPI), β_1 is the parameter estimate for the NCCPI, β_0 is the intercept, and ε is the random error or residual.

12.2.6 Sample Data

For the model development, we used simulated model results as inputs instead of using field scale data because field scale data for perennial grasses, particularly for miscanthus, is very limited and the utilization of limited sample data would not yield a reliable empirical model that can be used over larger regions. In this study, we used simulated biomass yields in southwest Michigan including nine counties (Allegan, Barry, Branch, Calhoun, Cass, Eaton, Kalamazoo, St. Joseph, and Van Buren) as input data.

12.2.7 Regional Model Simulations

We used the Spatially Explicit Modeling Framework (SEIMF) [37], in which the EPIC model is a central component, to simulate biomass yields for switchgrass and miscanthus. The EPIC model is a biophysical and biogeochemical model that can simulate numerous crop yields and environmental variables (*e.g.*, crop biomass, primary productivity, soil carbon, soil erosion) that are influenced by different growing conditions (*e.g.*, climate, soil, and management). The SEIMF was used to simulate potential switchgrass and miscanthus biomass yields on agricultural lands (both croplands and non-croplands) in nine counties of southern Michigan comprising. The management protocol used to parameterize the EPIC model is described in Table 12.1. The input data, including climate, soil, and topographic characteristics, was obtained from the North American Land Data Assimilation System (NLDAS) (http://ldas.gsfc.nasa.gov/nldas/NLDAS2 forcing_download.php), the SSURGO database (http://soils.usda.gov/survey/geography/ssurgo/), and the United States Geological Survey (USGS) digital elevation model (DEM) (http://nationalmap.gov/viewer.html), respectively. Firstly, homogenous modeling units were created by combining the SSURGO soil, CDL land cover, and DEM elevation maps. Each homogenous modeling unit represents a unique combination of elevation, soil, and land use and the spatial resolution of the modeling unit is 56 m, which is commensurate with the resolution of the CDL raster. The simulations were employed for 24 years for both switchgrass and miscanthus. The model was calibrated and validated using experimental data from the Kellogg Biological Station (KBS) at Michigan State University, Michigan [12, 37].

Tab. 12.1 Management protocol for switchgrass and miscanthus cropping systems as reported in the EPIC model.

Cropping system inputs	Switchgrass	Miscanthus
Seeding rate (kg ha^{-1})	5.6	17290[?]
Herbicides (kg ha^{-1})	none*	none*
Fertilizers (kg ha^{-1})		
N	60	60
P (as P_2O_5)	0	0
K (as K_2O)	0	0

* A mid-season mowing operation was assumed as a weed control strategy in the EPIC model for switchgrass and miscanthus during establishment only.

? Miscanthus seeding rate refers to rhizome number per ha.

12.2.8 Data Selection

To avoid over-fitting the OLS models, we randomly selected 250 samples from model simulations for each crop and divided them into two sub-samples: (i) a training data set (150 samples) for parameter estimation and (ii) a testing data set (100 samples) for evaluating goodness-of-fit. Later, we extracted representative NCCPI values of selected samples.

12.2.9 Model Development and Validation

Initially, we fitted the models for switchgrass and miscanthus using all training datasets and evaluated the fit diagnostic statistics for identifying outliers.
Based on Cook's d and R student values, we identified 20 and 21 observations as
outliers in the switchgrass and miscanthus training datasets, respectively, and
removed those observations from the training dataset to improve the overall fit
of the models. We again fitted the models using the training dataset without
the outliers. We evaluated the performances of the fitted models using the testing dataset and published the experimental data. Predicted biomass values for
the testing dataset were estimated using model parameters and were then were
compared with corresponding biomass values. In addition to this, we collected
experimental data to validate the models. We obtained the NCCPI values of
experimental sites and estimated biomass yields based on our models and then
compared them with the corresponding observed biomass values at experimental
sites. For switchgrass, we chose seven marginal land sites from three states (*i.e.*,
two in Oklahoma, two in North Dakota, and four in Texas). For miscanthus,
the experimental data was only available in Illinois, where we used data from
six different sites, and all miscanthus sites are located on productive croplands.
In addition to validation, for each model, we assessed the distributions of residuals to examine whether they are evenly distributed and also reported residual
standard errors, root mean square errors (RMSEs), and Pearson correlations.

12.3   Results and Discussion

12.3.1   USA Marginal Lands

The spatial distributions of the USA's marginal lands are shown in Figure 12.1.
The total area under marginal lands was estimated to be 74.26 million ha, most of
which predominantly stretched across the Great Plains (54.23%) and the eastern
parts of the mountain region (30.1%); the rest of the area is sparsely spread across
the Pacific (6.3%), northeast (2.7%), lake state (2.3%), Appalachian (2.3%), and
southeast regions (1.44%). The locations of marginal lands are clearly reflected
by the regional soil, topography, and climate features. The Great Plains are
generally considered transition regions between the eastern humid climate and
western arid climate and they are typically characterized by natural vegetation (*e.g.*, tall, mixed, and short grasses). Due to their transitional nature, the
marginal lands in the Great Plains region have proven to be sensitive to changes
in land use. Recent studies have found that marginal lands in the eastern Great

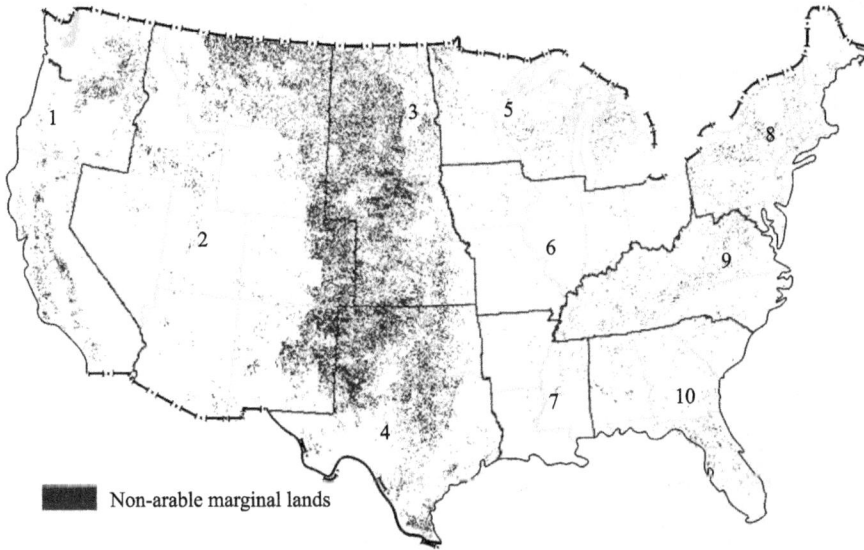

1—Pacific; 2—Mountain; 3—Northern Plains; 4—Southern Plains; 5—Lake States;
6—Corn Belt; 7—Delta States; 8—Northeast; 9—Appalachia; 10—Southeast

Fig. 12.1 Non-arable marginal lands in the contiguous USA. Light and dark lines represent state boundaries and USDA NASS production region boundaries, respectively.

Plains were converted to corn cultivation because of increases in corn market prices caused by increased corn demand [15].

12.3.2   Model Developments and Validations

Simulated biomass yields for switchgrass and miscanthus showed positive correlations with land productivity. Biomass yields were low for soils with low land productivity, but they were high for highly productive soils (Figs. 12.2a and 12.2b). Pearson correlations between biomass yields and the NCCPI were highly significant (Tab. 12.2). Switchgrass yields ($R = 0.75$) showed stronger positive responses to land productivity than miscanthus yields ($R = 0.66$) and this could be attributed to the higher bias in the simulated miscanthus yields. Due to the lack of adequate experimental data for miscanthus, the EPIC model could not be satisfactorily calibrated for miscanthus as it was for switchgrass. The OLS regression models of both switchgrass ($R^2 = 0.57$) and miscanthus ($R^2 = 0.43$) agree well with biomass yield values (Figs. 12.2a and 12.2b). Residual values are well within the acceptable range and showed more or less normal distribution.

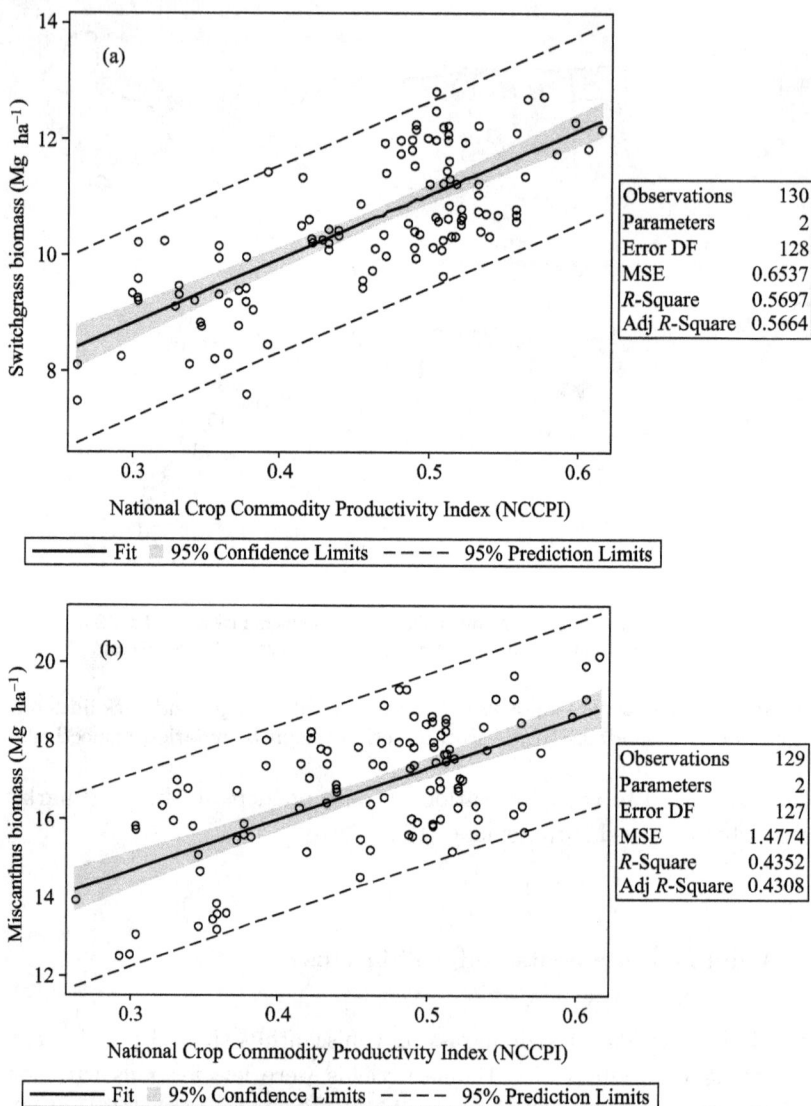

Fig. 12.2 Scatter plots of simulated biomass and the National Commodity Crop Productivity Index (NCCPI) along with best-fit ordinary least squares (OLS) regressions for switchgrass (a) and miscanthus (b).

Validation results indicated that models were able to explain variability in biomass yields reasonably well. When predicted yields of the testing data were compared with the EPIC model's simulated values of switchgrass and miscanthus, the respective models for these crops were able to explain 59% and 31% of the variability of biomass (Fig. 12.3a), respectively. Similarly, predicted biomass values at experimental sites were closely correlated with observed values for both

Tab. 12.2 Model statistics for each crop.

Model variable	Switchgrass		Miscanthus	
	Estimate	P	Estimate	P
Intercept	5.224	< 0.0001	8.971	< 0.0001
NCCPI	11.036	< 0.0001	13.034	< 0.0001
R	0.75		0.66	
R^2	0.57		0.43	
RMSE	0.808		1.215	

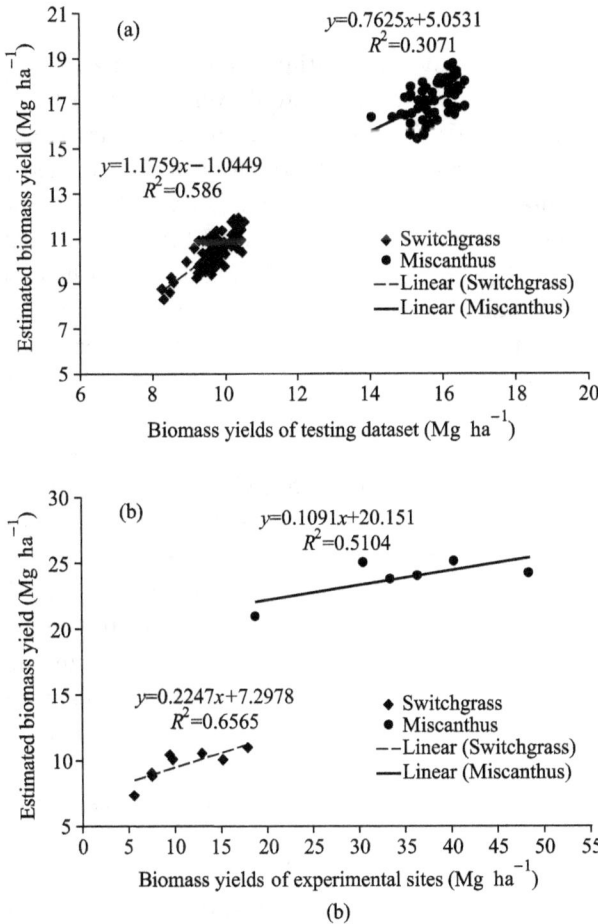

Fig. 12.3 Comparison of predicted biomass yields with the EPIC model's simulated biomass yields of the testing data set (a) and with observed biomass yields of experimental sites (b) for switchgrass and miscnathus.

switchgrass ($R^2 = 0.66$) and miscanthus ($R^2 = 0.51$) (Fig. 12.3b). These results implied that the empirical models based on the NCCPI can capture variability in biomass potentials reasonably well. However, it is worth noting that the NCCPI is a generalized index that represents multiple soil, climate, and topographic features in terms of land productivity for cultivation. Even though the NCCPI captures general variability in productivity, it may be lacking specific information that could better explain variability of biomass yields.

12.3.3   Biomass Estimates of Switchgrass and Miscanthus

The predicted yields of switchgrass and miscanthus ranged from 6.77 to 9.86 Mg ha^{-1} and 10.79 to 15.01 Mg ha^{-1}, with spatial averages of 8.12 and 12.89 Mg ha^{-1}, respectively (Fig. 12.4). The predicted biomass estimates and their spatial patterns showed good agreements with recently published model results. A published study simulated the switchgrass biomass potential on non-arable marginal lands over a ten-state region in the Midwest USA and reported an average biomass value of 8.0 Mg ha^{-1} for the study region [12], which is consistent with the average yield value (8.4 Mg ha^{-1}) from our study over the same region. To the best of our knowledge, there are no prior studies reporting miscanthus biomass values on marginal lands in the USA, except for the recent regional simulation study [4]. However, since we used the simulation results from this study to develop our empirical models, direct comparison was not appropriate. In general, our biomass yields for miscanthus were within the range of field and modeled results. For instance, an average simulated miscanthus biomass yield value (13 Mg ha^{-1}) for USA croplands [38] compared favorably with our average value. Biomass yields of switchgrass and miscanthus showed an expected decreased trend when moving from eastern to western states, which reflected general USA rainfall patterns. Higher yields were predicted in the Corn Belt, Delta, and Appalachian states and the lowest predicted yields were observed in the mountain states. In the remaining regions, moderate yields were predicted. Similar spatial patterns were reported in earlier studies [38–39].

The total estimated biomass potential for switchgrass and miscanthus on the USA's marginal lands was estimated to be 602.99 and 957.21 Tg yr^{-1}, respectively. Our total biomass estimates (143.95 Tg yr^{-1}) for switchgrass correlated well with the recent estimates (136 Tg yr^{-1}) over the ten-state midwest region [12]. There were no regional estimates on marginal lands to compare with our miscanthus estimates. Total projected biomass of a specific region may not necessarily provide an estimate of the actual biomass used for biofuel production and the location and size of a bio-refinery and its collection radius play critical roles in how much biomass will actually be produced to supply a bio-refinery.

(a)

(b)

Biomass (Mg ha^{-1} yr^{-1})

Min:6.77   Max:15.01

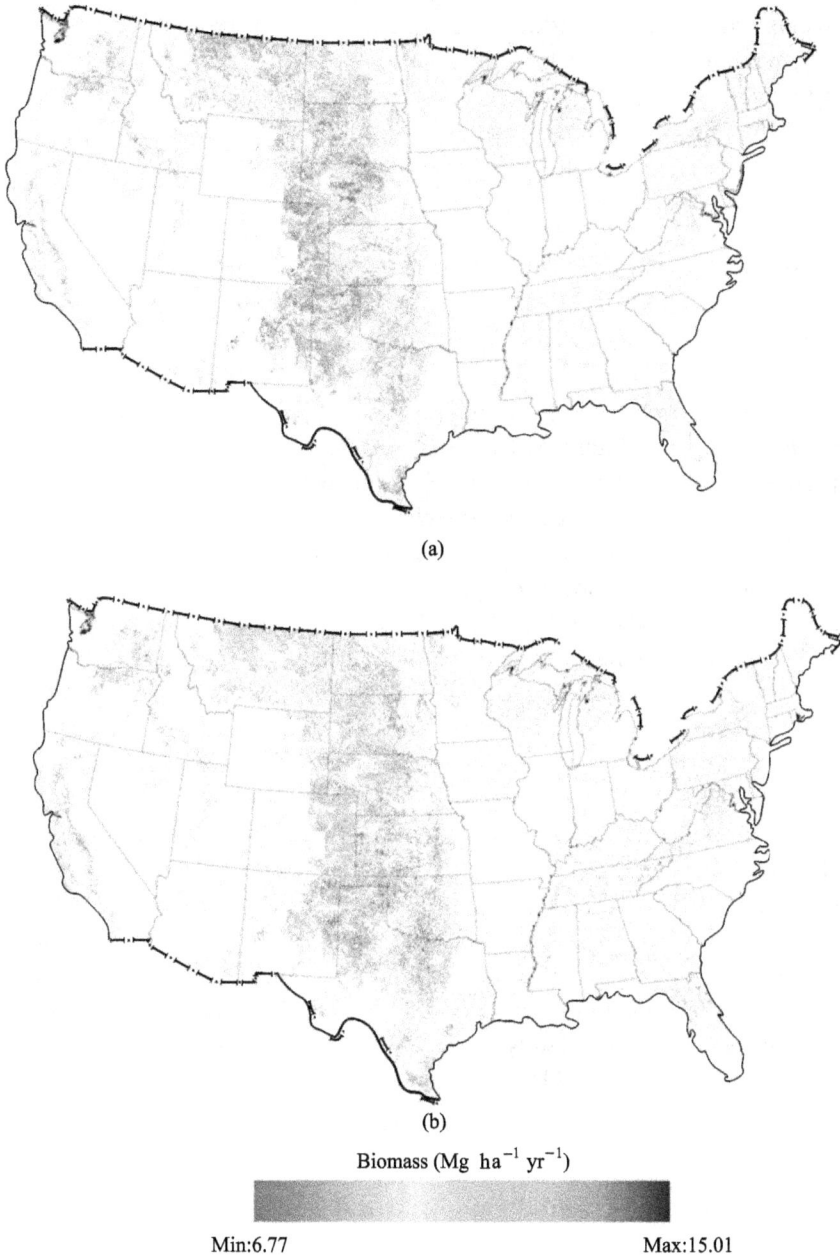

Fig. 12.4 Comparisons of biomass yields of switchgrass (a) and miscanthus (b) on the contiguous USA's marginal lands.

An earlier study [12] found that 58% of the total estimated biomass in the ten-state region was produced within an 80 km radius of projected bio-refineries, each with a minimum feedstock requirement of 0.65 Tg yr^{-1}. If we assume that this value (58%) holds the same for the entire USA, approximately 349.73 and 555.18 Tg yr^{-1} of the biomass could be supplied using switchgrass and miscanthus, respectively, for biofuel production. Based on the current biomass-to-ethanol conversion efficiency (0.30 L kg^{-1} biomass), this biomass would approximately produce 104.92 and 166.55 billion liters, respectively, which represent approximately 184% and 292% of the EISA-mandated cellulosic biofuel target (57 billion liters). This indicates that the USA's marginal lands have the potential to produce more than the amount of biomass required to meet the cellulosic biofuel targets.

Currently, the USA's biofuels are mainly derived from corn and other grains and recent studies have found that the demand for corn led to the expansion of corn production on marginal lands, which could result in detrimental impacts on the environment [15]. Substituting cellulosic biofuels for grain-based biofuels would minimize the negative consequences on the environment and would also avoid conflicts with food production. Projected energy from switchgrass and miscanthus could represent 77% and 122% of the total biofuel (both grain- and non-grain-based biofuels) target (136 billion liters), respectively. This implies that using miscanthus alone, the USA's biofuel targets can be met without utilizing croplands.

It is important to note that in recent times, new technologies are still evolving such that cellulosic biomass can be preprocessed, thus facilitating low-cost transportation to bio-refineries, which would increase biomass supply substantially for cellulosic biofuel production. In this situation, all of the available marginal lands may not need to be utilized for meeting long-term biofuel targets. It is also worth noting that all of the marginal lands identified here may not be available for biofuel production. For example, some of the marginal lands in the Great Plains region are primarily used for grazing purposes. For farmers to be willing to convert these lands for biofuel production will depend on several social and economic factors [15].

12.3.4 Comparison of Switchgrass and Miscanthus

Land use planning is an important component for achieving long-term biofuel targets sustainably. To reduce the pressure on natural resources, cropping systems with high energy potentials that use minimum resources should be recommended for biofuel production. Even though miscanthus and switchgrass are C4 warm season grasses with efficient N recycling mechanisms, miscanthus was shown to perform better than switchgrass in terms of biofuel potentials and environmental benefits. Our results indicated that miscanthus would produce, on average, 77.8% more biomass relative to switchgrass, which means that miscanthus requires less land (0.56 ha) for producing the same energy that can be produced by switchgrass in one ha of land. In addition, earlier studies reported many other benefits associated with the use of miscanthus for biofuel production. Due to higher plant biomass potential, miscanthus was reported to have higher C sequestration potential than switchgrass [6, 33]. Miscanthus was also reported to have higher water use efficiency than switchgrass [40–41]. However, the major concern with miscanthus is its adaptability to diverse USA environments, since it is not an exotic species and has a very narrow genetic base. To date, very few field scale studies have been carried out in the USA to evaluate the performance of miscanthus. Field studies carried out in Illinois reported promising results [42]. However, more recent field scale studies in Texas were found to be discouraging as the crop was not fully established due to problems with stand establishment [personal communication: James Kiniry, 2011].

12.3.5   Limitations and Future Study

We used the best available data for identifying marginal lands. However, there would be uncertainty by some extent in the spatial locations of marginal lands and this uncertainty can mainly be attributed to the errors in the classifications of grass and pasture lands in land cover data. Even though empirical models captured regional variability in biomass reasonably well, the model performance and confidence in the model results can be improved if the EPIC-simulated results are available over various regions across the nation or more field scale data is available for validation. The main goal of this study is to provide approximate biomass estimates on marginal lands based on land productivity so as to advance the understanding of the USA's marginal lands for cellulosic biofuel production. However, in addition to land productivity, local agronomic and environmental factors would influence the performances of switchgrass and miscanthus on marginal lands. A good example is the crop failure of miscanthus in Texas.

12.4   Conclusions

Following the recent concerns over available marginal lands and their potential to supply feedstocks for large-scale biofuel production, here, we determined the total available non-arable marginal lands in the USA and provided approximate estimates of total cellulosic biomass for switchgrass and miscanthus on the USA's marginal lands.

We identified 74.26 million ha of marginal lands in the USA and the results indicated that we could supply 349.73 to 555.18 Tg yr^{-1} of biomass for cellulosic biofuel production. This biomass can fulfill 120%–188% of the EISA-mandated cellulosic biofuel target (*i.e.*, 57 billion liters), which means that more marginal lands are available than required to meet the USA's long-term biofuel demands.

For large-scale, sustainable biofuel production, it is important to recommend highly productive energy crops that maximize energy with the use of limited natural resources. Among the two energy sources studied here, we found that miscanthus can produce approximately 60%–80% more biomass than switchgrass. This implies that miscanthus would produce the same energy as switchgrass using less land. Nevertheless, the adaptability of miscanthus to some of the USA's climate zones may be limited and, therefore, extensive field scale studies are required to evaluate its suitability to different USA environments. Our biomass estimates can be used to locate potential areas for designing field scale experiments for evaluating the performances of switchgrass and miscanthus under different management strategies.

References

[1] Righelato R, Spracklen DV. Carbon mitigation by biofuels or by saving and restoring forests? Science 2007; 317: 902.
[2] Njakou Djomo SN, Kasmioui OE, Ceulemans R. Energy and greenhouse gas balance of bioenergy production from poplar and willow: a review. GCB Bioenergy 2010; 3: 181–197.
[3] Wu Y, Liu S, Li Z. Identifying potential areas for biofuel production and evaluating the environmental effects: a case study of the James River Basin in the Midwestern United States. GCB Bioenergy 2012; 4: 875–888.
[4] Varaprasad. B, Izaurralde RC, Manowitz D, Link R, Zhang X, Post WM. Soil carbon change and net energy associated with biofuel production on marginal lands: a regional modeling perspective. Journal of Environmental Quality 2013; 42.6: 1802-1814.
[5] Heaton EA, Dohleman FG, Long SP. Meeting U.S. biofuel goals with less land: the potential of Miscanthus. Global Change Biol 2008; 14: 1–15.
[6] Qin Z, Zhuang Q, Chen M. Impacts of land use change due to biofuel crops on carbon balance, bioenergy production, and agricultural yield, in the conterminous United States. GCB Bioenergy 2012; 4: 277–288.
[7] Searchinger T, Heimlich R, Houghton RA, et al. Use of U.S. croplands for biofuels increase greenhouse gasses through land-use change. Science 2008; 319: 1238–1240.

[8] Fargione JE, Plevin RJ, Hill JD. The ecological impact of biofuels. Ann Rev Ecol Evol Sys 2010; 41: 351–377.

[9] Tilman D, Hill J, Lehman C. Carbon-negative biofuels from low-input high-diversity grassland biomass. Science 2006; 314: 1598–1600.

[10] Campbell E, Lobell DB, Genova RC, Field CB. The global potential of bioenergy on abandoned agriculture lands. Environ. Sci. Technol 2008; 42: 5791–5794.

[11] Bhardwaj AK, Zenone T, Jasrotia P, Roberson GP, Chen J, Hamilton SK. Water and energy footprints of bioenergy crop production on marginal lands. GCB Bioenergy 2011; 3: 208–222.

[12] Gelfand I, Sahajpal R, Zhang X, Izaurralde RC, Gross KL, Robertson GP. Sustainable bioenergy production from marginal lands in the US Midwest. Nature 2013; 493: 514–517.

[13] Kang S, Post W, Wang D, Nichols J, Bandaru V, West T. Hierarchical marginal land assessment for land use planning. Land U. Policy 2013; 30: 106–113.

[14] Reger B, Otte A, Waldhardt R. Identifying patterns of land-cover change and their physical attributes in a marginal European landscape. Landsc Urb Planning 2007; 81: 104–113.

[15] Swinton SM, Babcock BA, James LK, Bandaru V. Higher US crop prices trigger little area expansion so marginal land for biofuel crops is limited. Energy Policy 2011; 39: 5254–5258.

[16] Wright CK, Wimberly MC. Recent land use change in the Western Corn Belt threatens grasslands and wetlands. Proc Natl Acad Sci USA 2013; 110: 4134–4139.

[17] Wood S, Sebastian K, Scherr SJ. Pilot Analysis of Global Ecosystems: agroecosystems. International Food Policy Research Institute and World Resources Institute, Washington, DC, 2000 (Accessed March 23, 2012, at http://www.ifpri.org/sites/default/files/publications/agroeco.pdf).

[18] Wiegmann K, Hennenberg KJ, Fritsche UR. Degraded land and sustainable bioenergy feedstock production. Joint International Workshop on High Nature Value Criteria and Potential for Sustainable Use of Degraded Lands, Paris, 2008. (Accessed August 3, 2012, at http://www.unep.fr/energy/activities/mapping/pdf/degraded.pdf).

[19] Dale VH, Lowrance R, Mulholland PJ, Robertson GP. Bioenergy sustainability at the regional scale. Ecological Society 15, 2010. (Accessed July 23, 2012, at http://www.ecologyandsociety.org/vol15/iss4/art23).

[20] Schmer MR, Vogel KP, Mitchell RB, Perrin RK. Net energy of cellulosic ethanol from switchgrass. Proc Natl Acad Sci USA 2008; 105: 464–469.

[21] Angelini LG, Ceccarini L, Nicoletta NDNN, Bonari E. Comparison of Arundo donax L. and Miscanthus x giganteus in a long-term field experiment in Central Italy: analysis of productive characteristics and energy balance. Biomass and Bioenergy 2009; 33: 635–643.

[22] Gopalakrishnan G, Negri MC, Wang M, Wu M, Snyder SW, Lafreniere L. Biofuels, land, and water: a systems approach to sustainability. Environ Sci Technol 2009; 43: 6094–6100.

[23] Debolt S, Campbell JE, Smith Jr R, Montross M, Strok J. Life cycle assessment of native plants and marginal lands for bioenergy agriculture in Kentucky as a model for south-eastern USA. GCB Bioenergy 2009; 1: 308–316.

[24] Connor DJ, Loomis RS, Cassman KG. Crop Ecology: Productivity and Management in Agricultural Systems. Cambridge, UK: Cambridge University Press, 2011.

[25] Hamdar B. An efficiency approach to managing Mississippi's marginal land based on the Conservation Reserve Program (CRP). Resour Conserv Recylg 1999; 26: 15–24.

[26] Gopalakrishnan G, Negri MC, and Snyder SW. A novel framework to classify marginal land for sustainable biomass feedstock production. J Environ Qual 2011; 40: 1593–1600.

[27] Classen R, Fernando C, Joseph CC, Daniel H, Kohei U. Grassland to cropland conversion in the Northern Plains: the role of crop insurance, commodity, and disaster programs, ERR-120. USDA-Economic Research Service, 2011. (Accessed September 8, 2012, at http://www.ers.usda.gov/media/128019/err120.pdf).

[28] McLaughlin SB, Kszos LA. Development of switchgrass (Panicum virgatum) as a bioenergy feedstock in the United States. Biomass and Bioenergy 2005; 28: 515–535.

[29] Wright L, Turhollow A. Switchgrass selection as a "model" bioenergy crop: a history of the process. Biomass and Bioenergy 2010; 34: 851–868.

[30] Clifton-Brown JC, Stampfl PF, Jones MB. Miscanthus biomass production for energy in Europe and its potential contribution to decreasing fossil fuel carbon emissions. Glob Chang Biol 2004; 10: 509–518.

[31] Christian DG, Riche AB, Yates NE. Growth, yield and mineral content of Miscanthus × giganteus grown as a biofuel for 14 successive harvests. Ind Crop Prod 2008; 28: 320–327.

[32] Kerckhoffs H, Renquist R. Biofuel from plant biomass. Agron sustain dev 2013; 33: 1–19.

[33] Anderson-Teixeira KJ, Davis SC, Masters MA, DeLucia EH. Changes in soil organic carbon under biofuel crops. GCB Bioenergy 2009; 1: 75–96.

[34] Robertson GP, Hamilton SK, Parton WJ, Del Grosso SJ. The biogeochemistry of bioenergy landscapes: carbon, nitrogen, and water considerations. Ecol Appl 2011; 21: 1055–1067.

[35] Johnson DM, Mueller R. The 2009 cropland data layer. Photogramm Eng and Remote Sens 2010; 76: 1201–1205.

[36] Dobos RR, Sinclair HR, Jr Hipple KW Jr. User Guide National Commodity Crop Productivity Index (NCCPI). USDA-Natural Resources Conservation Service, 2008. (Accessed April 18, 2013, at ftp://ftpfc.sc.egov.usda.gov/NSSC/NCCPI/NCCPI_user_guide.pdf).

[37] Zhang X, Izaurralde RC, Manowitz D, et al. An integrative modeling framework to evaluate the productivity and sustainability of biofuel crop production systems. GCB Bioenergy 2010; 2: 258–277.

[38] Mishra U, Torn MS, Fingerman K. Miscanthus biomass productivity within US croplands and its potential impact on soil organic carbon. GCB Bioenergy 2013; 5(4): 391–399.

[39] Thomson AM, Izaurralde RC, West TO, Parrish DJ, Tyler DD, Williams JR. Simulating Potential Switchgrass Production in the United States. PNNL-19072, Pacific Northwest National Laboratory, Richland, WA, 2009. (Accessed January 08, 2013, at http://www.pnl.gov/main/publications/external/technical_reports/PNNL-19072.pdf).

[40] VanLoocke A, Twine TE, Zeri M, Bernacchi CJ. A regional comparison of water use efficiency for miscanthus, switchgrass and maize. Agri Forest Meteorol 2012; 164: 82–95.

[41] Zhuang Q, Zhangcai Q, Min C. Biofuel, land and water: maize, switchgrass or Miscanthus ? Environ Res Letters 2013; 8: 1–6.

[42] Jain AK, Khanna M, Erickson M, Huang H. An integrated biogeochemical and economic analysis of bioenergy crops in the Midwestern United States. GCB Bioenergy 2010; 2: 217–234.

Chapter 13

Global Agro-ecological Challenges in Commercial Biodiesel Production from *Jatropha curcas*: Seed Productivity to Disease Incidence

Bajrang Singh, Kripal Singh, Nidhi Raghuvanshi, and Vimal Chandra Pandey

13.1   Introduction

Jatropha curcas L. (*Jatropha*) has received considerable attention over the last few years in the biofuel sector, but plantations on an industrial scale have suffered with poor economic returns during the early growth stages (5–10 years) in most of the countries we have investigated so far. There are several constraints to make it a successful biodiesel crop; therefore, large investments and rapid expansion throughout the tropics warrant the review of its current status and future prospects. *Jatropha* is still a wild undomesticated plant with a wide variation in growth, production, and quality characteristics in different environmental conditions [1]. In this chapter, we are looking for the actual seed yield potential per unit of area in various biogeographic regions to evaluate several claims and facts of *Jatropha* for use in commercial biodiesel production. As long as *Jatropha* is considered a long-rotation small tree or shrub, the seed yield per hectare would be limiting to industrial biodiesel production. An international conference on *Jatropha* was held at Wageningen in the Netherlands (March, 2007) where experts of different countries agreed to ensure an annual yield of 4–5 tonnes of seed per ha for commercial viability. It has become clear that the positive claims on *Jatropha* are numerous, but only few of them could be scientifically sustained. Therefore, it is essential to develop a sound agro-technology to produce a reasonable seed yield per unit area annually. The environmental benefits, like the rehabilitation of barren land, soil reclamation, and C sequestration in terms of

C credit are not generally accounted for in the techno-commercial economics of biodiesel production.

Biodiesel nowadays is produced from edible crop plants or first-generation bio-fuels like wheat, maize, soybean, sunflower, safflower, sugarcane, Brassica, palm oil, Pongamia, *etc.* A challenge for biodiesel production is to use feedstock that would not compete with human food [2]. The biodiesel production from veg-etable oils during 2004–2005 was estimated to be 2.36 million tonnes globally [3]. Amongst the several plants, *Jatropha* has received wide acceptance as a 2nd-generation biofuel due to some of its merits over the other crops that are currently used in the biofuel sector (maize, soybean, rapeseed, *etc.*) Fluctuating oil prices and increasing concerns about climate change have led to a global boom of investments and enthusiasm for liquid biofuels [4]. Popular claims on drought tolerance, low nutrient requirements, pest and disease resistance, and assumed exaggerated yields [5] have promoted *Jatropha* plantations without considering their actual seed productivity per unit of area [4, 6]. Besides, wasteland recla-mation, CO_2 fixation, poverty alleviation, and soil and water conservation were advised as additional benefits on the investments [7]. However, many of these claims are yet to be supported by scientific evidence [4, 8]. Major knowledge gaps concerning basic ecological and agronomic properties (growth conditions, nutrient requirements, seed setting, oil content, and species genetics) make the seed yield poorly predictable [4, 6]. The seed productivity in native countries has not been assessed well and several assumptions and unauthentic extrapola-tions have created high expectations from the plant. *Jatropha* produces seeds rich in toxic oil varying from 27% to 40% [9]. Biodiesel is derived from the crude oil extracted from the seeds via a transesterification process of different sophistication levels [4]. *Jatropha* was identified as the most promising oil seed crop for biodiesel production in terms of oil yield as well as the desired fatty acid composition of the oil [10].

Ghosh *et al.* [11] proposed that it is difficult to get high seed yield from *Jatropha* shrubs on wastelands, which determine its success for biodiesel. The ambitious plan of India's government for producing sufficient biodiesel by the end of 2011–2012 has failed to meet its mandate of 20% blending with petroleum diesel due to the non-availability of sufficient feedstock for biodiesel production [12], it could not achieve even 5% against the projected target. Low productiv-ity is inherent to many *Jatropha* germplasm and raising large-scale plantations using such untested planting material can lead to wasteful expenditures. Unre-liable flowering and poor fruiting are generally responsible for low productivity with this species. The ratio of male to female flowers is quite high; pollinators, stigma receptivity, fertilization, and development of embryo to a viable seed are the main bottlenecks in good seed settings. As the *Jatropha* is gaining enough momentum to become a popular biofuel, genetic improvement, selection, breed-ing for elites, and domestication processes are important to ensure its economic productivity in large-scale monoculture plantations. The seed yield of *Jatropha* is generally poor and insufficient for the biodiesel industry in most of the countries [3, 13]. One of the major reasons is the lack of high yielding varieties with high oil contents. Apart from agronomic, socioeconomic, and entrepreneurial constraints, planned crop improvement programs are lacking globally [3]. Multi-location trial plantations launched in South Africa, Brazil, Nicaragua, and India indicated that the crop productivity is far below what is expected to be commercialized. In extreme cases, the plantations failed to produce fruits. Current literature and news reports from all over the world are increasingly documenting a growing disappointment regarding the crop performance. Presently, the major biodiesel feedstock accounts for over 80% of biodiesel production costs, which are limiting to the economic viability of the biodiesel industry [14]. In many trials, viral infestations, insects, and pest attacks led to extremely low yields. In recent years, a number of projects have been launched for the research and development of *Jatropha* but there has been no significant achievement toward the yield improvement or oil recovery. There are several reviews on *Jatropha* dealing with different aspects [3–5, 15–23] but none has composed the seed yield potential or oil content globally, on which its commercialization and scope depend for biodiesel production at the industrial level. In this review, an attempt was made to evaluate the yield status globally and to address issues related to yield improvement through selections, breeding (conventional and molecular), and standardization of agro-technology, including disease management.

13.2   Standardization of Agro-technology

Instead of harvesting the seeds from wild plants, it is essential to develop a standardized cultivation package from either seeds or cuttings. This includes the development of planting stock, planting techniques, optimum spacing, fertilizers and irrigation, disease management, and seed harvesting and processing. *Jatropha* is a cross-pollinated plant and therefore seed characteristics do not remain stable. As a consequence, propagation from stem cuttings is preferred from the elite plants.

13.2.1   Propagation Techniques

Jatropha can be propagated from seeds as well as cuttings. Cuttings are typically prepared with one-year-old terminal branches of 25–30 cm. It is good practice to inoculate cuttings with mycorrhizal fungi when establishing them into the nursery. This treatment improves the quality of the plant-fungal symbiosis in field conditions, especially in soil with poor fertility, as endo-mycorrhizal fungi were demonstrated to be commonly found in association with *Jatropha* in natural conditions. Seeds or cuttings of stems or twigs, at least one year old, can be directly planted in the field. The seedlings grown in PVC bags are generally transplanted to the field. The proper season for good root induction in the cuttings is the spring (February to April), however in a polyhouse, rooting can be developed throughout the year by managing an adequate temperature and humidity (Fig. 13.1). The lengths and basal areas of the cuttings determine the growth and development of the plants [24]. Short cuttings favor early sprouting, but long and thick cuttings promote more shoot and root growth. Plants that originated from stem cuttings obtained from the base of the branch grew more shoot structures (buds, stems, and leaves) than the stem cuttings from the middle and apex of the branch. The most vigorous root systems were observed in the plants that originated from direct seeding, without any transplanting. Propagation using bags or root plugs interferes with the formation of a normal

Fig. 13.1 Plant propagation under poly-house and hardening in open fields: (a) cuttings planted in thermacol trays, (b) cuttings shifted in PVC bags after one month, (c) three-month-old nursery plants, and (d) an overview of nursery plants ready for field planting.

root system [24]. Propagation through stem cuttings resulted in very different root structures, with the predominance of superficial and thin roots. Seeds are pre-soaked for 24 hrs in water and, after sowing, germinate in 5–10 days at 27–30°C with humidity saturation. Efforts have been made to propagate elite plants vegetatively as well as through micro-propagation [24]. However, the seed yields of such plantations have not been compared with parent plantations. Similarly, yields of hybrid plants are still unknown in comparison to parent accessions. The selected germplasm has been clonally multiplied through both cuttings and micro-propagation. Micro-propagation protocols have been developed and the ramets were planted in the field to assess their performance under adverse conditions. Since all of these efforts have been initiated recently, within the last 2–3 years, seed yields are yet to be assessed.

13.2.1.1 Direct planting

The lands should be ploughed once or twice depending on the nature of soil. In the case of heavy soil, deep tillage is required, whereas in light soils, shallow tillage is enough. The seeds/cuttings should be planted in the main field with the onset of monsoon season. Two seeds should be dibbled at each spot at the spacing indicated above. When the seedlings are 4 weeks old, weaker seedlings should be removed to retain one healthy seeding in each spot and the seedlings so removed could be used for gap-filling [25].

13.2.1.2 Transplanting

The main field was prepared by digging small pits of 30 cm × 30 cm × 30 cm at specified spacing (2 m × 2 m – 3 m × 3 m). Pits are refilled with soil and compost or organic manure at variable rates depending on soil degradation status. Two seeds should be sown around 6 cm deep in each poly bag and watering should be done regularly. When the seedlings are around four weeks, the weaker of the two seedlings should be removed and used for gap-filling [25].

13.2.2 Planting Material

Around 5 to 6 kg of seeds is required for 1 ha planting; otherwise, the rooted cuttings, nurtured in PVC bags for about four to six months, may be planted. At 2 m × 2 m spacing, 2,500 plants ha^{-1} shall be required under irrigated or partially irrigated conditions. On rain-fed wastelands, high-density plantations at 2 m × 1 m or 1.5 m × 1.5 m, accommodating 5,000 or 4,444 plants ha^{-1}, respectively, have been reported [1, 25–26]. Plants from seeds each develop a taproot and four lateral roots, whereas it has been reported that cuttings do not develop a taproot [27].

13.2.3 Nursery Management

Irrespective of plants raised by seeds or cuttings, the potting mixture is an important medium for plant growth at the juvenile stage. Generally, soil, sand, and compost manure or leaf mold in a 2 : 1 : 1 proportion is used. However, in a protected nursery environment, vermiculite, peat, sand, and vermi-compost are also used. The potting mixture should be light and porous for the proper development of the root. Thermocol trays, cones, and plastic root trainers on stands are also used in modern nurseries for root induction and, after one month, these plants are transferred in PVC bags. Regular weeding, irrigation, and monthly shifting of the seedlings/ramets should be carried out. Hardening of poly-house-raised plants is done in a net house for 15–30 days and thereafter under the shade of trees until the field plantings. Hardy plants can be developed in the nursery through extreme water stress, where the root/shoot ratio significantly increases [18].

13.2.4 Field Planting

Plantings should be in pits of 30^3 cm size, duly refilled with soil, compost manure, or local soil amenders. The soil of the dug-out pits should be solarised at least 15 days before filling the pits to control root infections. The plants of 50 cm height and at least six months old are preferred for plantation on degraded soil sites. The best time for planting is in the warm season before or at the onset of the rains. In the former case, watering of the plants is required for initial establishment. Direct seed sowing under field conditions is not advised as it leads to poor germination and high mortality of the seedlings under suboptimal/marginal conditions. High plantation densities, such as 2,500 plants per hectare, are possible only under good soil and water conditions while in rain-fed conditions on marginal soils, density may be low or high depending on use of soil amendments or without, respectively. The resulting effect is that the production per hectare is likely to be lower on such lands. Due to the wide variation in key economic parameters and the lack of standardized, genetically superior planting material, adequate cultivation practices call for intensive research and development prior to a large-scale industrial plantation based on incomplete information.

In order to develop high-yielding plant varieties, the best suitable germplasm has to be identified from various locations. This implies characterization of provenances with broader geographical backgrounds in order to widen the genetic base of *Jatropha*. Recently, some accessions have been screened out through large-scale plantations with high yields and oil contents [26, 28]; however, their evaluations in different geographical regions have not yet been carried out.

About 2.5 million ha have been planted with *Jatropha* in India and China alone, with plans for an additional 10 million ha by 2010 [6]. Despite the several advantages associated with *Jatropha* plants, some negative lesions were also observed in India [1, 26], which indicate a low genetic diversity among the material collected from different geographical regions of India. Many trials have shown the exaggerated yield extrapolated to per hectare area on the basis of per plant fruit or seed yield obtained with a few experimental plants. About 30%–50% of the plants in one hectare area do not bear fruit or seed, so in a yearly cycle, we get a reduced empirical yield from a unit area. This situation is further compounded by the production of kernel less seeds or seeds with rudimentary kernels that adversely affect the oil percent of the bulk lot obtained from the unit area in which the commercial oil percent is generally extracted not more than 20%–22%.

In Zimbabwe, various land types as well as agro-ecological conditions were examined for the production of *Jatropha* by Jingura *et al.* [32–33]. They emphasized the need for elite planting materials to optimize seed yield and seed quality following suitable agro-techniques and also recommended the judicious application of fertilizers based on other related crops, such as castor beans, to realize biological yield potential of above 5 t ha^{-1}. Macro-nutrients such as nitrogen (N), phosphorus (P), and potassium (K) are essential for optimization of seed yield in *Jatropha* [31]. *Jatropha* has been found to respond better to organic than inorganic fertilizers in terms of seed yield [1, 7]. Both inorganic and organic fertilizers tend to promote more vegetative growth in *Jatropha*. Mohapatra and Panda [34] studied the effects of N : P : K fertilizers on the growth and yield of *Jatropha* in an aeric Tropaquept soil of eastern India. Based on the results of growth and yield attributes, the application of N fertilizer proved to be beneficial for *Jatropha* under tropical agro-climatic conditions. Kumar *et al.* [35] investigated the effect of FYM and vermi-compost on the biomass yield of vegetatively propagated *Jatropha*. According to them, vermi-compost performed better over FYM. It has increased the oil content by improving the physical characteristics (seed length, breadth, and thickness) of *Jatropha* seeds as compared to FYM and control, which directly influence the oil contents of seeds.

The agro-technique for raising a plantation has been standardized to some extent as spacing, pruning, fertilizer, and irrigation requirements would be variable for different types of soils and climatic conditions as well as the age of the *Jatropha* plantations on the degraded land in India [1]. Rajaona *et al.* [36] investigated the pruning effects on growth, canopy size, and leaf area density. Pruning of *Jatropha* promotes the initiation of primary branching and higher productivity. However, the pruning effects on seed yield have been mostly negative [1, 11, 26]. The complete agro-technology of *Jatropha* has not been standardized, yet; however, some fragmentary information on spacing and pruning treatments has been recently published [28, 36], which concluded that pruning of large lateral branches, with shorter main stems, rendered significantly high growth and biomass in comparison to plants with small laterals and longer main stems; however, their effect on yield is yet to be determined. The pruning of apical buds of the main stem of one-year-old plants can increase the number of main and secondary branches [37]. Proper pruning of *Jatropha* helps in producing more branches with healthy inflorescences. This enhances flowering and fruit set, which ultimately increases yield [31]. However, in some studies, pruning of one-year-old plants did not affect branching [26]. Pruning in *Jatropha* plantations has shown controversial results due to variation in climatic and edaphic factors as well as time of pruning, pruning height (top and bottom), and age of the plant. Therefore, the uniform effect of pruning is not expected.

Studies of exogenous applications of various plant growth regulators (PGRs) and analysis of endogenous phytohormones showed that PGRs play important roles in floral development [38–40]. Exogenous cytokinin application has been shown to increase inflorescence meristem activity and promote floral initiation in several species [41–43]. This plant is propagated from seed or vegetatively from stem cuttings. It is still a wild plant and has not yet undergone much selection or many improvement processes for agricultural use [7]. The plant improvement program should be mainly focused on three traits, such as seed yield, oil content, and oil quality [44]. The spacing is an important factor along with the pruning management for long-term cultivation.

13.3   Global Seed Productivity

Jatropha is well adapted to marginal and degraded lands, as a fence or protection hedge of cultivated lands from animals and erosion. Recent studies show the potential of approximately 30 million ha of land on which *Jatropha* could be grown, especially in South America, Africa, and Asian countries such as China, India, and Indonesia. Seed maturity is asynchronous in the crop; therefore, one-time harvesting is not possible. Periodical harvesting by picking the yellow fruits is better than the dry, black fruits. Drying of the yellow fruits after detachment from the plant provides good seeds with healthy kernels; whereas, several black fruits collected from the plant had rudimentary kernels without any oil contents. Therefore, the harvesting of yellow fruits and their drying on the floor is recommended for the extraction of potential seeds from the fruits. In the crop, several plants do not bear fruits; therefore, extrapolation of seed yield from a few plants will lead to an exaggerated yield per unit of area. It is therefore suggested that extrapolation should be done from the seed yield of a sample plot area. Yield depends on site characteristics such as rainfall, temperature, and soil fertility [5, 7], genetics [29], plant age [27, 30], and management like the propagation method, spacing, pruning, fertilizer applications, irrigation, etc. [27, 31], as summarized in Table 13.1. As a consequence of the picking of dry, black fruits from the plants, nearly 25%–30% seeds do not bear good kernels that can be referred to as pseudo-seeds. Mexico, Nicaragua, and India have identified superior selections on the basis of seed yield. However, their multi-location trials have not yet been performed. In India, two selections of NBPGR from Gujarat,

Tab. 13.1 Global status of seed yield of *Jatropha curcas*.

Country	Latitude	Longitude	Annual rainfall (mm)	Age	Seed Yield (kg ha^{-1} yr^{-1})	Oil (%)	References
India	26° 55′	80° 59′	1,027	5	298	22–33	[1]
India	21° 46′	72° 11′	695	5	534.67	36.8	[1]
India	30° 19′	78° 04′	1,965	5	5.10	n.d.	[1]
India	25° 19′	68° 21′	786	5	415.6	n.d.	[1]
India	26° 45′	94° 13′	1,605	5	120	28.33	[1]
India	27° 06′	93° 00′	3,106	5	150	n.d.	[1]
India	25° 28′	81° 54′	1,250	4	4,000	n.d.	[4]
India	26° 55′	75° 52′	220	2.5	313	n.d.	[4]
Thailand	13° 92′	101° 01′	1,400	n.d.	2146	n.d.	[27]
Paraguay	24° 92′	57° 37′	665	9	4,000	n.d.	[27]
Nicaragua	12° 59′	85° 87′	n.d.	n.d.	5,000	n.d.	[27]
South Africa	30° 24′	29° 24′	680	3–4	348.8	n.d.	[55]
Mali	14° 81′	5° 50′	2,322	n.d.	3,000	n.d.	[74]
Mali	10° 95′	−7° 63′	978	2	550	n.d.	[75]
Nicaragua	12° 59′	85° 87′	n.d.	4	3,484	n.d.	[76]
India	20° 02′	73° 50′	1,108	n.d.	1,200	n.d.	[77]
India	28° 5′	76° 35′	n.d.	3	208	n.d.	[77]
India	25°19′	68° 21′	n.d.	3	911	n.d.	[78]
India	21°46′	72° 11′	4,000	2	1,270	n.d.	[79]
India	16°00′	80° 00′	n.d.	2.5	1,000	n.d.	[80]
India	10° 87′	76° 61′	1,085	5	1,420	35	[81]
India	11° 00′	78° 00′	n.d.	3	1,573	n.d.	[82]
Indonesia	−4° 04′	105° 65′	n.d.	2	1,000	n.d.	[82]
Zambia	−13° 86′	28° 66′	n.d.	2.5	500	n.d.	[82]
India	8° 44′	77° 44′	n.d.	3	2,000	n.d.	[82]
India	11° 20′	77° 46′	n.d.	2.5	350	n.d.	[82]

n.d. = not determined.

namely Urlikanchan and Chhatrapati, have been examined in a wide range of habitats. In most of the locations, they have outscored the other accessions, indicating their bright futures.

A trend analysis with rainfall, latitude, and age was carried out on the available data, but significant correlations were not found (Fig. 13.2). However, there

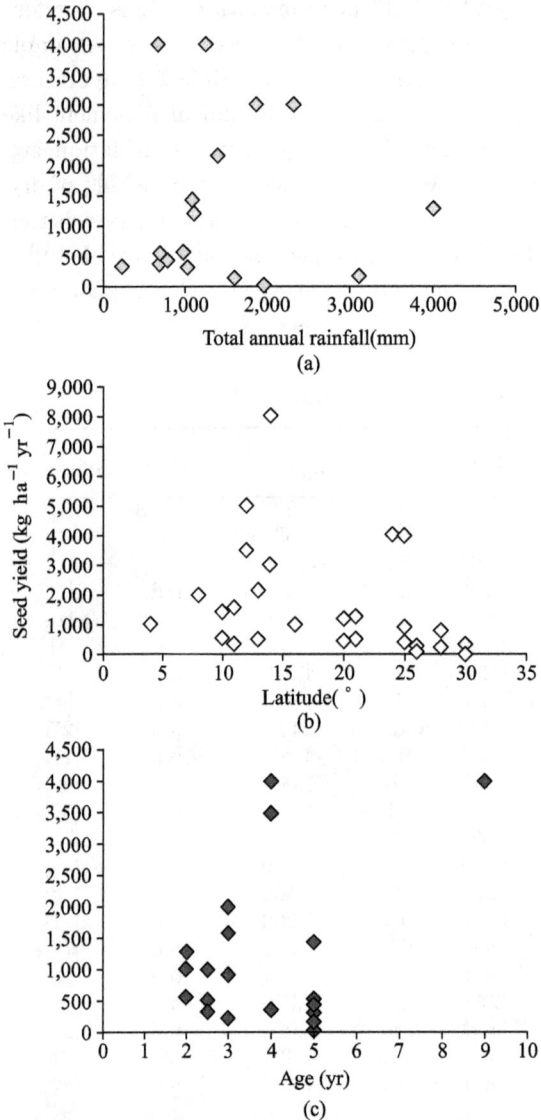

Fig. 13.2 Relation of seed yield to total annual rainfall (a), latitude (b), and age (c) of the *Jatropha*.

For the source of the data, see Table 13.2.

is an indication to get a high yield at 800–1,200 mm of annual rainfall in the latitude between 10° and 15°. The yield was not consistent with plant age due to unavailable yield records at high ages (>5 years). Seeds developed below three years of age have been found to be mostly kernelless or have only rudimentary kernels.

Various researchers recommended climatic conditions suitable for *Jatropha* cultivation (Tab. 13.2). They recommended an average annual rainfall of 300–3,000 mm, temperatures of 18–40°C, and well-drained, sandy soils with pH <9 as suitable for the cultivation of the *Jatropha* plant, but even these preferred conditions may not ensure a good seed yield.

Tab. 13.2 Climatic conditions recommended for *Jatropha* cultivation.

Average annual rainfall (mm)	Temperature (°C)	Soil condition	Source
300–1,000	20–28	Well drained, sandy soils with pH <9	[27]
250–1,000	18–40	Grows on wide range of soils including degraded land, saline and sandy soils, needs soil depth at least 45 cm	[31]
250–3,000	20–28	Well drained, sandy soils with pH <9	[4]
1,000–2,000	12.7–33.3	Well drained, sandy soils with pH <9	[5]
400–1,000	26–28	Well drained, porous, low bulk density, saline and sodic not suitable, thrives well even at 20 cm soil depth in Gujarat	[1]

13.4   Techno-commercial Economics

Seeds are harvested at maturity and the total average global seed yield potential of *Jatropha* is 1625 kg ha^{-1} or 1.6 t ha^{-1} (Tab. 13.1). However, assessment of its seed yield is still a difficult issue [4]. Actually, the mature seed yield per ha per year is not known, since systematic yield monitoring only recently started [1, 26, 45]. Earlier reported figures exhibit a very wide range (0.4–12 t ha^{-1} yr^{-1}) and are not coherent, mainly because of inadequate extrapolation of annual yields of individual plants to ha^{-1} yr^{-1}. Gopinathan and Sudhakaran [84] estimated that with an average seed yield of 3.75 t ha^{-1}, oil content of 30%–35%, and oil yield 1,200 kg ha^{-1}, *Jatropha* is superior to the other tree-born oil seeds and even the potential crops such as soybeans (USA) and rapeseed (Europe), which produce 375 kg and 1,000 kg of oil ha^{-1}, respectively. Since the average seed yield compiled globally herewith is determined as 1,625 kg ha^{-1}, which would turn over the oil yield of 504 kg ha^{-1} yr^{-1} (equivalent to 475 kg ha^{-1} yr^{-1} biodiesel on 5% loss in the transesterification process) against the anticipation. The cost of *Jatropha* oil production was assessed as 72.89 cents L^{-1} [5], which was 4.5 times higher than the cost of petroleum diesel at that time. Therefore, biodiesel production and its blending in petroleum diesel would not be possible unless the cost of biodiesel production is much more reduced than the prevailing prices of the diesel in the market. It is only possible when the seed yield per unit of area is increased to that much extent (4–5 t ha^{-1} equivalents to 1,200–1,500 kg ha^{-1} oil). Presently, the benefit/cost ratio is highly negative for widely adopting this venture. The major challenges to achieve this goal lie in overcoming its several demerits through proper research and development (Tab. 13.3).

Tab. 13.3 Prospect and retrospect of *Jatropha* cultivation as an energy crop.

Merits	Demerits
• *Jatropha* yields high-quality oil suitable for use in diesel engines.	• *Jatropha* is a wild species, not yet domesticated as a competitive energy crop.
• *Jatropha* is anticipated to yield more than 2 tons of oil per hectare per year.	• Yield expectations are very uncertain due to exaggerated extrapolation and lack of improved seed material.
• *Jatropha* can grow on poor soils of low fertility that are not suitable for food production.	• *Jatropha* does not produce good yields in poor conditions; there are trade-offs between rehabilitation of wastelands and maximization of oil production.
• It is suited for the rehabilitation of waste lands and erosion control.	• Harvesting is very labor-intensive and may jeopardize the economic viability.
• It grows well in semi-arid regions not suited for oil palms.	• Asynchronous maturity of seeds
• *Jatropha* seeds do not have to be processed immediately (unlike palm oil); therefore remote areas can be included in the production schemes.	• Being an exotic large scale plantations may heavily distort local social and ecosystems services
• *Jatropha* can be planted as a hedge around fields which offers small holders an opportunity to create additional revenues.	• *Jatropha* seed contains toxic substance (curcin), so the seed cake cannot be used as fodder for animals.
• It is an asset to rural development.	• Does not provide any fuel-wood or fodder

Merits	Demerits
• It is claimed as drought tolerant /resistant.	• Yield is poor in rain-fed conditions, modest irrigation is required
• Its seed consists of 30%–40% oil.	• Commercial extraction of oil from seed is quite less than that reported.
• *Jatropha* may be propagated through seeds as well as stem cuttings.	• Due to cross pollinations seeds differ from parent
• *Jatropha* may be developed as an energy crop without any competition with food crops.	• High yielding variety is not available yet; selections from wild accessions do not perform well on other sites even in isoclimatic regions.
• *Jatropha* plantations do not need any biotic protection as it is not browsed by cattle.	• There is very poor social acceptability of plants without fodder value in rural areas in the developing countries
• It is reported as disease resistant.	• Virus infestation, insect-pest attack and fungal diseases are quite common in many plantations.
• It does not require any specific agro-technology for plantation.	• Sound agro-technology is desired if domesticated.

13.5   Scope for Improvements

Selections from the wild germplasm for the desired traits and their integration in one plant are the primary processes for domestication, which could be achieved by breeding and multiplications. Nowadays, complex molecular techniques are also followed to form genetically modified crops. The objectives should aim at higher seed yields and oil contents, early maturity, reduced vegetative growth, and resistance to pests and diseases, drought tolerance, higher ratio of female to male flowers, and improved fuel properties [16–17]. Few researchers have strongly suggested evolving a high-yielding variety through breeding before its massive plantation on a farmer's field [26, 46]. Conventional breeding is suitable for selecting traits, such as seed yield, seed size, and oil yield, but also for developing planting material that has adapted to local environmental conditions. Plant tissue cultures and molecular techniques are powerful tools of biotechnology that can complement conventional breeding, expedite crop improvement, and meet the demand for the availability of uniform clones in large numbers [22]. The breeding strategies for commercial plantations will require a breeding program with crossings between selected genotypes, testing of offspring from the crosses, and, finally, development of superior offspring through either clonal propagation or seed propagation in seed orchards [47]. To secure long-term genetic gains, the breeding program could be organized using the concept of multiple breeding populations [48]. *Jatropha* closely resembles Ricinus and has attracted interest because it possesses various beneficial traits not found in castor [49–51].

There are several strategies that can be used to improve the morphometric traits associated with seed yield such as the number of branches and number of fruits per branch. These include traditional breeding, mutation breeding, and alien gene transfer through inter-specific hybridization and genetic engineering [52]. King *et al.* [46] also recommended the latest breeding techniques: the creation of inter-specific hybrids, molecular breeding, and *Agrobacterium tumefaciens*-mediated plant transformations for pathway engineering to create the new plant varieties. The seed yield can be enhanced by increasing the female-to-male flower ratio and through the modification of plant architecture, *i.e.*, increasing the number of branches [46]. The oil contents in seeds can be increased by altering the expression levels of enzymes in the triacyl glycerol biosynthetic pathway [46]. Inter-specific and inter-generic crossing is needed in cases where there is little variation for economically important traits. The hybridization improved vegetative, flowering, and fruiting traits of *Jatropha* [53]. Techniques in tissue culture such as *in vitro* fertilization, somatic hybridization, and gene transfer can be used to facilitate such crosses [44]. Mutation breeding techniques have also been used for the improvement of seed yields in *Jatropha* [44], which are more efficient and cheaper [52]. Induced mutation can be used to improve the quality of *Jatropha* in terms of seed production, oil content, and days to maturity [52].

Mutations occur in natural populations at low rates and are generally recessive. Random mutation breeding studies in *Jatropha* were carried out in Thailand using fast neutrons and isolated dwarfs or early flowering mutants from the M3 generation, but the potential productivity of these variants under intensive cultivation conditions was not proven [54]. Induced mutations were used, which identified mutant plants with early maturity, 100 seed weights 30% over that of the control, and better branch growth, as carried out by Dwimahyani and Ishak [52]. Several techniques are available for the improvement of the seed yield of *Jatropha*. The first step is gene mapping to establish an elite population with desirable characteristics that breeds true. The elite mapping population data could then be used to screen the germplasm of *Jatropha* in order to identify DNA markers or major quantitative trait loci associated with high yield and the information would be used to begin the genetic improvement for yield characteristics [44]. The development of high-yielding crop varieties through plant breeding has significantly increased agricultural productivity, especially in the latter half of the 20th century [55]. *Jatropha* is monoecious and has a male-to-female flower ratio of around 29 : 1. Efforts are underway to increase the female flowers through hormone treatments at the National Botanical Research Institute (NBRI), Lucknow, but they have only achieved mild success so far. The plant is insect-pollinated, although self-pollination is possible via geitonogamy [56]. Its flowers are very small and its nectar is not attractive to pollinators. Only a few insects and ants visit the flowers. This is one of the drawbacks for the efficient pollination and fertilization for viable seed formation. A correlation between the male-to-female flower ratio and yield has been observed, which was a highly heritable trait [57]. Yield increases in a number of plant species have also been obtained through the modifications of plant architecture [58]. Increasing the number of branches may lead to an increased number of inflorescences, which results in an increase of fruits.

Jatropha has few female flowers, which is one of the most important reasons for its poor seed yield. Pan and Xu [59] studied the effects of the plant growth regulator 6-benzyladenine (BA) on floral development and floral sex determination. According to their study, the seed yield of Jatropha can be increased by the manipulation of the floral development and floral sex expression. BA treatments induced bisexual flowers, which were not found in control inflorescences, and a substantial increase in the female-to-male flower ratio provides the scope for greater fruiting. Each Jatropha inflorescence is composed of 100–300 flowers and yields approximately 10 or more ovoid fruits [15, 57]. Rarely, four-seed fruits may be noticed, although they have been observed in some Mexican genotypes [60–61]. Inbreeding depression reduces individual fitness, survival, and growth variables [62] and raises the possibilities of population and/or species extinction [63–65]. It aims to evaluate the wide genetic resources for developing as an energy crop that is suitable for the marginal/degraded lands of the tropics and to establish a corresponding breeding program along with the establishment of good agronomic practices for crop management. Outstanding accessions that can be readily cloned offer an improved yield and drought tolerance, but plant breeding would allow for the continuous increase and release of evermore productive varieties. Varieties with higher oil contents in percents of seed dry weights will also provide increased revenue per working hour for the farmers. The development of non-toxic varieties will allow farmers to have additional markets for other products (not just biodiesel), like protein-rich Jatropha cake meal, making it a highly attractive animal feed.

13.6   Disease Incidence

Unlikely to the presumptions for disease resistance, Jatropha is susceptible to many insects, pests, and viral and fungal diseases (Fig. 13.3). Severe damage to plants by viral infections has been observed at Lucknow in India [66]. Root rot, stem borers, and fruit damage by Webber (spider) at Hyderabad, fruit sap suckers at Lucknow, and bark-eating rodents at Jodhpur are quite common problems at different locations in India [1]. The major problems reported are caused by the scutellarid bug *Scutellera nobilis* Fabr., the capsule borer *Pempelia morosalis* Saalm and Uller, *Pachycoris klugii* Burmeister (Scutelleridae), *Leptoglossus zonatus* Dallas (Coreidae), the blister miner *Stomphastis thraustica* Meyrick (Acrocercops), the semi-looper *Achaea janata* L., and the flower beetle *Oxycetonia versicolor* Fabr. [67–70]. They also studied different types of insects and pests in the *Jatropha* plantations. According to their studies, mealy bugs, aphids, and crocuses were the most frequent insects in the *Jatropha* plantation. In the meantime, leaf spots and fungus infections were the most severe in many areas in Thailand. They proposed harmless chemicals such as sodium lauryl sulphate and consumable products (toothpaste, shampoo, etc.) and biological treatments such as the natural predator and green lace wing to control the insects.

Fig. 13.3 Occurrences of various diseases in *Jatropha curcas*: (a) Healthy fresh leaf, (b) leaf infected with Cucumber Mosaic Virus, (c) leaf infected with Gemini Virus closely related to Indian Cassava Mosaic Virus, (d) virus infestation in a pruned plant, (e) dead plant as a result of severe virus infestation, (f) leaf necrosis by fungal infection, and (g–i) fruit damaged by sap-sucking insect *Scutellera perplexa*.
Source: Figs. g–i [67]

13.7   Soil Amelioration

Jatropha is known to reclaim degraded lands with potentially positive impacts on biodiversity and soil resources through the building of soil organic matter and its root symmetry controlling soil erosion [4, 7, 71–72]. Singh *et al.* [26] indicated that *Jatropha* had a modest ability to reclaim the sodic soils and considerable changes in soil properties were found after five years for *Jatropha* plantations but the seed yield potential was extremely low. The reduction in soil pH and electrical conductivity, with a parallel increase in organic C, N, microbial biomass, and dehydrogenase activity, ameliorate the degraded soil by a significant extent. A recent study on the impact of the cultivation of *Jatropha* on the structural stability and C-N content of degraded Indian Entisol reported that the cultivation of *Jatropha* resulted in an 11%, on average, increase in mean weight diameter of the soil and a 2% increase in soil macro aggregate turnover [73]. It has been observed that a 3- to 5-year-old plantation can sequester 305 kg ha^{-1} yr^{-1} of C in the soil and nearly 400 kg ha^{-1} yr^{-1} of C in the plant biomass [74].

13.8   Conclusions

Jatropha curcas has not yet achieved enough momentum in biodiesel production because of the poor yield and considerable genetic improvement is required in the existing germplasm. Unless ~5 tonnes per ha per year good seed yield (equivalent to ~3.8 kg per plant per year) is not ensured, no plantation would be commercially viable. As of today, only a limited number of plants with such yields can be found in the population of organized plantations in ideal geographical/environmental conditions. Genetic improvement using various approaches for breeding should be the main focus for developing a high-yielding variety as the standardization of agro-technology can only increase about 15%–20% of their genetic potential. Development of techniques such as somaclonal variants, mutations, doubled haploids, and gene transfers, which support plant breeding activities, should be emphasized. Similar to other crops, heritability of seed traits is the most common predictor of genetic gains for different breeding methods in *Jatropha*. Information on breeding and domestication of *Jatropha* is scarce as it has recently acquired its importance in biofuel. There is a need to develop this suitable domesticated variety that incorporates the traits of being high-yield with an oil content for wide implementation of *Jatropha*-based biodiesel programs. Some researchers claim to have high-yielding plants through natural selections, but the consistency of the results of such research has not yet proven a commercially viable biofuel. It is unclear how long it might take for such efforts to be realized. So far, it cannot be recommended universally in all localities and should be restricted in specific areas where high yields are ensured. A general recommendation for the entire tropical region of the world would be dangerous, particularly for the small holding farmers or entrepreneurs. We have tried to address the myths and facts for using the plants carefully in energy plantations so that everyone can realize the true prospects and scopes for its exploitations in sustainable biodiesel production. If we were able to develop a disease-resistant, high-yielding variety as an annual crop similar to that of castor (*Ricinus communis*), which has recently become a potential lubricant crop in the tropics, it would be an incredible achievement. We must take lessons learned from castor and follow a similar road map for the improvement of present germplasm to an economic crop.

References

[1] Singh B, Singh K, Rao GR, *et al.* Agro-technology of Jatropha curcas for diverse environmental conditions in India. Biomass Bioenerg 2013; 48: 191–202.

[2] Azam MM, Waris A, Nahar NM. Prospects and potential of fatty acid methyl esters of some non-traditional seed oils for use as biodiesel in India. Biomass and Bioenerg 2005; 29: 293–302.

[3] Divakara BN, Upadhyaya HD, Wani SP, Gowda CLL. Biology and genetic improvement of Jatropha curcas L.: A review. Appl Energ 2010; 87: 732–742.

[4] Achten WMJ, Verchot L, Franken YJ, *et al. Jatropha* bio-diesel production and use. Biomass Bioenerg 2008; 32: 1063–1084.

[5] Openshaw K. A review of Jatropha curcas: an oil plant of unfulfilled promise. Biomass Bioenerg 2000; 19: 1–15.

[6] Fairless D. Biofuel: the little shrub that could – maybe. Nature 2007; 449(7163): 652–655.

[7] Francis G, Edinger R, Becker K. A concept for simultaneous wasteland reclamation, fuel production, and socio-economic development in degraded areas in India: need, potential and perspectives of Jatropha plantations. Nat Resour Forum 2005; 29: 12–24.

[8] Jongschaap REE, Corré WJ, Bindraban PS, Brandenburg WA. Claims and Facts on *Jatropha curcas* L. Plant Research International, BV Wageningen, The Netherlands 2007.

[9] Achten WMJ, Mathijs E, Verchot L, Singh VP, Aerts R, Muys B. *Jatropha* biodiesel fueling sustainability? Biofuel Bioprod Bior 2007; 1: 283–291.

[10] Martín C, Moure A, Martín G, Carrillo E, Domínguez H, Parajó JC. Fractional characterisation of *Jatropha*, neem, moringa, trisperma, castor and candlenut

seeds as potential feedstocks for biodiesel production in Cuba. Biomass Bioenerg 2010; 34: 533–538.

[11] Ghosh A, Jitendra C, Chaudhary DR, Prakash AR, Boricha G, Zala A. Paclobutrazol arrests vegetative growth and unveils unexpressed yield potential of *Jatropha curcas*. J Plant Growth Regul 2010; 29: 307–315.

[12] Kumar S, Chaube A, Jain, SK. Critical review of *Jatropha* biodiesel promotion policies in India. Energy Policy 2012; 41: 775–781.

[13] Sanderson K. Wonder weed plans fail to flourish. Nature 2009; 461: 328–329.

[14] Knothe G. Analytical methods used in the production and fuel quality assessment of biodiesel. Trans ASAE 2001; 44: 193–200.

[15] Kumar A, Sharma S. An evaluation of multipurpose oil seed crop for industrial uses (*Jatropha curcas* L.): a review. Ind Crop Prod 2008; 28: 1–10.

[16] Sujatha M. Genetic improvement of *Jatropha curcas* L.: possibilities and prospects. Indian J. Agroforestry 2006; 8: 58–65.

[17] Sujatha M, Reddy TP, Mahasi MJ. Role of biotechnological interventions in the improvement of castor (*Ricinus communis* L.) and *Jatropha curcas* L. Biotech Adv 2008; 26: 424–435.

[18] Achten WMJ, Maes WH, Aerts R, *et al.* Jatropha. From global hype to local opportunity. J Arid Environ 2010; 74: 164–165.

[19] Parawira W. Biodiesel production from Jatropha curcas: a review. Sci Res Essay 2010; 4: 1796–1808.

[20] Koh MY, Ghazi TIM. A review of biodiesel production from *Jatropha curcas* L. oil. Renew Sustain Energ Review 2011; 15: 2240–2251.

[21] Ong HC, Mahlia TMI, Masjuki HH, Norhasyima RS. Comparison of palm oil, *Jatropha curcas* and *Calophyllum inophyllum* for biodiesel: a review. Renew Sustain Energ Review 2011; 15(8): 3501–3515.

[22] Mukherjee P, Varshney A, Johnson TS, Jha TB. *Jatropha curcas*: a review on biotechnological status and challenges. Plant Biotech Rep 2011; 5: 197–215.

[23] Pandey VC, Singh K, Singh JS, Kumar A, Singh B, Singh RP. *Jatropha curcas*: a potential biofuel plant for sustainable environmental development. Renew Sustain Energ Review 2012; 16(5): 2870–2883.

[24] Severino LS, Lima RLS, Lucena AMA, *et al.* Propagation by stem cuttings and root system structure of *Jatropha curcas*. Biomass Bioenerg 2011; 35: 3160–3166.

[25] Gubitz GM, Mittelbach M, Trabi M. Exploitation of the tropical oil seed plant *Jatropha curcas* L. Bioresour Technol 1999; 67(1): 73–82.

[26] Singh B, Singh K, Shukla G, *et al.* Field performance of some accessions of *Jatropha curcas* L. (biodiesel plant) on degraded sodic land in north India. Int J Green Energ 2013; 10: 1026–1040.

[27] Heller J. Physic Nut. *Jatropha curcas* L. Promoting the Conservation and Use of Underutilized and Neglected Crops. Rome: Institute of Plant Genetics and Crop Plant Research 1996.

[28] Srivastava P, Behra SK, Gupta J, Jamil S, Singh N, Sharma YK. Growth performance, variability in yield traits and oil content of selected accessions of *Jatropha curcas* L. growing in a large scale plantation site. Biomass Bioenerg 2011; 35: 3936–3942.

[29] Ginwal HS, Rawat PS, Srivastava RL. Seed source variation in growth performance and oil yield of *Jatropha curcas* Linn. in Central India. Silvae Genet 2004; 53: 186–192.

[30] Sharma GD, Gupta SN, Khabiruddin M. Cultivation of *Jatropha curcas* as a future source of hydrocarbon and other industrial products. In: Giibitz GM, Mittelbach M, Trabi M (Eds.). Biofuels and Industrial Products from *Jatropha Curcas*. DBV Graz, 1997; 19–21.

[31] Gour VK. Production practices including post harvest management of *Jatropha curcas*. Biodiesel Conference Towards Energy Independence—Focus on *Jatropha*. 2006.

[32] Jingura RM, Matengaifa R, Musademba, D, Musiyiwa K. Characterisation of land types and agro-ecological conditions for production of *Jatropha* as a feedstock for biofuels in Zimbabwe. Biomass Bioenerg 2011; 35: 2080–2086.

[33] Jingura RM. Technical options for optimization of production of *Jatropha* as a biofuel feedstock in arid and semi-arid areas of Zimbabwe. Biomass Bioenergy 2011; 35: 2127–2132.

[34] Mohapatra SA and Prasanna Kumar P. Effects of fertilizer application on growth and yield of *Jatropha curcas* L. in an aeric tropaquept of Eastern India. Notulae Scientia Biologicae 3.1 2011; 95–100.

[35] Kumar A, Sharma S, Mishra S. Application of farmyard manure and vermicompost on vegetative and generative characteristics of *Jatropha curcas*. J Phytol 2009; 1: 206–212.

[36] Rajaona AM, Asch HBF. Effect of pruning history on growth and dry mass partitioning of *Jatropha* on a plantation site in Madagascar. Biomass Bioenerg 2011; 35: 4892–4900.

[37] Kureel RS. Prospects and potential of *Jatropha curcas* for biodiesel production. In: Singh B, Swaminathan R, Ponraj V (Eds.). Biodiesel Conference Toward Energy Independence-focus of *Jatropha*, Hyderabad, India, June 9–10. Rashtrapati Bhawan, New Delhi, India, 2006; 43–74.

[38] Krizek B, Fletcher J. Molecular mechanisms of flower development: an armchair guide. Nat Rev Genet 2005; 6: 688–698.

[39] Irish V. The flowering of Arabidopsis flower development. Plant J 2009; 61: 1014–1028.

[40] Santner A, Calderon-Villalobos LIA, Estelle M. Plant hormones are versatile chemical regulators of plant growth. Nat Chem Biol 2009; 5: 301–307.

[41] Wang YH, Li JY. Molecular basis of plant architecture. Annu Rev Plant Biol 2008; 59: 253–279.

[42] Werner T, Schmulling T. Cytokinin action in plant development. Curr Opinion Plant Biol 2009; 12: 527–538.

[43] Kiba T, Sakakibara H. Role of cytokinin in the regulation of plant development. In: Pua EC, Davey MR (Eds.). Plant Developmental Biology-biotechnological Perspectives. Springer, New York, 2010; 237–254.

[44] Nambisan P. Biotechnological intervention in jatropha for biodiesel production. Curr Sci 2007; 93: 13–47.

[45] Everson CS, Mengistu MG, Gush MB. A field assessment of the agronomic performance and water use of *Jatropha curcas* in South Africa. Biomass Bioenerg 20132.; 59: 59–69. http://dx.doi.org/10.1016/j.biombioe.2012.03.013.

[46] King AJ, He W, Cuevas JA, Freudenberger M, Ramiaramanana D, Graham IA. Potential of *Jatropha curcas* as source of renewable oil and animal feed. J Exp Enviorn Botany 2009; 60: 2897–2905.

[47] Achten WM, Nielsen LR, Aerts R, *et al.* Towards domestication of *Jatropha curcas.* Biofuels 2010; 1: 91–107.

[48] Namkoong G, Lewontin RC, Yanchuk AD. Plant genetic resource management. The next investments in quantitative and qualitative genetics. Genet Resour Crop Evol 2004; 51: 853–862.

[49] SathaiahV, Reddy TP. Seed protein profiles of castor (*Ricinus communis* L.) and some *Jatropha* species. Genet Agr 1985; 39: 35–43.

[50] Reddy KRK, Ramaswamy N, Bahadur B. Crossing compatibility between *Ricinus* and *Jatropha.* Plant Cell Incomp Newslet 1987; 19: 60–65.

[51] Sujatha M, Mukta N. Morphogenesis and plant regeneration from tissue cultures of *Jatropha curcas.* Plant Cell, Tissue Organ Culture 1996; 44: 135–141.

[52] Dwimahyani I, Ishak. Induced mutation on *Jatropha* (*Jatropha curcas* L.) for improvement of agronomic characters variability. 2004; 53–60. Accessed on February, 23, 2013 at http://digilib.batan.go.id/atom-indonesia/fulltex/v30-n2-7-2004/Ita-Dwimahyani-Ishak.pdf.

[53] Paramathma M, Reeja S, Parthiban KT, *et al.* Development of interspecific hybrids in *Jatropha.* Proceedings of the biodiesel conference toward energy independence—focus on *Jatropha,* June 2006; 9: 10.

[54] Sakaguchi S, Somabhi M. Exploitation of Promising Crops of North East Thailand Siriphan Offset. Thailand: Khon Kaen Press 1987.

[55] Evenson RE, Gollin D. Assessing the impact of the green revolution, 1960 to 2000. Science 2003; 300: 758–762.

[56] Raju AJS, Ezradanam V. Pollination ecology and fruiting behaviour in a monoecious species, *Jatropha curcas* L. (Euphorbiaceae). Curr Sci 2002; 13: 1395–1398.

[57] Rao G, Korwar G, Shanker A, Ramakrishna Y. Genetic associations, variability and diversity in seed characters, growth, reproductive phenology and yield in *Jatropha curcas* (L.) accessions. Trees-Struct Funct 2009; 22: 697–709.

[58] Sakamoto T, Matsuoka M. Generating high-yielding varieties by genetic manipulation of plant architecture. Current Opinion in Biotechnology 2004; 15: 144–147.

[59] Pan BZ, Xu ZF. Benzyladenine treatment significantly increases the seed yield of the biofuel plant *Jatropha curcas.* J Plant Growth Regu 2011; 30: 166–174.

[60] Makkar H, Martinez-Herrera J, Becker K. Variations in seed number per fruit, seed physical parameters and contents of oil. J Plant Growth Regul 2008; 30: 166–174.

[61] Makkar HRS, Becker K. *Jatropha curcas,* a promising crop for the generation of biodiesel and value-added coproducts. Eur J Lipid Sci Technol 2009; 111: 773–787.

[62] Nielsen LR, Siegismund HR, Hansen T. Inbreeding depression in the partially self-incompatible endemic plant species *Scalesia affinis* (Asteraceae) from Galápagos islands. Evol. Ecol. 2007; 21: 1–12.

[63] Hansson B, Westerberg L. On the correlation between heterozygosity and fitness in natural populations. Mol Ecol 2002; 11: 2467–2474.

[64] Charlesworth D, Charlesworth B. Inbreeding depression and its evolutionary consequences. Annu Rev Ecol Syst 1987; 18: 237–268.

[65] Reed DH, Frankham R. Correlation between fitness and genetic diversity. Conserv Biol 2003; 17: 230–237.

[66] Raj SK, Snehi SK, Kumar S, Khan MS, Pathre U. First molecular identification of a begomovirus in India that is closely related to Cassava mosaic virus and causes mosaic and stunting of *Jatropha curcas* L. Aust Plant Disease Note 2008; 3: 69–72.

[67] Sahai K, Srivastava V, Rawat KK. Impact assessment of fruit predation by *Scutellera perplexa* Westwood on the reproductive allocation of Jatropha. Biomass Bioenerg 2011; 35(11): 4684–4689.

[68] Banjo AD, Lawal OA, Aina SA. The entomofauna of two medicinal Euphorbiacae in southwestern Nigeria. J Appl Sci Res 2006; 2(11): 858–863.

[69] Regupathy A, Ayyasamy R. Need for generating baseline data for monitoring insecticide resistance in leaf Webber cum fruit borer, *Pempelia morosalis* (Saalm Uller), the key pest of biofuel crop, *Jatropha curcas*. RPM Newsletter 2006; 16: 1–5.

[70] Shanker C, Dhyani SK. Insect pests of Jatropha curcas L. and the potential for their management. Curr Sci 2006; 91(2): 162–163.

[71] Abhilash PC, Srivastava P, Jamil S, Singh N. Revisited Jatropha curcas as an oil plant of multiple benefits: critical research needs and prospects for the future. Environ Sci Pollut Res 2011; 18: 127–131.

[72] Achten WMJ, Akinnifesi FK, Maes WH, *et al. Jatropha* integrated agroforestry systems—biodiesel pathways towards sustainable rural development. In: Ponterio C, Ferra C (Eds.). *Jatropha curcas* as a Premier Biofuel: Cost, Growing and Management. Nova Science Publishers, Hauppauge, NY, USA, 2010; 85–102.

[73] Ogunwole JO, Chaudhary DR, Ghosh A, Daudu CK, Chikara J, Patolia JS. Contribution of *Jatropha curcas* to soil quality improvement in a degraded Indian entisol. Acta Agric Scand Sect B Soil Plant Sci 2008; 58: 245–251.

[74] Wani SP, Chander G, Sahrawat KL, *et al.* Carbon sequestration and land rehabilitation through *Jatropha curcas* (L.) plantation in degraded lands. Agric Ecosyst Environ 2012; 161: 112–120.

[75] Henning R. Combating desertification: The *Jatropha* project of Mali, West Africa. Aridlands Newsletter 1996; 40.

[76] Wijgerse I. The electricity system for a rural village in Mali. Msc Thesis, Delft University of Technology., Delft, 2008; 104.

[77] Foidl N, Foidl G, Sanchez M, Mittelbach M, Hackel S. *Jatropha curcas* L. as a source for the production of biofuel in Nicaragua. Bioresour Technol 1996; 58: 77–82.

[78] Wani SP, Osman M, Silva ED, Sreedevi TK. Improved livelihoods and environmental protection through biodiesel plantations in Asia. Asian Biotech Develop Review 2006; 8: 11–29.

[79] Rao VR. The Jatropha hype: promise and performance. Boidiesel Conference Towards Energy Independence—Focus on *Jatropha*. Papers presented at the Conference Rashtrapati Nilayam, Bolaram, Hyderabad 2006.

[80] Daey Ouwens K, Francis G, Franken YJ, *et al*. Position paper on *Jatropha curcas*. State of the art, small and large scale project development. Fact Foundation 2007.

[81] GEXSI. Global market study on *Jatropha*—final report. GEXSI LLP, Berlin, Germany, 2008.

[82] Gunaseelan VN. Biomass estimates, characteristics, biochemical methane potential, kinetics and energy flow from *Jatropha curcas* on dry lands. Biomass Bioenerg 2009; 33: 589–596.

[83] Trabucco A, Achten WMJ, Bowez C, *et al*. Global mapping of *Jatropha curcas* yield based on response of fitness to present and future climate. GCB Bioenergy 2010; 2: 139–151.

[84] GEXSI. New Feedstocks for Biofuels—Global market study on *Jatropha*. Kraftstoffe der Zukunft. Berlin, December 2, 2008.

[85] Gopinathan MC, Sudhakaran R. Biofuels: opportunities and challenges in India. Biofuel 2011; 173–209.

Subject Index

www.ingramcontent.com/pod-product-compliance
Lightning Source LLC
Chambersburg PA
CBHW051333200326
41519CB00026B/7408